Popular Geopolitics and Nation Branding in the Post-Soviet Realm

This seminal book explores the complex relationship between popular geopolitics and nation branding among the Newly Independent States of Eurasia, and their combined role in shaping contemporary national image and statecraft within and beyond the region. It provides critical perspectives on international relations, nationalism, and national identity through the use of innovative approaches focusing on popular culture, new media, public diplomacy, and alternative "narrators" of the nation. By positing popular geopolitics and nation branding as contentious forces and complementary flows, the study explores the tensions and elisions between national self-image and external perceptions of the nation, and how this complex interplay has become integral to contemporary global affairs.

Robert A. Saunders is Professor in the Department of History, Politics, and Geography at Farmingdale State College, a campus of the State University of New York, where he teaches courses on Russia, Central Asia, and world religions. His research interests include post-totalitarian states, geopolitics, popular culture, and the mass-mediation of national identity.

Routledge Research in Place, Space and Politics Series
Series edited by Professor Clive Barnett, Professor of
Geography and Social Theory, *University of Exeter, UK*

This series offers a forum for original and innovative research that explores the
changing geographies of political life. The series engages with a series of key
debates about innovative political forms and addresses key concepts of political
analysis such as scale, territory and public space. It brings into focus emerging
interdisciplinary conversations about the spaces through which power is exer-
cised, legitimized and contested. Titles within the series range from empirical
investigations to theoretical engagements and authors comprise scholars working
in overlapping fields including political geography, political theory, development
studies, political sociology, international relations and urban politics.

Popular Geopolitics and Nation Branding in the Post-Soviet Realm

Robert A. Saunders

LONDON AND NEW YORK

First published 2017 by Routledge

2 Park Square, Milton Park, Abingdon, Oxfordshire OX14 4RN
52 Vanderbilt Avenue, New York, NY 10017

Routledge is an imprint of the Taylor & Francis Group, an informa business

First issued in paperback 2020

British Library Cataloguing in Publication Data
A catalogue record for this book is available from the British Library

Library of Congress Cataloging in Publication Data
Names: Saunders, Robert A., 1973- author.
Title: Popular geopolitics and nation branding in the post-Soviet realm / Robert A. Saunders.
Description: New York, NY : Routledge, 2016. | Series: Routledge research in place, space and politics series | Includes bibliographical references and index.
Identifiers: LCCN 2016007450| ISBN 9781138830172 (hardback) | ISBN 9781315737386 (e-book)
Subjects: LCSH: Geopolitics—Former Soviet republics. | Nation-building—Former Soviet republics. | Branding (Marketing)—Political aspects—Former Soviet republics. | Popular culture—Former Soviet republics. | Former Soviet republics—Relations.
Classification: LCC DK293 .S38 2016 | DDC 327.47—dc23
LC record available at https://lccn.loc.gov/2016007450

ISBN: 978-1-138-83017-2 (hbk)
ISBN: 978-0-367-66823-5 (pbk)

Typeset in Times New Roman
by Swales & Willis Ltd, Exeter, Devon, UK

"It is said that evil men have no songs. How is it that the Russians have songs?"

— Friedrich Nietzsche

Contents

Figures

Tables

Acknowledgments

I am convinced that all academic works are, at least in some way, autobiographical. The book you are about to read is unabashedly so. Growing in up in a military town during the last decade of the so-called Second Cold War, I was a lab rat for understanding the power of popular culture on geopolitical understanding. My friends (many of whom were the sons of military officers or civilian contractors) and I would play with G.I. Joe action figures, read *Captain America* comic books, listen to politically infused alternative music, watch *The A-Team* (1983–1987) and *Airwolf* (1984–1986), and rush out to see the premieres of films like *Red Dawn* (1984) and *Dreamscape* (1984). Cold War pop-cultural production was even a part of my junior high school curriculum. In seventh grade, my class watched *The Day After* (1983) television mini-series in class (on VCR), a rather traumatic experience for a pre-teen; however, the experience would come in handy a few years later when I was tasked with being the "group leader" in a semester-long project in my freshman English course which centered on surviving a limited nuclear strike. My high school years were purportedly a time when the US–Soviet hostilities were abating, yet for me, things only seemed to be getting hotter with movies like *Top Gun* (1986), *Russkies* (1987), and *Rambo III* (1988) on the big screen and the mini-series *Amerika* (1987) on television, just as I was being required to study a six-week course in "Americanism versus Communism," as was mandated by the State of Florida from 1962 until the early 1980s (despite its repeal in 1983, my "old school" World History instructor continued to observe the [defunct] requirement). Yet, change was in the air and I watched in utter disbelief as the Wall came down in Berlin only a few short months after I had started studying the Russian language at Booker T. Washington High School in Pensacola, Florida.

My instructor was a man without a shadow, in that he did not have a nation to call his own. Born into a Cossack family in a World War II refugee camp in Austria, he grew up in a Spanish-speaking neighborhood in Rio de Janeiro before moving to a Ukrainian neighborhood in Brooklyn. He spoke five languages, but none with genuine fluency. His Cold War credentials were above reproach. To prove his loyalty to his adopted country of the United States, he volunteered for the Vietnam War, and was tasked with placing markers in the jungles of Southeast Asia for subsequent Agent Orange defoliation operations. While I did not learn much Russian in the pivotal period from August 1989 until November 1990 (when

I moved back to Panama City, Florida), I learned a lot about the outside world from Comrade Golovko. He was always connecting the popular with the political, especially through anecdotes of his own life, such as when the actor Yul Brynner threatened to throw him out of an Orthodox Church for snickering during a speech by Alexander Kerensky, the leader of the February Revolution. Leaving behind my toys, comics, and youthful enthusiasm for "red-blooded American" TV fare, I started university as the Soviet Union dissolved in 1991, being lucky enough to study under Maria Todorova, who was working on her manuscript for *Imagining the Balkans* (1997) at the time. Through her tutelage and that of other professors at UF, I came to understand that the popular and the political are inextricably imbricated. However, I think I always knew that deep down, and I hope that this work will reflect that life-long understanding.

This book has been a long time coming, and as a result the list of people to thank is not a short one and will likely fail to include a few who are more than deserving of being listed herein. First and foremost, I wish to extend my gratitude to Vlad Strukov, who read and commented on the following chapters, and who arranged for me to spend a semester as a Visiting Senior Research Fellow at the School of Languages, Cultures, and Societies' Leeds Russian Centre at the University of Leeds (Fall 2014). My thanks also go out to Sarah Hudspith, Paul Cooke, and Frank Finlay for making me feel so welcome at the University of Leeds, which provided substantive financial support to this project which I hereby acknowledge. While there, Vlad and I co-taught a module entitled "Imagining the Post-Soviet Realm: Popular Culture and Representation of Russia and the Newly Independent States since 1991." Despite its unwieldy title, the class was a great success and I wish to thank my students for their insights, comments, and contributions to this project. I also want to thank Jason Dittmer, Kyle Grayson, and Nick Robinson for contributing to our seminar "Interdisciplinarity of Popular Geopolitics: Popular Culture and the Making of Space and Place," hosted by the Leeds Russian Centre. In addition to hearing them present their own groundbreaking research and receiving their support for our seminar series, I was lucky enough to pick their brains for my own project.

My deep appreciation also goes out to Tine Roesen for inviting me to teach a masterclass on "Popular Geopolitics: The Role of Film, TV, and Other Media in Contemporary International Relations" at the Department of Cross-Cultural and Regional Studies, University of Copenhagen, in late 2014. I was also lucky enough to be asked to speak about my long-running experiences with Sacha Baron Cohen's "Borat," which prompted me to reconsider and reflect on the initial catalyst for the project you hold in your hands. Looking beyond institutional support for this project, I am particularly indebted to Federica Caso for her suggestions on key chapters, as well as serving as a sounding board on a variety of theoretical issues that I grappled with in the final stages of writing. To Joel Vessels, the Lucien Wilbanks to my Jake Brigance, I remain in awe of your ability to transform the banal into the lofty and the rarefied into the intelligible; I hope I remain worthy of continued guidance on that score. Throughout this process, I have been lucky to benefit from all types of support from other scholars in the field, from invitations to present or publish my work to feedback on case studies, theory, and methodology. In this regard, I wish to thank Peter van Ham, Iver Neumann, Lara Ryazanova-Clark, Vera Zvereva, Nadia Kaneva, Chris Homewood, Wolf Schäfer, Jeremy Morris, and Natalia Rulyova, in particular.

For their part in helping me understand reality in a world swamped by and mired in illusions, I want to give thanks to Vasyl Myroshnychenko, Lila Osaulenko, Dalia Bankauskaitė, Roman Vassilenko, and Jeremy Faro. For their assistance in securing the rights to publish images and other content included in the volume, I offer my gratitude to Kadi Nõmmela, Inta Briede, Edita Gaigalienė, Arto Halonen, Pauline Hirsch, Faith Stein, and Elizabeth Sheehan. I wish to thank my students at Farmingdale State College for their great ideas and contemporary examples. Moreover, this book would not have come to fruition were it not for the continued support of the following people: Tamara Sooknauth, Jeff Gaab, Lou Reinisch, Lucia Cepriano, and W. Hubert Keen. At Routledge, I wish to recognize the excellent work and infectious enthusiasm of my editor Faye Leerink, as well as Clive Barnett for bringing into being a series on three topics near and dear to my heart.

Additionally, I want to acknowledge all those pub-keepers and bartenders who played roles both large and small in the long and winding road of bringing this idea from its genesis into its final form. Special shout-outs go to the following: Alex Fitton (Kim Marie's Eat-n-Drink Away, Asbury Park, New Jersey) who inspired me with his knowledge of the former Soviet republics, as well as his creation of a new one called "Tzatzikistan"; Carol Mendez (Black Forest Brew Haus, Farmingdale, New York) for keeping my thirsty mug at the ready; John Merklin, owner of Beach Haus Brewery (Belmar, New Jersey) for providing a space and place for finishing the last chapter and my final edits (as well as great chats about Russia); and all the folks at the Pack Horse Pub (Leeds, England) for making sure the Wi-Fi was always on and the music on the jukebox superb. Lastly, I want to thank my wife and son for their enduring patience and constant support during this lengthy process.

Part of Chapter 3 appeared in my 2012 article "Undead spaces: Fear, globalisation, and the popular geopolitics of zombiism" in the Routledge journal *Geopolitics*. Chapter 5 is partially adapted from my chapter "Brand interrupted: The impact of alternative narrators on nation branding in the former Second World," which appeared in Nadia Kaneva's edited volume *Branding post-communist nations: Marketizing national identities in the "new" Europe* (2012) and is reproduced by permission of Taylor and Francis Group, LLC, a division of Informa plc. Some passages in Chapters 2 and 7 appeared in my 2008 article "Buying into brand Borat: Kazakhstan's cautious embrace of its unwanted 'son'," which appeared in the Association for Slavic, East European, and Eurasian Studies' journal *Slavic Review*. Some parts of Chapter 7 were also previously published in "The geopolitics of Russophonia: The problems and prospects of post-Soviet 'global Russian'," in *Globality Studies Journal* (2014); all are reproduced with the permission of the publisher. All maps in the volume were produced using open source software provided by DIVA-GIS, Global Administrative Areas, and Natural Earth; my deepest thanks to Emily A. Fogarty for her time and skill in crafting these.

Introduction: [Not] made in the USSR

We live in a world characterized by what the Germans call *Fremdbebilderung* or the state of being "totally engulfed by foreign images." Turn on a TV or log on to the Internet and you will be bombarded by images that were generated far from your home by people you will never know. Even those living in the Hollywood neighborhood of central Los Angeles can expect a barrage of images of distant origin at any given point in the day, and a significant portion of these representations will undoubtedly be geopolitical in nature: maps, depictions of "foreigners," representations of distant lands. As one scholar notes, "There can be no doubt that mass media influence the way a country's people form their images of the people and governments of other countries because it is the mass media that disseminate the greater part of information about foreign countries" (Kunczik 1997, 7). Using popular culture itself as a tool for navigating the new world order, we can look to the powerful and cogent maxim that the "Producer" Christof (Ed Harris) uttered in the tragicomedy *The Truman Show* (1998), which followed the life of a single individual who was literally born to star in his own television show and did so unknowingly for nearly 40 years: "We accept the reality of the world with which we are presented; it's as simple as that."

Certain images tell us who our friends are, while others show us whom to fear. In the contemporary realm of international relations, images have been weaponized in an unending war of ideas, and while they may not cause actual physical damage they are nonetheless capable of producing tangible, often destructive outcomes (Mirzoeff 2015). We may laugh at some of these images and retract in horror at others. Some images are carefully designed to make us visit, buy, or try something, while others teach us what to shun, hate, or seek to destroy. Many images are simply "background," filler for the structured realities of the everyday lived experience; yet, these too may serve the political interests of agents operating in the name of this or that state or nation. In the words of Stuart Aitken and Leo E. Zonn, "The very heart of geography—the search for our sense of place and self in the world—is constituted by the practice of looking and is, in effect, a study of images" (1994, 7). Some images are simply the flotsam and jetsam of the artistic mind, yet these too convey subtle meanings and influence our mental mapping of the world around us. Accepting the fact that we are "constantly bombarded by the visual" (Särmä 2015, 117) and that the "contemporary dominance of the visual has a decidedly social and political dimension" (Stocchetti and Kukkonen 2011, 1), this book interrogates a narrow slice of these images (or more accurately, imaginaries), specifically those Western-generated representations of the people, places, and spaces that once composed the Soviet Union.

In 1991, the Union of Soviet Socialist Republics (USSR) dissolved, bringing an end to a seven-decade-long geopolitical experiment of creating a worker-state in northern Eurasia. This seminal event triggered the birth of fifteen independent countries across two continents. Out of a single state—albeit it a vast and diverse one—four new world regions came into being almost overnight: the Baltic States (Estonia, Latvia, and Lithuania); the Western Republics (Ukraine, Belarus, and Moldova); Transcaucasia or the South Caucasus (Georgia, Armenia, and Azerbaijan); and Central Asia (Kazakhstan, Uzbekistan, Turkmenistan, Kyrgyzstan, and Tajikistan). The largest of the constitutive units of the USSR—the Russian Soviet Federative Socialist Republic (RSFSR)—emerged as the renamed Russian Federation, a multifaceted geographical and political entity in its own right, comprised of European Russia, the North Caucasus, Siberia, the Russian Far North, and the Russian Far East. With the act of disbanding the Soviet Union, Russian president Boris Yeltsin (and his counterparts in the non-Russian republics) wrought a geopolitical transformation the likes of which history has rarely—if ever—witnessed.[1] The "Red Giant"—a source of geopolitical fear, fascination, and sociological, technological, and ideological inspiration for nearly seventy years—was not only humbled, it hacked itself to pieces and remade the world map in the process, a fact that Russia's long-serving president Vladimir Putin once lamented as the "greatest geopolitical tragedy of the last century" (Allen 2005).

As a geographical space in intense political, economic, and socio-cultural flux and one containing more than a dozen new countries, i.e., states which have come into existence under new names since 1991, it is not surprising that myriad struggles for control over national image rage across the post-Soviet realm. Complicating this set of affairs is the fact that each of these states had grappled for decades with the politics of national identity production within the Soviet structure, a phenomenon fraught with myriad conflicts and contradictions (see Martin 2001). According to Vladimir Lebedenko, former Deputy Director of Russia's Ministry of Foreign Affairs, "Most of these newly independent states were faced with the need for self-identification and assertion of their image in the international arena" (2004, 71). These states debuted at a rather complicated juncture in history, one in which globally networked new media and information and communication technologies (ICTs) have become nearly ubiquitous, global economic interdependence is a daily reality, and cultural hybridization and transnationalism are the norm (Saunders 2006). As national formations, the peoples of the region enjoy long and robust histories; however, due to the post-Cold War abandonment of the one-party system, the transition from state socialism to a market economy, and the fracturing of the USSR's socialist federation, these countries face a crisis of (self-)representation. Moored in the geopolitical imaginary of the "Other Europe" (Wolff 1994), stigmatized by the "post-Communist" moniker (Kovačević 2008), and easily conflated or ignored in the "supermarket" of nation brands (Anholt 2002), many of the post-Soviet republics have adopted the tools and techniques of brand management with unbridled enthusiasm in an effort to burnish their respective national images.

Despite expensive, sustained, and often inventive efforts at establishing and enhancing their respective national images (see, for instance, Dzenovska 2005; Marat 2009; Kaneva 2012), the countries of post-Soviet Eurasia remain the targets of novelists, filmmakers, satirists, and other cultural producers who readily and often thoughtlessly besmirch the "brand" of the post-Soviet world through their own works of popular culture and artistic imagination (see Saunders 2012). In the current era of postmodern geopolitics defined by globalization, deterritorialization, and cultural fragmentation, mass media's role in shaping geographical imagination and making sense of the geopolitical order is steadily increasing, making *popular geopolitics* as important in international relations as its *elite* and *academic* counterparts. This study will argue that the former Soviet republics sit at the bleeding edge of this trend, and, consequently, are important sites of socioeconomic and cultural change which will serve as a harbinger for other regions of the world in the twenty-first century.

Recognizing the realities of this new era of public, corporate, and para-diplomacy, the community of nations has come to function as a marketplace where states must define and differentiate themselves in order to attract foreign direct investment (FDI), add value to their export products, promote tourism, and develop their diplomatic and strategic alliances. This set of activities has been identified as nation branding. In many post-totalitarian countries, political elites have taken up this mantle with pride and purpose. However, in a world where CNN has displaced the diplomatic pouch, where Twitter trumps *The New York Times*, and where celebrities like Angelina Jolie often wield greater influence than the prime minister of Namibia, it is painfully evident that any country's branding strategy must anticipate, negate, and (occasionally) accommodate its unwanted "branding" by alternative actors. The challenges are particularly acute as a result of the growing importance of mass-mediated images, which are transmitted around the world via movie screens, satellite TV, smart phones, and other digitally networked devices. Reflecting a widening trend in the field of critical international relations theory associated with the recognition of the "popular culture–world politics continuum" (Grayson, Davies, and Philpott 2009), it is vital to explore the important role of conceptualizing territory through popular media. Likewise, it is incumbent on scholars to understand how "imagined geographies" (Said 1979) influence power, ideology, and identity in the contemporary world.

Focusing on the competition between governmental efforts to manage their national images and external forms of geographical imagination via popular culture, *Popular Geopolitics and Nation Branding in the Post-Soviet Realm* explores the complexities of national image among the post-Soviet republics. This text uses two comparatively new fields of analysis—popular geopolitics and nation branding—to explore how the Russian Federation, the Baltic states, the Central Asian republics, and other post-Soviet countries have been portrayed in Western popular culture and how they have sought to represent themselves to the outside world. By positing popular geopolitics and nation branding as oppositional, even countervailing forces (though ones which employ many of the same tools), my analysis explores the tension between national self-image and external/international perceptions of the nation. Furthermore, I am interested in how countries' international reputations have come to function as a key determiner of national identity in the current era of globalism. Grounded in current history, this book attempts to expand the field of nationalism studies by utilizing new approaches to the study of the nation through what I refer to as *digital nationality*, i.e., the shift of national identity production from analog to digital modes of transmission which in turn allow for greater geographical reach and more diverse forms of transmission, replication, manipulation, and distortion (see Saunders 2010). Through a critical examination of twenty-first century phenomena, including viral marketing of the nation, computer-mediated and manipulated national images, and the colonization of cyberspace by national(ist) forces (see Saunders 2014), I suggest that the nation now primarily functions as a easily reproducible and networked commodity product.[2]

Taking Eric Hobsbawm's (1992) notion of "invented traditions" to its logical (and digitally enhanced) extremes, I contend that what we currently refer to as national identity is something that is produced, consumed, recycled, and in a constant state of rebranding. This is done with the aim of attracting the greatest number of nation-state "fans" (Dittmer and Dodds 2008), thus reifying what Peter van Ham (2001) has called the "brand state." Popular culture is and will remain an important site of such activity (see Edensor 2002). I differentiate popular culture from so-called "high culture" based on a threshold related to the ease of consumption.[3] Accessing and enjoying artefacts of popular culture require much less in the way of pecuniary and/or interpretive resources. A graphic novel based on Fyodor Dostoyevsky's *Crime and Punishment* (1866) might cost the same as a reprint of the original novel, but requires less intellectual work for the reader, while streaming a YouTube video of Lady Gaga's latest hit demands less time, money, and acumen than attending a performance of *Swan Lake*. Admittedly, the advent of the Internet has proved transformative to both high and low culture, making both more "popular" in form; however, this is not the appropriate venue for a discussion of this important and ongoing phenomenon.

In interpolating (Western-produced) popular culture with (post-Soviet) governmental attempts at branding the state, this text aims to expand Jean Baudrillard's contention that "everything in politics is faked" (qtd. in Butler 2010, 24) to the realm of contemporary, postmodern geopolitics. In doing so, I hope to illuminate the day-to-day realties of the "geopolitical staging" (Gagnon 2007) in order to increase a state's relative power in the world system, with the aim of demonstrating how widespread this phenomenon has become. While I am not willing to go so far as to suggest that reality no longer has any meaning, I contend that representation via mediation is paramount in contemporary geopolitics; representation is thus the ultimate tool of *world-building* (Graham 2002), an assertion that is buttressed by recent work in the area of geopolitical assemblage (see Dittmer and Dodds 2013; Dittmer 2014; Bleiker 2015) and further supported by critical IR treatments of statecraft as dependent on "identity-related practices at a multiplicity of levels and sites" (Auchter 2014, 170). With a critical focus on images produced in popular culture and as part of nation branding campaigns, this work is also a preliminary response to the criticism that "the *study* of images seems to have been unable to engage the *politics* of images," thus leaving a large lacuna in the academic study of (geo)political imagination (Stocchetti 2011, 22). Through the application of the aforementioned conceptual tools to fifteen states with a common history (all were part of the USSR from at least 1944 until 1991),[4] this study seeks to provide a multicomponent case study of the challenges and opportunities presented to the newly independent states in the twenty-first century, while simultaneously investigating the impact of the Soviet legacy on geographical imagination and national identity projects. In doing so, I anticipate that this monograph will speak to a number of issues associated with the seemingly unending process of "making the West," that is, the visual-discursive production of what it means to be part of Western civilization. This process has historically relied on the creation of imagined Others at the shifting borders of this protean zone of inclusion/exclusion, with Russia almost always playing a critical and sometimes even defining role (see Makarychev 2014). My larger, though admittedly tangential, goal is to address certain emerging issues in the current "crisis in conceptualization" (Lewis and Wigen 1997) that afflicts global geography in the post-Cold War era, particularly those related to quotidian views of the world and how it *really* works (see Dittmer 2010), especially in the Anglophone world.

The first several chapters of this text are extended interrogations of the three concepts that gird the subsequent analysis: national image, nation branding, and geographical imagination/popular geopolitics. Each of these is treated to an in-depth interrogation, with an accompanying literature review and brief case studies being provided. The first chapter, "Of idols and idylls: The question of national image," explores the rather scattershot literature of national image, focusing on how such representations have been historically constructed and altered over time. An in-depth discussion of the distinction between nations' images of themselves and how they conceive of other nations is provided. I pay particular attention to the issues of the Other in international relations, using national image as one of several optics for understanding and framing nations' actions within the global system. Subsequently, the chapter distinguishes between national image and related concepts, including nationalism, national identity, and stereotypes, with the aim of demonstrating the unique power of national image production in the realm of international relations. The chapter concludes with an extended analysis of Russia's image in the West dating back to the foundation of Kievan Rus over 1,000 years ago.

Chapter 2, entitled "The supermarket of nations: Competitive identity and the brand state," focuses on the field of nation branding, providing an introduction to the most important theoretical works and contextualizing current practices within a larger historical framework. An exploration of the strategies, difficulties, and successes of nation branding projects is provided, as well as a discussion of the merits and pitfalls of becoming a "brand state." The chapter distinguishes state branding from public diplomacy and soft power, while delineating the similarities across all three undertakings. A discussion of the changes in the world system that have made brand management of national identity so critical is also included. Lastly, this chapter argues the importance of distinguishing and burnishing national images in the "global supermarket" of nation brands through brief case studies of some of the most successful examples of nation branding (New Zealand, Qatar, and Spain).

With the third chapter, "The mind's eye: Popular culture, geographical imagination, and international relations," I provide an introduction to the relatively new field of popular geopolitics and the work of key scholars in the discipline. After discussing the marriage of critical geopolitics with cultural studies, the chapter explains how geographies are constructed and maintained in our *Fremdbebilderung*-bound world where television programs, videogames, comic books, motion pictures, and web sites have emerged as important, even dominant sources of everyday geopolitical "knowledge." In an effort to provide a comparative component, the analysis attempts to acquaint the reader with the complex interplay between popular culture, geographical imagination, political ideologies, and international relations through the use of other examples of popular geopolitics, including pop-culture mediations of "Africa," the "Orient," and "Latin America," with special attention paid to the Arab world/Middle East as a geopolitical imaginary.

Chapter 4 is provided as a segue between conceptual considerations and the subsequent case studies. In this section of the book, I present an introductory geopolitical analysis of the post-Soviet realm, focusing on the complex interplay between domestic politics, international affairs, political culture, demographics, and physical geography. A caveat is in order here; like any work of this nature, ideology inevitably colors any sustained description of places, populations, and politics, and this chapter is no different, being a markedly "Western" depiction of post-Soviet Eurasia. However, as the reader will undoubtedly discover, this analysis, while seeking neutrality, tends to suffer from some of the very same geopoliticized orientations that will be critiqued in the ensuing chapters. That being said, any text on geopolitics (popular or otherwise) would be incomplete without a cursory account of the places, spaces, and peoples that make up the former Soviet Union, or as I will call it, "A brand new Eurasia."

Over the next few chapters, I interrogate pivotal examples of popular culture that have contributed to the construction of post-Soviet space in the collective/popular "Western mind," and specifically the Anglophone world. Through close readings of several key texts and analysis of their reception and (re)presentation as works of politics, I explore the role of popular media as a critical nexus where power, ideology, and identity manifest. The fifth chapter, "The post-Soviet bogeyman: A guide to the dangerous personae of the former USSR," looks at Russian/post-Soviet villains in contemporary cinema and television, focusing on a series of post-1991 media representations that portrayed the Russian Federation and other post-Soviet republics as redoubts of anti-Westernism, peopled by "feral" anarchists," "ruthless" mafiosi, "corrupt" politicians, and "revanchist" military commanders. The post-1991 James Bond films, *Air Force One* (1997), the television drama *24* (2001–2010), and other popular media depictions of post-Soviet Russia and Russians (including revisionist and/or wholly imagined histories of the USSR) are explored. This chapter explains how Cold War fears associated with the Soviet Union (nuclear war, KGB agents, and gulags) were seamlessly transformed into new stereotypes with a global reach (radical ideologues with "loose nukes," mad scientists bent on bringing down Western capitalism, Central Asian warlords, and gangland criminals infiltrating the American "Heartland" and London's leafy suburbs). This chapter also provides a baseline exploration of the impact of popular geopolitics on the construction of everyday opinions about Russian/post-Soviet space and its inhabitants after the collapse of the USSR, as well as exploring the highly gendered notions of threats that emanate from the post-Soviet realm.

In Chapter 6, "Laughable nations: Parodying the post-Soviet republics," I assess Western media's tendency to treat the post-Soviet republics as an undifferentiated and ridiculous morass, which can be used as a handy geopolitical setting for corruption, poverty, backwardness, intolerance, and—most importantly—self-righteous laughter. Through a close reading of three lampooning "geographies" of the post-socialist "East"—entries on post-Soviet nations from *The Onion*'s recent atlas of world geography, JetLag's fictive travel guide of Molvanîa, and Sacha Baron Cohen's farcical primer on Kazakhstan (as well as several shorter supplemental case studies)—this chapter interrogates the lingering Cold War-era prejudices of the First World towards the nations of the former Second World, as well as the generation of new ones based on the transitological paradigm.[5] The focus is on the political dynamics of geopolitically informed cultural production in the highly mediatized environment of contemporary international relations, how parodic "travel writing" impacts geographical imagination in the Anglophone West, and how satirical mapping influences geopolitical "realities" in the neoliberal world system.

The seventh chapter of the book, "Mapping Trashcanistan: The post-Soviet badlands in popular culture, news media, and academe," utilizes the concepts of geopolitical aesthetics (Bleiker 2001, 2012, 2015) combined with the notion of "tabloid geopolitics" (Debrix 2008), as a mechanism for interrogating the ways in which systemic prejudices towards the former Soviet Union are visually—or more accurately, ideographically—maintained through popular culture, travelogues, nightly news reports, political cartoons, talk shows, and tabloid periodicals. Returning to several previously explored artefacts, as well as extending the analysis of common themes in Chapter 4, this section interrogates how Russia's physical geography has come to serve as an adjunct "enemy agent" of Westerners and their interests. Coverage of international sporting events like the Olympic Games—which tend to revel in "ruin porn"—and quasi-academic analyses of the region by public intellectuals such as Zbigniew Brzezinski, Stephen Kotkin, and Robert D. Kaplan will also be considered, as will travelogues, recent works of popular literature, and graphic novels set in the former USSR. Themes include "geo-graphic" framing of the region as irredeemably corrupt and in a perpetual state of decay. By focusing on how various public understandings of international relations are formulated, this chapter provides a real-world/highbrow complement to the fictional/lowbrow content analyzed in the previous two chapters. My focus is on how textual information shapes geographical imaginaries of post-Soviet space, and how such representation often perpetuates stereotypes that are increasingly untethered to lived realities in the region.

In the last full chapter of the text, "Branded! Marketing Eurasia's new nations to the (Western) world," the nation branding efforts of the Newly Independent States are explored, with a focus on how each of these countries has either failed or succeeded in their efforts to become brand states. In the first section, the focus is on the stewardship of "Brand Russia" under Vladimir Putin, a leader who has attempted to rehabilitate Russia's national image after the Yeltsin era. This section examines Russia's use of Western PR firms, Russia Today (RT), Russia's cultural patrimony (particularly through the *Russkiy Mir* program), and world-class sporting events to promote the country as a bastion of civilization and culture, while also exploring Moscow's mixed results in its re-branding of the country's land mass, economy, corporations, and even military actions (particularly the 2008 Russo-Georgian War and recent incursions into the Arctic Basin and Ukraine) as a vehicle for rehabilitating the country in geopolitical terms. The second part of the chapter explores the "Baltic Wunderkinds" of Estonia, Latvia, and Lithuania, analyzing how these small countries mixed tourism investment, economic reform, and high-priced image makeovers (including quirky logos, Facebook pages, and other tools of the trade) to craft recognizable and marketable images in Europe and the West, ultimately winning admission to NATO and the European Union. In the final part of the chapter, attention turns to the lesser-known (and thus effectively "un-branded") states of the region. This section investigates the often haphazard and occasionally counterproductive efforts of Kazakhstan, Ukraine, Georgia, and Belarus (among others) to present positive images in the global supermarket of nation brands. From well-funded Eurovision entries to educational campaigns on YouTube, my focus is on the difficulties presented to new states trying to market themselves while attempting to craft "new histories" (Watson 1999) that are distinct from USSR/Russia, while holding their tenuous polities together through other national identity projects.

The concluding section, "Post-Soviet Eurasia: The once-and-future geopolitical imaginary," seeks to reconcile the competing forces of popular geopolitics and nation branding, relating both to older forms of national image construction, maintenance, and change. In the end, the aim of this undertaking is to provide value to scholars of geopolitics, nationalism, mass media, and globalization, particularly those interested in transitional and post-totalitarian societies.

Popular Geopolitics and Nation Branding in the Post-Soviet Realm is grounded in literature and analytic methodologies drawn from the fields of critical geopolitics and international relations theory, as well as cultural studies. I employ a constructivist approach to international relations and the concept of nationhood and attempt to add to the developing field of postmodern diplomacy, with a particular focus on how relations between states are adapting to new information and communication paradigms that allow for non-governmental organizations, individual actors, and transnational communities to make their voices heard. I question the proposition that national brands can be tightly controlled and managed by national governments, arguing instead that narratives of nationhood—whether they are thought of as "brands" or not—are continually negotiated and perpetually "staged." As Vladislav Surkov, former deputy prime minister of the Russian Federation, bluntly frames the issue: "If a nation does not create its own images and ideas, and does not communicate them to other nations, it does not exist in a political and cultural sense" (qtd. in Simons 2010, 6). Keeping such ideals in mind, I contend that in the era of globalized media flows, these negotiations have come to include new, extra-national actors and narrators, whose ultimate impact on national identity has yet to be realized. Reflecting the importance of Peter Brandt's trenchant assertion that "the image which one nation has of another, or which is being presented to one nation of another, plays a role in relations between peoples which can hardly be exaggerated" (2003, 39), my ultimate aim is shed light on how the radically changed state of affairs in today's interlinked world of digital information, images, and videos is reflected in the realm of national image. This undertaking is particularly important given 2012 Republican presidential nominee Mitt Romney's assertion that Russia is America's "number one geopolitical foe" (qtd. in Willis 2012) and the growing conceptualization of a brewing "New Cold War" (see, for instance, Lucas 2008; Ciută and Klinke 2010) between the US and the Russian Federation, as each and every one of the countries covered in the volume plays a pivotal role as an interlocutor, ally, and/ or adversary in this prophesied—though unrealized—world conflict.

Notes

1 Perhaps the only comparison is the breakup of the Ottoman Empire, which was composed of territories on three continents; however, the Ottoman imperial disintegration was a multi-century process, whereas the total dissolution of the USSR occurred in less than two years.

2 It is important to keep in mind that the state and nation remain separate entities in the current global era. In the ensuing analysis, I use "state branding" and "nation branding" interchangeably; however, I do not mean to imply that nations without states are incapable of branding, nor do I mean to suggest that states maintain an unquestioned monopoly on representing the nation from which they purportedly take their authority.

3 I eschew linking the field of class to definitions of high and popular culture, although I do recognize intersections between class and cultural consumption and that this has long been a staple in attempts to conceptually differentiate the two.

4 It should be noted that the three Baltic Republics were not part of this political continuum from the end of World War I until 1940, when they were annexed by the Soviet Union; likewise, the greater part of Moldova, which was only raised to the status of a republic during World War II, belonged to Romania during the interwar period.

5 Since the end of the one-party system and the shift towards market-based economies in the former Eastern Bloc, Western scholarship has tended to focus on the raft of

socio-political problems faced by polities in the region rather than accentuating the economic successes of these states. Due to its often triumphalist vantage point and (pre)
deterministic lenses, critics within and outside of the region have labelled transitology's
mode of analysis as "flawed and hegemonic" (Sajjad 2013, 6).

References

Aitken, Stuart, and Leo E. Zonn. 1994. "Re-Presenting the Place Pastiche." In *Place,
Power, Situation, and Spectacle: A Geography of Film*, edited by Stuart Aitken and
Leo E. Zonn, 3–25. Lanham, MD: Rowman & Littlefield.
Allen, Nick. 2005. "Soviet Break-up was Geopolitical Disaster, Says Putin." *Telegraph*,
26 April.
Anholt, Simon. 2002. "Forward." *Brand Management* no. 9 (4–5):229–239.
Auchter, Jessica. 2014. *The Politics of Haunting and Memory in International Relations*.
London: Routledge.
Bleiker, Roland. 2001. "The Aesthetic Turn in International Political Theory." *Millennium:
Journal of International Studies* no. 30 (3):509–533.
——. 2012. *Aesthetics and World Politics*. New York: Palgrave Macmillan.
——. 2015. "Visual Assemblages: From Causality to Conditions of Possibility." In
Reassembling International Theory: Assemblage Thinking and International Relations,
edited by Michele Acuto and Simon Curtis, 75–81. Houndsmills, UK: Palgrave Macmillan.
Brandt, Peter. 2003. "German Perceptions of Russia and the Russians in Modern History."
Debatte no. 11 (1):39–59.
Butler, Andrew M. 2010. "Jean Baudrillard (1929–2007)." In *Fifty Key Figures in Science
Fiction*, edited by Mark Bould, Andrew M. Butler, Adam Roberts, and Sherryl Vint,
22–27. London: Routledge.
Ciută, Felix, and Ian Klinke. 2010. "Lost in Conceptualization: Reading the 'New Cold
War' with Critical Geopolitics." *Political Geography* no. 29:323–332.
Debrix, François. 2008. *Tabloid Terror: War, Culture, and Geopolitics*. London and New
York: Routledge.
Dittmer, Jason. 2010. *Popular Culture, Geopolitics, and Identity*. Lanham, MD: Rowman &
Littlefield.
——. 2014. "Geopolitical Assemblages and Complexity." *Progress in Human Geography*
no. 38 (3):385–401.
Dittmer, Jason, and Klaus Dodds. 2008. "Popular Geopolitics Past and Future: Fandom,
Identities and Audiences." *Geopolitics* no. 13 (3):437–445.
——. 2013. "The Geopolitical Audience: Watching *Quantum of Solace* (2008) in London."
Popular Communication no. 11:76–91.
Dzenovska, Dace. 2005. "Remaking the Nation of Latvia: Anthropological Perspectives on
Nation Branding." *Place Branding* no. 1 (2):173–186.
Edensor, Tim. 2002. *National Identity, Popular Culture and Everyday Life*. Oxford: Berg.
Gagnon, Serge. 2007. "L'intervention de l'État québécois dans le tourisme entre 1920 et
1940." *Hérodote* no. 4 (127):151–166.
Graham, Elaine L. 2002. *Representations of the Post/Human: Monsters, Aliens and Others
in Popular Culture*. New Brunswick: Rutgers University Press.
Grayson, Kyle, Matt Davies, and Simon Philpott. 2009. "Pop Goes IR? Researching the
Popular Culture–World Politics Continuum." *Politics* no. 29 (3):155–163.
Hobsbawm, Eric J. 1992. "Introduction: Inventing Traditions." In *The Invention of
Tradition*, edited by Eric J. Hobsbawm and Terence Ranger, 1–13. Cambridge, UK:
Cambridge University Press.

Kaneva, Nadia. 2012. "Nation Branding in Post-Communist Europe: Identities, Market, and Democracy." In *Branding Post-Communist Nations: Marketizing National Identities in the "New" Europe*, edited by Nadia Kaneva, 3–22. New York and London: Routledge.

Kovačević, Nataša. 2008. *Narrating Post/Communism: Colonial Discourse and Europe's Borderline Civilization*. London and New York: Routledge.

Kunczik, Michael. 1997. *Images of Nations in International Public Relations*. Mahwah, NJ: Lawrence Erlbaum Associates.

Lebedenko, Vladimir. 2004. "Russia's National Identity and Image-Building." *International Affairs: A Russian Journal of World Politics, Diplomacy & International Relations* no. 50 (4):71–77.

Lewis, Martin W., and Kären E. Wigen. 1997. *The Myth of Continents: A Critique of Metageography*. Berkeley: University of California Press.

Lucas, Edward. 2008. *The New Cold War: Putin's Russia and the Threat to the West*. New York: Palgrave Macmillan.

Makarychev, Andrey. 2014. *Russia and the EU in a Multipolar World: Discourses, Identities, Norms*. Frankfurt: Verlag.

Marat, Erica. 2009. "Nation Branding in Central Asia: A New Campaign to Present Ideas about the State and the Nation." *Europe–Asia Studies* no. 61 (7):1123–1136.

Martin, Terry. 2001. *The Affirmative Action Empire: Nations and Nationalism in the Soviet Union,1923–1939*. Ithaca, NY, and London: Cornell University Press.

Mirzoeff, Nicholas. 2015. *How to See the World*. London: Pelican.

Said, Edward. 1979. *Orientalism*. New York: Vintage Books.

Sajjad, Tazreena. 2013. *Transitional Justice in South Asia: A Study of Afghanistan and Nepal*. London and New York: Routledge.

Särmä, Saara. 2015. "Collage: An Art-inspired Methodology for Studying Laughter in World Politics." In *Popular Culture and World Politics: Theories, Methods, Pedagogies*, edited by Federica Caso and Caitlin Hamilton, 110–119. Bristol, UK: E-International Relations.

Saunders, Robert A. 2006. "Denationalized Digerati in the Virtual Near Abroad: The Paradoxical Impact of the Internet on National Identity among Minority Russians." *Global Media and Communication* no. 2 (1):43–69.

———. 2010. *Ethnopolitics in Cyberspace: The Internet, Minority Nationalism, and the Web of Identity*. Lanham, MD: Lexington Books.

———. 2012. "Brand Interrupted: The Impact of Alternative Narrators on Nation Branding in the Former Second World." In *Branding Post-Communist Nations: Marketizing National Identities in the "New" Europe*, edited by Nadia Kaneva, 49–78. New York and London: Routledge.

———. 2014. "Mediating New Europe–Asia: Branding the Post-Socialist World via the Internet." In *New Media in New Europe–Asia*, edited by Jeremy Morris, Natalya Rulyova, and Vlad Strukov, 143–166. London: Routledge.

Simons, Greg. 2010. *Mass Media and Modern Warfare: Reporting the Russian War on Terrorism*. Farnham, UK: Ashgate.

Stocchetti, Matteo. 2011. "Images: Who Gets What, When, and How?" In *Images in Use: Towards the Critical Analysis of Visual Communication*, edited by Matteo Stocchetti and Karin Kukkonen, 11–38. Amsterdam and Philadelphia, PA: John Benjamin Publishing Co.

Stocchetti, Matteo, and Karin Kukkonen. 2011. *Images in Use: Towards the Critical Analysis of Visual Communication*. Amsterdam and Philadelphia, PA: John Benjamin Publishing Co.

van Ham, Peter. 2001. "The Rise of the Brand State: The Postmodern Politics of Image and Reputation." *Foreign Affairs* no. 80 (5):2–7.

Watson, Rubie S. 1999. *Memory, History and Opposition: Under State Socialism.* Martlesham , UK: Boydell & Brewer.

Willis, Amy. 2012. "Mitt Romney: Russia is America's 'Number One Geopolitical Foe'." *Telegraph*, 27 March.

Wolff, Larry. 1994. *Inventing Eastern Europe: The Map of Civilization on the Mind of the Enlightenment.* Stanford, CA: Stanford University Press.

1 Of idols and idylls

The question of national image

What is national image? Like many seemingly straightforward questions associated with concepts in the social sciences, the answer becomes increasingly difficult as one tries to articulate it, particularly given that defining the term necessarily leads one into the treacherous conceptual thickets of *national identity* and *nationalism*. Is national image the mental picture conjured up when a country or its people are referenced? In part, the answer is yes. Is it the reputation that a country trades on in the international marketplace? Again, the answer is yes. Is it the raw material from which jokes and satire are formed? But of course—it is all these things and more. Simply put, national image is a fluid, socially constructed view of a nation, and it exists on both the domestic and foreign levels. When one thinks of Switzerland, one undoubtedly conjures up visions of snow-covered Alps, Emmental cheese, discreet bankers, the distinctive white cross on a red field, and Heidi (among other things). These fairly arbitrary examples represent physical geography, local products, emblematic occupations, symbols, and literary/artistic referents. A similar exercise could be conducted for many countries in the world: the Great Plains, hot dogs, the cowboy, the stars-and-stripes, and Superman (United States); the Cotswold Hills, beans on toast, the London bobby, the Union Jack, and Sherlock Holmes (United Kingdom); the Loire Valley, the baguette, the winemaker, the *tricoleur*, and Marianne (France). However, despite the ease with which we can explain national image in a commonsensical fashion, there is a surprising lacuna in analysis of the concept.[1]

Towards a holistic understanding of national image

National image is a human construct that exists only when a significant portion of humanity can agree on a basic set of markers that distinguish one nation from another (and—ideally—all others).[2] Being a "multidimensional construct," its sources include both "discursive and non-discursive elements" (Wang 2008, 9). Like other images, national image can exist in a variety of forms: graphic (pictures, symbols, designs, statues); optical (mirrors, distortions, reflections, projections); perceptual (sense, data, appearances); mental (dreams, fantasies, ideas, memories); and verbal/textual (descriptions, idylls, epics, metaphors) (see Mitchell 1986; Beller 2007a). To be more precise, national image functions as a synecdoche for vast, complex, and ultimately unknowable (yet still imaginable) congeries of places, things, peoples, experiences, and ideas. Moreover, we should not forget that such visions of the nation carry with them requisite affective components, i.e., intangibles that provoke a set of feelings that can come to function as a form of intelligence for navigating the world (see Thien 2005; Thrift 2008; Pile 2010).

Geography is a central component of any national image, as a nation without a (current or historical) "place" is an entity which has yet to come into being (though certainly, we can conceive of such a phenomenon in a post-national future). Most states have statuary or architectural icons which are immediately recognizable: Russia's onion-domed St. Basil's Cathedral, Greece's Parthenon, India's Taj Mahal, and Brazil's Christ the Redeemer. Each of these iconic man-made monuments instantly conveys information. Other countries lay claim to landscapes that possess such powerful evocations that they literally "own" certain types of geographies: fjords (Norway); precipitous karst islands (Vietnam); polders (Netherlands); or the 45-year-old burning gas crater labeled the "Door to Hell" (Turkmenistan; see Figure 1.1). However, uniqueness is not necessarily required to evoke national image, particularly in the instances of agricultural products and foodstuffs. In certain cases, a country's image is so deeply intertwined with a given thing that it functions as a mnemonic device for calling up larger associations: sushi (Japan), pad Thai (Thailand), Guinness stout (Ireland), or kielbasa (Poland). Symbols are perhaps the most straightforward communicator of national image; the maple leaf (Canada), the black, double-headed eagled (Albania), the Star of David (Israel), and the cross of the Knights Hospitaller (Malta) are just a few examples of how a signifier can come to represent the nation-state in a "condensed form" (Jourdan 2007, 436).[3]

Traditional occupations and lifestyles can also evince the nation, e.g., the Argentine gaucho, the Burmese monk, the Italian gondolier, and the Iranian bazaari. Likewise, famous individuals—living or dead—can elevate a country's prestige and serve as powerful agents of national image: Mustafa Kemal Atatürk (Turkey), Toussaint L'Ouverture (Haiti), Nelson Mandela (South Africa), or Fidel

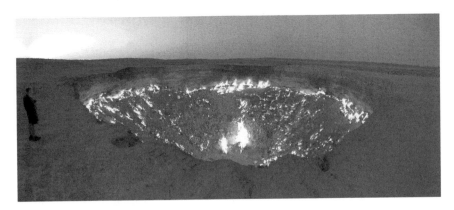

Figure 1.1 Door to Hell, Derweze, Turkmenistan (Tormod Sandtrov/Creative Commons)

Castro (Cuba). Imaginary entities or quasi-historical personages—forged in myth, literature, or popular culture—enjoy similar power, from Don Quixote (Spain) to Hayk (Armenia) to Count Dracula (Romania) to the Queen of Sheba (Ethiopia). Systems and structures of government, as well as political culture, are also key elements of any country's national image, both for its residents and for foreigners. While this is not the forum for interrogating each and every nation's quiddity, social scientists have not been shy about articulating the unique set of characteristics, otherwise known as stereotypes, that define a given nation's (mythical) character. The Swedes, for example, are rational, effective, predictable, harmonious, and independent (O'Dell 2011), the Uzbek national character is defined by duty, hospitality, courtesy, musicality, and a poetic nature (Allworth 1990), Mexicans are defined by machismo, a "different sense of time," and a socio-cultural uniqueness associated with *mestizaje* (Castañeda 2012), while the Latvians are nature-loving, self-reliant, risk-averse, reserved, and modest (Mežs 2010).

National images can be ancient, e.g., that of Egypt, or nascent, such as is the case with Namibia (1990), Kyrgyzstan (1991), Timor-Leste (2002), and Kosovo (2008). In the globalized world of fast-moving digital data, national images are increasingly influenced by more transient icons. New idols—be they celluloid, televisual, or YouTube-based—can easily trump venerated national symbols rooted in literature or history: Crocodile Dundee serves as the face of Australia, Borat speaks for Kazakhstan, Shakira personifies Colombia, and Mel Gibson's "Braveheart" (William Wallace) is the apotheosis of an independent Scotland. Corporate brands can now lend their power to national brands, feeding off and feeding into international recognition: LG is synonymous with South Korea; Nokia means Finland; BMW connotes Germany; and Al Jazeera cannot be disentangled from Qatar. Sports figures and teams also add to national image in the contemporary world, with Novak Djokovic influencing how Serbia is viewed abroad, just as Yao Ming redefines the role of China in the world, cricketer Imran Khan shapes Pakistan's international image, and Didier Drogba makes the Ivory Coast a known quotient around the globe.

Mega-events can also lend power to national image in the era of deterritorialized media, from royal weddings to the hosting of environmental or peace summits to major international sporting events (Giffard and Rivenburgh 2000). Taken together, these impressions often provide what one scholar calls a "quasi-statistical appraisal of the country's domestic and international performance" (Amienyi 2005, 73). However, it is important to note that all national images of foreign peoples and places are derived through a highly selective process based on value judgments which are informed by the state, travel writers, cultural producers, and other preceptors of "knowledge" (Beller 2007a). These and other contemporary inputs into national image will be explored in greater depth in the following chapter.

While there is little in the way of an established body of literature on the concept of "national image," much ink has been spilled on the origins, nature, and outcomes of nationalism as well as its compositional essence: national identity; however, less has been written on the idea of national image. Yet, it is impossible for one to exist without the others. Consequently, it is appropriate to provide an aperçu of these concepts before moving on to a more nuanced analysis of national image.

Related concepts: Nationalism, national identity, and national character

To fully understand national image, it is necessary to explore a variety of linked concepts, including nationalism, national identity, national character, and national stereotypes. Each of these fields of inquiry impacts national image, shaping how a given national polity is viewed by its own members and—more importantly from the perspective of this text—how other nations view that particular national grouping. As Nicholas J. O'Shaughnessy reminds us:

> A national image is not just for external consumption, as it can be used to infuse a nation with a sense of pride that helps unite it. The promotion of national image can help generate a sense of solidarity with others. In some cases, this leads to extreme nationalism and chauvinism. (2003, 196)[4]

According to another author, "In each country the creation of a national image was undertaken by those with power in reference to other national images, because only through this globalization did these images make sense. A relatively clear 'standard' of nationalism thus emerged" (Tenorio-Trillo 1996, 243). Taken together, these "pulls" on national image suggest a deep and abiding link to the concept of nationalism, which— unlike national image—is a perennial topic of study across multiple disciplines.

Nationalism is, at its root, an ideology that espouses a perilously simple maxim: each nation should maintain political, social, cultural, and economic control over the territory in which its peoples reside (see, for instance, Deutsch 1953; Gellner 1983; Hutchinson and Smith 1994; Guibernau 1996).[5] The ideology of nationalism is constructed and maintained through a complex system, or—using a more contemporary analogy—a web, of competing and complementary power dynamics. As ideologies with a life of their own, particular nationalisms are educed over time, the product of much work by political, economic, and cultural elites. Nationalism, once formed, is reinforced and sometimes transformed through three principal forums: politics, education, and popular/high culture (see Kohn 1944; Motyl 1992; Breuilly 1994; Kramer 2011). Nationalism is impossible without the existence of an "Other," and more specifically a host of various national Others, including internal Others of a regional, ethnic, linguistic, and/or religious persuasion (Neumann 1999; Kligman 2001; Wingfield 2003; Rash 2012; Alami 2013; Burton 2013). From a historical perspective, the modern form of nationalism spread across Europe and then outwards to all corners of the globe through political structures, which—ironically—undid rather than buttressed imperialism, the primary medium for the spread of nationalism. At the risk of oversimplifying the types of nationalism, it is possible to divide the various nationalisms into two varieties: ethno-nationalism and civic nationalism. The former is rooted in imagined ties of a common ethnogenesis, or blood ties (*jus sanguinis*). The latter is forged through allegiances and experiences related to a national community determined primarily by birth in a particular state (*jus soli*) and adherence to a creed associated with that state.[6] While there has been a surfeit of academic analysis on the emergence of postmodern nationalism in recent decades (see, for instance, Buell 1994; Ong 1999; Croucher 2004; Calhoun 2008), these two forms of nationalism remain an important binary in the field of study.

National identity links the internal and the external senses of self, producing a "fellow feeling" between members of the same nation, combined with an awareness of difference from other nations (see Smith 1991; Anderson 1991; Cameron 1999; Edensor 2002; Gottlieb 2007). In short, national identity is a "state of mind" (Beller 2007a, 12). Basic elements required for a distinct national identity are (at a minimum): association with a particular territory or homeland (or lack thereof); the political will to be a single people; a belief in a set of common values; and a mass culture (Smith 1991). Similar to nationalism, national identity is reliant on the triad of the military, public school system, and mass media for its fecundity; however, national identity can also flower outside and even in opposition to state support where other avenues of expression come to the fore (sport, mutual aid, religious ties, cultural production, etc.). National identity is produced through a variety of means: visual/performative (music, theatre, art, ceremony, athletic competition); narrative (oral traditions, epic songs, myths, discourses, histories); affect (emotional work, sense of belonging, memory); and association (pledges, affirmations, communal events, collective service). A myth of common descent ("ethnies") is a typical, though not requisite, element of many nationalities (Smith 1988). The borders of national identity are informed by a variety of markers, often jealously policed by society to maintain distinctions between "in" and "out" groups (see, for instance, Joireman 2003); these include language/dialect, religion/sectarian divisions, dress/appearance/mannerisms, cuisine, sport, celebrations/holidays/veneration of heroes, and other forms of customs and habitus. Economic activity and everyday geographies further shape and inform national identity.

Both of these concepts rely on a clear understanding of what constitutes the "nation." As I have written elsewhere, the nation is:

> [a] named group of people who imagine themselves to be part of a community based on common history, language, and a set of traditions/culture, and who strive towards common legal rights, political responsibilities and rewards and economic interaction for all members of the group. (Saunders 2010, 10)[7]

Nationalism without national identity is thus impossible. Likewise, national identity must be articulated and negotiated within an international discursive space that recognizes the existence of multiple nations, each with its own image (Rusciano and Ebo 1998). This discursive space, once dominated by diplomats and political elites, is increasingly influenced by popular culture. As Michael Kunczik, a historian of public relations, states, "The mass media are, in fact, continually offering images of nations" (1997, 5). Elucidating this assertion is one the principal foci of this text.

The politically laden and extremely problematic notion of "national character," a term made popular by the esteemed IR theoretician Hans J. Morgenthau, is an additional aspect of the matrix that informs and influences ideas about national image. According to Morgenthau, certain qualities set one nation apart from others and these tend to be fixed or at least "highly resilient to change" over time (Morgenthau 1985 [1948], 124). Morgenthau links character to prestige and reputation, as if such sentiments only existed in an individual mind; however, it is important to remember that prestige is "an effect produced upon the international imagination" (Wight qtd. in Koski 2011, 93). Crises tend to accentuate elements of this national character, which in times of peace and prosperity may remain submerged or at least subdued. The concept of particular national characters is rather controversial in the social sciences (Inkeles 1997), owing to the fact that such notions are often applied from the outside, and attempt to "yoke national differences based on a wide variety of experience to a few key psychological traits to which those national characteristics may have no connection" (Mandler 2006, 2). As will be discussed shortly, the study of national character emerged as a national security prerogative during World War II and the Cold War, with specific focus on the so-called "Russian national character," which Morgenthau and other IR theorists often interrogated, hoping to predict Soviet behavior on the international stage through a curious marriage of anthropology and psychology (see Smith 2008). Most emblematic of this trend were undoubtedly John Fischer's *Why They Behave Like Russians* (1947) and George F. Kennan's "The Sources of Soviet Conduct" (1947) in *Foreign Affairs*;[8] however, there are countless treatises of the same ilk up to the present day. In his seminal work *Politics among Nations*, Morgenthau associates "obedience to the authority of the government" and a "traditional fear of the foreigner" as basic traits of the Russian national character, which, when combined with the "elementary force and persistence" of the Russian people, privileges this nation in terms in the long-term "struggle for power" (Morgenthau 1985 [1948], 127–128). According to one author, "National character . . . is a much simpler, more malleable, and superficial notion than national identity, but it is also one that can be more easily described, surveyed, probed, and quantified"(Castañeda 2012, xxi). What then distinguishes national character from national image? The answer is rather straightforward: one's ideological vantage point.

National image in history, or the politics of stereotypes

As stated in the Introduction, national image exists in two forms: self-image (*Selbstbild*), or how a given polity views itself and its country versus others (overlapping the aforementioned concept of national identity); and how a nation is perceived and viewed by others (*Fremdbild*) (see Rusciano and Ebo 1998). The latter can be made even more complex, by breaking down external representation into more refined subsets, namely: (1) external self-representation (Country A's presentation of itself to Country B); (2) mirrored or "refracted" self-representation (how Country A views its representation in Country B); (3) hetero-representation according to the presenter (how Country A sees itself represented to Country B);

and (4) hetero-representation according to the consumer (how Country B views its representation of itself in Country A) (see Metzeltin 1998; Westphal 2011).[9] *Fremdbild*—in its many varieties—is the focus of this study, though it is important to understand that internal and external national images are inextricably linked and constantly feed off one another, a trend that has only quickened and deepened in the era of cyberspatial nationalism (see, for instance, Bieber 2000; Bernal 2006; Eriksen 2007; Brinkerhoff 2009; Saunders 2010; Gaufman and Wałasek 2014).

How the French see themselves is an important part of how they are perceived by the British, while the reverse is equally true. As one scholar notes, "National stereotypes . . . are not as a rule unambiguous in the judgement but constructed ambivalently, even if self-images are usually more positive images and images of others more negative in hue" (Brandt 2003, 40). While not often explored in the social science literature, national stereotypes are nonetheless important to the formulation of national image. In the words of one scholar, "National stereotypes . . . continue to be highly recognizable, and many people, while conceding that these stereotype are generalizations, stubbornly contend that there might be a core of truth substantiating the basic allegation" (Chew 2006, 182). Janus-faced in its essence, the notion of national image functions as a mechanism for reinforcing hidebound "truths" about a given people while simultaneously providing fodder for ethnic humor. The oft-told "Heaven and Hell" joke is a quotidian example of how well-established key "national character" traits are understood:

> Heaven is where the police are British, the lovers French, the mechanics German, the chefs Italian, and it is all organized by the Swiss.
> Hell is where the police are German, the lovers Swiss, the mechanics French, the chefs British, and it is all organized by the Italians.

At the risk of being pedantic, an unpacking of this joke relays information about the British (forthright and upstanding, but a disaster in the kitchen), French (amorous to a fault, yet lacking in technical acumen), Germans (precise and skillful though prone to ruthlessness when in power), Italians (adept at the culinary arts, whereas abysmal at management), and the Swiss (master bureaucrats, yet cold and unapproachable). Jocularly reflecting perceived strengths and weakness of various nations in turn, such jests are assumed to be politically harmless (particularly in the context of postmodern Europe where the supranational integration project has made war nigh unimaginable). However, at the same time, there are significant data to be mined here as salient forms of what Iver Neumann calls "geographical folk models of understanding" (Neumann 1999, 207); in essence, this refers to the place-based corroboration of stereotypes based on the creation of knowledges formed by pre-judging "other" peoples whom one encounters. As Michael Kunczik points out, "[The] images of nations at least partly can be understood as hardened prejudices. These are not suddenly there, but often have grown in long historical processes" (Kunczik 1997, 39). This process thus produces the various national stereotypes which feed national image; however, as the subsequent chapter on nation branding will argue, this is a two-way street where discourses can be altered to abolish (negative) stereotypes, often in favor of the creation of new (positive) ones.

The term "stereotype" owes its creation to American writer and commentator Walter Lippmann (1922), who made the notion of "pictures in our heads" part of the everyday vernacular.[10] According to Lippmann, a stereotype "precedes the use of reason; [it] is a form of perception, imposes a certain character on the data of our senses before the data reach the intelligence" (1922, 96); in other words, we *define* before we *see*. Put in more straightforward terms, stereotypes demonstrate the ever-present disconnect between *construed* and *actual* images associated with the nation (see Simons 2013). While most social scientists emphasize the inaccuracy of stereotypes—be they national, ethnic, religious, socio-economic, gendered, or age-based—in fact, "basic cognitive processes have been identi- fied that lead people to exaggerate real differences between groups, ignore or misremember stereotype-inconsistent information, and develop false beliefs to justify injustice" (McCrae 2013, 832). These distorted "pictures" can also exist as auto-stereotypes wherein a given polity views itself in a particular way (often despite "reality"). One particularly succinct definition of national stereotypes is as follows:

> Stereotype refers to oversimplified representations of reality. Historical ste- reotypes are created in public discourse in order to strengthen the identity of a group, either the group of our own or the group of the other. Therefore there are autostereotypes and heterostereotypes. The most powerful heterostereo- types are enemy images. They are used to legitimate and provoke hostilities among groups. (Ahonen 2001, 25)

While malleable, national stereotypes do not tend to be frangible in nature, persisting even when the majority of data demands a revaluation. Stereotypes are fictions, but ones which create their own "problematic realities" (Beller 2007b, 430). Such mental pictures, whether affective or cognitive, flow across a wide variety of informational platforms. As a result, "National stereotyping is never a monomedial process because the creation and perpetuation of culturally nor- mative knowledge builds on the perpetual reaffirmation of cultural notions of self and other" (Neumann 2009, 276). In somewhat simpler language, national stereotypes do not exist in a vacuum. They are sustained through jokes, songs, novels, television, films, news reports, and other forms of textual and visual media, as well as a rich array of less concrete artefacts that engender nationalist affect (Mankekar 2015), ultimately delineating and reinforcing the diacritica, or, borrowing from Freud (1915 [1991]), the narcissism of small differences, that separate one nation from another.

In the field of literature, a number of scholars have investigated how notions of national character (and accompanying stereotypes) manifested in key works of "national" literature, therein framing the nation, as well as its internal and exter- nal Others, with a particular focus on the "tropes, patterns, and structural features that can be used in literature and other media to construct, naturalize, and render popular persuasive stereotypes" (Neumann 2009, 277). Such nuanced and academically acute analysis of national image has helped to create a methodology for examining the political and cultural meaning behind national image, particularly *Fremdbilder*. Birgit Neumann argues that the active promotion of national image—linked to domestic politics and international relations—first rose to prominence in the eighteenth century. For Great Britain, the Netherlands, France, and other nascent imperial powers, "representations of national others helped interpret the striking experience of cultural difference and satisfied the need for national self-assurance" (Neumann 2009, 276). Manfred Beller similarly argues that "increasing national awareness and its political instrumentalization . . . provided the themes and the characteristic typecasts for literary image formation" (Beller 2007a, 11). This field of study, which shares a number of topical (if not necessarily methodological) commonalities with popular geopolitics, is known as imagology, or image studies. Imagology primarily focuses on discourses surrounding ethnotypes or "stereotypical characterizations attributed ethnicities or nationalities, national images and commonplaces" in an effort to understand how cultural differences are processed and retooled (Beller and Leerssen 2007, xiv).

Popular in France and Germany but gaining ground elsewhere,[11] image studies provides an interdisciplinary approach to interrogating, both historically and diachronically, how important works of literature (and, by extension, other forms of popular culture) project and then reinforce certain characteristics associated with the national self and the Other (Leerssen 2000). Many national images are hundreds of years old, initially forged by sixteenth- and seventeenth-century travelers to the various regions of Europe (and farther afield), later to be delivered *en masse* via periodicals and dime novels during the 1800s as mass literacy became a reality. As imagology constantly reinforces, selective perception inevitably leads to selective observation, thus guaranteeing minimal changes over time (Beller 2007a). Going beyond simple exoticism, image studies delves into the circulation of national stereotypes as a pragmatic, audience-centric tool for achieving national cohesion, or, in the words of Iver Neumann (1999), these images of the Other are of "use." As in the field of popular geopolitics, the scholarly analysis of texts need not necessarily be concerned with the degree of truth of the representation; in the words of Leerssen (2014), "even though the belief is irrational, the impact of that belief is anything but unreal." In fact, such representations often show more about the relationship between elites and masses in the domestic national context, and provide little in the way of elucidating the "reality" of the Self–Other relationship of the time.

At the end of the nineteenth century, the emergent discipline of anthropology did much to promote the notion of national traits, behaviors, characters, and—as a byproduct—images. Coinciding with the heyday of high imperialism, such "scientific" treatment of the "races" (what we would today call ethnicities or nations) produced a variety of questionable offspring, including cultural determinism, Social Darwinism, and—most darkly—eugenics. French diplomat Arthur de Gobineau's *The Inequality of the Human Races* (1915 [1853–1855]) represents a key text in the evolution of scientific racism, and would serve a wellspring for

Nazi racial theories seventy-five years later. However, in the intervening period the Great War would tear Europe apart, resulting in an explosion of rabid nationalisms wherein large and small peoples would stand beside one another or fight to the death in the name of the nation, or what Benedict Anderson (1991) described as the "imagined community."

Popular culture was quickly yoked to this new and all-consuming conflict, with representations of the national "self" being cast in sharp contrast with the barbaric Other. While countless examples of such manipulation of national image exist, few surpass the raw power of the US Army's enlistment poster depicting a German imperial soldier as a rampaging ape carrying the half-naked body of a (white) woman (see Figure 1.2). Despite the fact that German-Americans made up one of the largest ethnic groups in the United States and that German was the second most-spoken language in the country at the time, the national image of

Figure 1.2 Destroy this mad brute (Library of Congress)

Germany was quickly and easily transformed into a bestial figure.[12] While the type of scientific racism and gross jingoism that characterized the early twentieth century faded into the background after the destruction of National Socialism in Germany, global national image production and consumption remain closely tied to ideas of ethnic origin. Moreover, it is important to note that the meaning of "race" is in a constant state of flux in the contemporary world which is defined by mobility, hybridity, contestation, and multidimensionality, thus complicating pre-existing notions of "us" and "them," while simultaneously creating new ones (see Kibria 2011). Interestingly, Joep Leerssen (2007) points out that Nazism, in its death, gave birth to the structured study of national images by stripping away the notion of any "real" national characters and debunking the ethnic myths, allowing for scholars to begin to see how "imagined truths" are constructed over time. It is notable that such a transition has generally failed to take place with regards to how the "West" *sees* the post-Soviet world, hence the rationale for this very book.

National image re-emerged as a field of academic analysis in the early years of the Cold War, a period when national governments became increasingly aware of the "significance of systematic image construction" (Li and Chitty 2009, 1) and the importance of "world opinion" in maintaining international order (Rusciano 1998). As the United States and the UK entered into a belligerent relationship with their erstwhile-ally the Soviet Union, IR specialists and geopoliticians began to apply a variety of methodological and analytical tools and techniques drawn from the behavioral sciences (particularly cognitive psychology) to understand and ultimately "predict" the behavior of states in the international arena. Kenneth Boulding, writing in 1959, assessed the notion of national image as such:

> The national image, however, is the last great stronghold of unsophistication. Not even the professional international relations experts have come very far toward seeing the system as a whole, and the ordinary citizen and the powerful statesman alike have naive, self-centered, and unsophisticated images of the world in which their nation moves. Nations are divided into "good" and "bad"—the enemy is all bad, one's own nation is of spotless virtue. (Boulding 1959, 131)[13]

As a number of scholars have demonstrated, out-group stereotypes, heterostereotypes, or xenostereotypes created dangerous conditions where enemy behaviors were often misinterpreted, producing flawed information-processing systems among political elites (McDermott 2004).[14] From a historical standpoint, this had much to do with the undeniable power and vitality of the "enemy image" or *Feindbild* (see Gienow-Hecht 1997; Neumann 1999; Rash 2012), as well as intellectual or ideological blinders which caused actors to maintain flawed—even false—national images of other states (see Jervis 1970). As will be explored in the coming chapters, the enemy image is constantly being refined by popular-cultural referents (see Kunczik 1997), and has become "omnipresent in modern societies" (Fiebig-von Hase 1997, 2). Enemy images are just part of a larger constellation that includes the self-image, images of friends and allies (*Freundbilden*), and images of neutral and "non-relevant" states/peoples. In the words of Ole Holsti:

> The relationship of national images is clear: decision-makers act upon their definition of the situation and their images of states—others as well as their own. These images are in turn dependent upon the decision-maker's belief system, and these may or may not be accurate representations of "reality." Thus it has been suggested that international conflict frequently is not between states, but rather distorted images of states. (1962, 244)

Holsti's trenchant analysis of key decision-makers suggested that rigid patterns develop wherein contra-indicators and even new trends in behavior or policy are rejected as irrelevant or simply as exercises in propaganda. In such ideologically suffused milieus, positive signals and indices were often ignored due to overriding negative *Fremdbilder*. The massive amount of propaganda that characterized the Cold War (and the preceding era of fascist–Bolshevik–Western conflict) was directly to blame for this situation.

First used by Pope Gregory X, the term "propaganda" dates to the mid-seventeenth century. Over time, the concept evolved from a mechanism for the supervision of missionary activities of the Roman Catholic Church to a systematic effort undertaken to influence specific target audiences through discursive tactics, psychological persuasion, and the use of symbols, which, collectively, are meant to influence emotions, alter mindsets, and most importantly affect behavior. The emergence of the modern nation-state in the first half of the twentieth century and the rise of one-party totalitarian regimes in the interwar period saw a vast expansion of state propaganda, which benefited from the explosion of new broadcast media such as the motion picture, radio, and television. Propaganda differs from diplomacy in that it does not focus on government-to-government relations, but instead looks to influence the masses, either outside of the country (foreign propaganda) or at home (domestic propaganda). According to certain scholars, propaganda is most easily recognized not by its content but as a technique of delivering information (Jowett and O'Donnell 2006); however, this produces certain problems as "propaganda works best when audiences are not aware they are being exposed to it" (Duncan and Smith 2009, 249). Propaganda, unlike other forms of government-to-people foreign policy (so-called "public diplomacy"),[15] is inherently self-centered and fails to take into account the interests of other nations (Chatterjee 2010). In common usage, the term carries a negative connotation, as it suggests the manipulation of the audience rather than their edification, thus differentiating it from "education" (Marlin 2002). Importantly, the term is almost never applied to one's own information management, only that of the Other.

In authoritarian and totalitarian states, identifying propaganda directed at the masses in other countries tends to be fairly simple; however, for pluralistic, democratic nations, the situation is less clear-cut.[16] For example, while Germany proudly used the term propaganda in its efforts to sway populations abroad, the US government eschewed the appellation and often attempted to veil its own efforts at political persuasion under the mantle of "cultural exchange." Antonio Pedro Tota's study of American public diplomacy, propaganda, and pop-culture political manipulation in Brazil provides a glimpse of this complicated situation, with the US funding a variety of media to promote its image in Latin America, most notably the serial *On Guard*. "The magazine propagated an image of the United States as the fortress of continental democracy, a stronghold from which the countries of the American continent could request support whenever necessary" (Tota 2010, 33). Such "soft propaganda," which made use of what Rob Kroes calls America's "modular culture"—a "Lego block" civilization that uses "empty signifiers" to construct "ever-changing meaningful structures" (Kroes 1996, 104–105)—functioned more effectively than did those political machinations of its competitors (Nazi Germany, USSR, and the People's Republic of China), particularly in the realm of national image where states used various media to broadcast their own *Selbstbild* to foreign populations while also representing their *Fremdbilder* of foreign states to third parties.

During the twentieth century, the construction—or more accurately, projection—of national image via mass media, particularly radio and film (and later, television) became ascendant.[17] For the United States, Hollywood was absolutely integral to establishing the country's national image in the minds of Europeans after World War II, particularly in those countries in Western Europe where the allure of socialism was strong (Falk 2011). Likewise, American cinema served as a handmaiden for the national interest by sculpting important countries as dangerous enemies or reliable friends (see McAlister 2005). As a host of scholars have demonstrated (see, for instance, Kuisel 1993; Gemünden 1998; Elsaesser 2005; Sari Karademir 2012), the "seductive" nature of cinema formed a central plank in the Americanization of Europe during the 1950s (and even earlier),[18] as well as making inroads in other parts of the world during the later decades of the Cold War. The conduct of foreign policy via "cultivation and exportation a national image abroad" inevitably impacts a nation's *Selbstbild*, influencing national identity at home as well overseas (Arnold 2012, 106). This resulted in a steady "Americanization" of the United States and a refining and/or hardening of its notions of others' "Otherness" within the global realm.[19] While many countries engaged in globally distributed national image production via popular culture (most notably Great Britain, France, and the USSR), there is little argument that the US culture industry maintained a dominant, even preeminent, position during the Cold War.[20]

As American comic books flooded European tabacs, Hollywood films shone on silver screens across the Americas, and Memphis-twanged rock 'n' roll filled the airwaves in places as disparate as the Philippines and Finland, many countries found it necessary to erect barriers to the deluge of American cultural production. In the case of Soviet-allied states, outright prohibition, censorship, and jamming of broadcasts were the norm. Other countries used less overt mechanisms to prevent the proliferation of American cultural products (and resultant consumerism) within their borders, from normative regulations meant to achieve decolonization (India and Ghana) to ones advocating the protection of national cultures and unique forms of artistic expression (France and Canada). Despite formal and informal attempts at stanching the flow of American (and British) media products,

Anglophone popular culture steadily seeped into the public consciousness of much of the world's population from the 1950s onwards, from Beatles albums illicitly transferred onto "x-rays" in the Soviet Union to the emergence of an American-style "beauty culture" in Brazil. Importantly, this barrage of cultural output carried with it a vast array of highly coded and subtly framed messaging about the US, its allies (particularly Great Britain), its enemies (especially the USSR), and various "internal" Others (American blacks, Latinos, Jews, immigrants, etc.), as well as promoting particular ideologies (neoliberal capitalism and democratic pluralism) versus "alternatives" (state socialism and authoritarianism/totalitarianism).[21]

While the vast majority of cultural production during the Cold War focused on "domestic themes" (as it continues to do today), films, comics, popular novels, and other forms of popular media invariably entered territory which resulted in the use, production, and manipulation of national image. By setting a motion picture in an "exotic locale," American and British filmmakers inevitably found themselves in a position to advance the national image of such a place and its people to its audience, whether it was Italy in *Roman Holiday* (1953), France in *To Catch a Thief* (1954), Israel in *Exodus* (1960), the Arab world in *Lawrence of Arabia* (1962), West Africa in *Ashanti* (1979), or South Asia in *A Passage to India* (1984). While I will explore this phenomenon in greater depth in Chapter 3, it is vital to understand the visual power of film and other forms of popular culture here, particularly in the context of the Cold War. Popular culture in its myriad forms functioned as a tool for projecting both American and British self-images to others as well as informing various other nations about American and British hetero-images of a variety of nations. Such "imagined discourses" (Beller and Leerssen 2007) proved powerful influencers on geographical imagination during the post-World War II era. With the deepening influence of complex economic interdependence following the Bretton Woods conference (1944)[22] and the so-called "coming of globalization" (Langhorne 2001), national image became deadly serious business.

The end of the Cold War in 1989 only fueled this trend, with a renewed focus on questions of "identity" (Fiebig-von Hase 1997), leading to the recent claim that national image is "one of the most salient concepts in the era of globalization" (Kinsey and Chung 2013, 5). World opinion—now thoroughly delinked from the ideological shackles of the US–Soviet struggle—has become "a more free floating phenomenon, whose meaning for each nation must be negotiated" (Rusciano and Ebo 1998, 65). Through a variety of mechanisms under the rubrics of public diplomacy, soft power, and state branding, countries—both large and small—have begun to actively manage their national images abroad, a process often called nation branding. As will be explored in detail in the next chapter, few if any countries are now content to let their national image be determined without their input. Faced with the challenge of attracting tourists and direct investment, advancing the national interest through strategic alliances and partnerships, maintaining trade relationships and promoting exports, and promoting patriotism at home, countries large and small now engage in multivalent efforts to burnish their images through outreach efforts, including advertising, political communication, sport, news programming, popular culture, people-to-people networks, investment in foreign educational and relief organizations, and a host of other activities, leading to what Simon Anholt (2002) has labeled the "giant supermarket" of nation brands.

Russia's national image: A case study

Russia possesses one of the world's oldest and most storied brands. Ever since Rurik the Dane carved a Viking kingdom of the Slavic lands between the forests of central Europe and the vast Eurasian steppe, Russia has been in the process of forming a variegated but undeniably robust national image that resonates both across space and through time. From Rurik to Vladimir I, Kievan Rus's great Christian monarch, the state that would become Russia grew in strength, wealth, and influence. While a peripheral power by European standards of the time, Kievan Rus did link into the great world trade network, which pushed goods, people, and ideas from Byzantium to Scandinavia, China, and South Asia. Unfortunately, Russia's fate was intimately tied to that of Constantinople, and when the Eastern Roman Empire waned under the pressure of Arabo-Muslim expansion, Turkic raiders, and rapacious Crusaders, Russia's fortunes languished in tandem with those of the *Imperium Romanum*. Economic weakness combined with geographical exposure left the country exposed on both flanks; while the immediate threat came from ascendant Catholic powers to the west (orders of Teutonic Knights and the Polish Piast dynasty), an existential danger arose in the east. The descendants of the world-conqueror Genghis Khan stormed into Russia, establishing the Mongols as nominal rulers of all the Eastern Slavic lands, and so began hundreds of years under the so-called Tatar Yoke. Turco-Mongol suzerainty left an indelible imprint on the image of Russia (if not the Russian "soul," as is so often claimed). Some historians have argued that the Golden Horde effectively preserved the Russian proto-nation by protecting the Eastern Slavs from subjugation by the Germanic "hordes" (see, for instance, Bulag 2010), a scenario that would have likely resulted in their forced conversion to Catholicism and perhaps even submersion into a combined Teutonic-Slavic ethnos. However, under Batu Khan (1207–1255) and his successors, the Slavs were relatively free to practice their Orthodox Christian faith, and to govern their own affairs (at least at the local level), assuming that they paid their tribute and maintained loyalty to the khan. The long interlude between Kievan Rus and the establishment of the proto-Russian state of Muscovy removed the country from the European mainstream, thus prompting the long-running, even quintessential, question of Russian identity: East or West/both or neither? (see Shlapentokh 2007).

Russia's sloughing off of the Tatar Yoke served as the initial catalyst for the West's construction of a *Feindbild* of the first order, beginning in the Middle Ages and reaching its apex in the modern era (see Halperin 2009). Not surprisingly, this national image has always been colored by the content and form of Russian leadership. As indigenous rule returned to Russia,[23] state builders were confronted with the host of thorny issues which ultimately came to be solved by harsh top-down policymaking. The primitive structures of Slavic democracy were all but wiped away by Viking, Mongol, and finally Romanov rulers. Under the reign of

Ivan the Great (1462–1505) and his successors, the early modern Russian state evolved into an exemplar of Early Modern autocracy. As Western Europe dispensed with the institution of serfdom following the ravages of the Black Death, Russian tsars clung to Russia's own form of feudal slavery into the nineteenth century. Yet, Russia was not monolithic in its retrograde tendencies vis-à-vis the West; under the rule of Peter the Great (1682–1725), the country shifted its geopolitical vision and political center towards Europe, establishing strong trade links and embarking on naval expansion that would ineluctably tie Russia to the oceans and the emerging world system. It was during this era that Russia became a member of the evolving European community of nations, though one which was often geopolitically situated alongside the Ottoman Empire as a "not quite" European, or at best a "quasi-European" state.

With increasing diplomatic and economic links to the West, travelers from England, Sweden, France, and the Germanies made their way to Russia, only to find a land that seemed locked in the past. Daniel Treisman summarizes the historical treatment of Russia in the "West":

> From sixteenth-century European travelogues, one learns that Russian peasants at that time were drunks, idolaters, and sodomites. Seventeenth-century travelers report that the country's northern forests were a breeding ground for witches. Then come the famous denunciations of the Marquis de Custine, along with the jeremiads of Chaadaev—a homegrown convert to the idiom—who, just as Pushkin was publishing *Eugene Onegin*, chastised Russia for failing to contribute anything to human civilization. Russia, he charged, was a "blank page in the intellectual order," which existed only to "teach the world some great lesson." (2011, x)

By way of example, William Coxe, upon crossing from Poland into Russia, opined, "The peasants seemed greedy of money . . . also in general much inclined to thieving" (1784, 256). Whereas as Frederick Burnaby, in his famous travelogue of Central Asia, noted, "It will take the Russians a long time to shake off from themselves the habits and way of thought inherited from their barbarous ancestry" (1876, 82).[24] Such literature, travel writing, and other forms of popular culture are instrumental in transforming "vague intuitions" about a place, space, or people into "establishedness" (Prieto 2011, 14).

Jaundiced accounts of Russian peasants filled travelogues and solidly established a lasting archetype of the Russian as backward, suspicious, and cold. The territory of Russia, situated as it was between Central Europe and East Asia, embodied a zone of transition between the Occident and Orient, more often manifesting elements of the latter (at least in the fantasies of the popular press and the contemporary writers). Highly attuned to the tastes and mores of such cultural producers, Catherine the Great (1762–1796) and her retainers, particularly Prince Grigorii Potemkin, worked assiduously to reposition the image of Russia in the eyes of the West (often to the point of absurdity in the case of erecting attractive facades for the benefit of visiting dignitaries—the so-called Potemkin Village, an example of over-the-top nation branding if there ever was one). However, despite such efforts, many in Western foreign policy circles continued to see Russia as a hotbed of "mongolism" bent on world supremacy (Brandt 2003, 43), a sort of Oriental power that could masquerade as a Western state.

Following Napoleon's failed invasion in 1812, Russia emerged as the single most powerful country in European affairs, forcing an almost instant rapprochement between England and France to counter Petersburg's influence. The tsar subsequently emerged as the principal bulwark against republicanism in Europe, earning Russia a quasi-permanent image as a force against "freedom" and "political evolution" (not only at home but now also in other parts of the world). As Russian power encroached on Western European interests, Russia's "brand" as a highly conservative, anti-progressive monolith of the East grew in the minds of Westerners (the result of popular as well as political framing of the country). In the mid-century, events came to a head with a short disastrous war involving Russia on one side and a motley array of unlikely allies on the other: Ottoman Turkey, Great Britain, France, and Sardinia. Russia's defeat in the Crimean War (1853–1856) had long-lasting repercussions on how Russia was perceived within and beyond its borders. Feeling betrayed by its "co-religionists" in Western Europe, Russia turned its gaze to the east, focusing its attention on expanding the empire in the South Caucasus, Turkestan, and the Pacific Basin. The inclusion of new (non-Slavic) populations significantly diversified an already multilingual, multiconfessional array, providing more fodder for the burgeoning notion of Russian exceptionalism as a uniquely "Eurasian" country, as well as a weakening of pro-Westernism (*zapadnichestvo*).

In the era of mass literacy, political cartoons emerged as visual complements to older textual works, which had painted Russians as haughty, xenophobic, rude, and prone to drunkenness and slovenliness (Cheauré 2010). However, whereas travelogues might be seen as somewhat innocent manifestations of personal prejudice, the populist media of the late 1800s took on an entirely new patina. Popular culture now began to more directly serve the interests of the state as images of the Other were re-tooled to inform national identities at home, with Russia often serving as convenient foil for this process in the US and Great Britain (see Zhuravleva 2012). Through popular culture and other forms of national image projection, Britain and the US engaged in the ideological Othering of Russia to reinforce their own credos and reify their "mutable self-images" (Ginzburg 2014, 606). We should remember that the term jingoism has its roots in pop-culture Russophobia, namely G. W. Hunt's "Macdermott's War Song" (1877), which included the chorus:

We don't want to fight but by Jingo if we do,
We've got the ships, we've got the men, we've got the money too,
We've fought the Bear before, and while we're Britons true,
The Russians shall not have Constantinople.

The American tabloid magazine *Puck* (1871–1918) exemplifies the ways in which cartoons reflected popular sentiments about national images, as well as how such conceptualizations could be mobilized for political purposes. A case in point: Russians were often depicted as violent drunks, as in the 1904 representation of a blood-thirsty and drunken Russian soldier (see Figure 1.3). Differing somewhat from its more reserved counterpart *Punch* (1941–2002), which tended to use animals as national symbols (the British lion, the American eagle, the Russian bear, etc.) or caricatures of the sitting heads of state, *Puck* was not above gross exaggerations of stereotypes to get the message across to its readers (or, alternatively for the less-than-literate, "seers"). Such imagery reinforced notions of Otherness which had been incubating in the public consciousness for centuries, specifically an "'Immutable Rus' . . . that remained alien and incomprehensible to the West" (Zhuravleva qtd. in Koshkin 2013).

Political instability and religious persecution tended to frame the Russian national image during this period, with Jewish emigration functioning as a double-edged sword. On the one hand, the popular press condemned the tsar for mistreatment of the Jewish population; on the other hand, portions of American and British society often reacted negatively to the influx of Jewish immigrants from Russia, whom they associated with anarchy, crime, disease, and poverty (see Lüthi 2013), thus reinforcing negative stereotypes about Russia and "Russianness."

Figure 1.3 Running amuck (Puck/Library of Congress)

As the nineteenth century progressed, Russia slowly turned back towards Europe, but the Romanovs' fear of political change translated into a predilection for legendarily violent repression, of radicals, Jews, and colonized peoples (especially the Poles). News reports of political crackdowns, alongside a steady flow of persecuted refugees, did "Brand Russia" no favors in the late 1800s, situating the empire as the "home of cruelty and barbarism" and an "enemy of all human culture" (Brandt 2003, 44). In fact, the stream of nihilists heading to foreign capitals did much to besmirch the "idea" of Russia abroad as the country became inextricably linked to the very terrorists the (tsarist) state sought to destroy.

Meanwhile, the Russian and British empires became more deeply ensconced in a "cold war" on countless fronts, from intrigues in the Black and Caspian Seas to proxy conflicts in Afghanistan to adventurism in Ethiopia to trade wars in East Asia. The British popular press kept pace with overseas developments, stirring up public sentiment to such a degree that had anyone asked the "man on the street" in Victorian London who was Britain's "number one geopolitical enemy," the answer would have undoubtedly been "The Russians." Meanwhile, in the US, Russia was increasingly framed as America's dark twin, "demonic" in its imperial acquisitions (as opposed to the "angelic" efforts of Washington to carry the "White Man's Burden") (see Koshkin 2013). This represented a rather sharp departure from earlier depictions of tsarist Russia, which were mainly positive; however, following the Russo-Japanese War (1904–1905), St. Petersburg was generally framed as a retrograde force in world affairs against the image of the "progressive" Japanese (see Zhuravleva 2012). Only the precipitous rise of an imperial Germany forced an end to this spiraling Anglophone popular-political enemy framing of Russia. With the start of the Great War Russians and British marched for the same cause, although many of the old stereotypes remained imbedded, being far less elastic than the geopolitical alignments of British High Command. Following the removal of the tsar in the February Revolution, hopes of a new Russia emerging were high; however, this optimism (and slipshod framing of "democratic Russia") was short-lived. Vladimir Lenin's victory over the pro-Western Provisional Government in the autumn of 1917 ushered in a new era of nation-image production in the West, linking together older negative tropes with new, ideologically-infused fears associated with Bolshevism and the "red menace."

The new regime was pilloried on every front, from aiding the Germans in the last days of the war to fomenting strikes and revolution in Birmingham and other British cities. Besieged by foreign invaders, engaged in a civil war on two continents, and economically isolated from the outside world, Soviet Russia (renamed the Union of Soviet Socialist Republics in 1922) saw its national image sink to a nadir in the interwar period. Treated as a pariah state, Russia was denied a place at the Paris peace negotiations in 1919, and saw an Allied-designed "cordon sanitaire" of anti-Soviet republics established from the Baltic to the Black Sea to contain Moscow's influence in the Eurasian steppes (a geopolitical stratagem that, at least from Moscow's perspective, seems to have been echoed since 1991).[25] Fear of international communism emerged as a palpable theme in contemporary popular culture, with the sinister hand of Moscow behind every plot. Joseph Stalin, upon assuming control of the Communist Party in late 1920s, sought to soften his nation's image, and lessened the USSR's rhetoric and overt support for world revolution. However, his disregard for the well-being of the Soviet Union's citizenry reinvigorated latent notions about Russia's unchanging orientation towards autocracy (ultimately in Stalin's historical rendering as the "Red Tsar"),[26] thus resurrecting connotations of the "immutability" of Rus.

Necessity bred a return of Soviet Russia to the ranks of the international community in 1941 as Nazi Germany invaded the USSR, followed shortly thereafter by a Japanese attack on the US naval base at Pearl Harbor, thus triggering a global conflict involving nearly every European country and dozens of other nations around the world (though it should be said that, at least for a time, Nazi representations of the USSR as a barbarous *untermenschen* steppe regime had purchase well outside of Germany [see Brandt 2003]). Increased religious freedom and Stalin's shift away from "communism" towards "patriotism" at home endowed Western propagandists with effective tools for a rebranding via popular culture, including the *Why We Fight* film series (1942–1945) which positively showcased the Soviet war effort and glossy spreads on the country in *Life* magazine (see, for instance, Leontief Alpers 2003).[27] Such representations buttressed positive images of the Soviet Union's "Great Experiment" espoused by American intellectuals and artists during the visits arranged by the All-Union Society for Cultural Ties Abroad (VOKS), including such luminaries as Paul Robeson and George Bernard Shaw (see David-Fox 2011). However, decades of anti-Bolshevism in the US and a century of populist Russophobia in Great Britain was not easily unmade.

Within a year of the Japanese surrender, the Anglo-American alliance with the USSR began to fray. Now out of power, Winston Churchill delivered a speech in Westminster College in Fulton, Missouri, on 5 March 1946 in which he declared: "From Stettin in the Baltic to Trieste in the Adriatic, an iron curtain has descended across the continent." This was one of the opening salvos in the awesome discursive conflict that would come to be known as the Cold War. Harry S. Truman, working with an anti-Soviet clique in the State Department, sculpted a new image for the USSR,[28] i.e., the center of "totalitarianism" and global threat to peace, security, and stability, starting in places like Greece, Turkey, and Iran (the geopolitical boundaries of this threat would soon be extended across Asia and into Africa). Churchill's framing of the coming conflict soon entered the public consciousness, supported by sophisticated foreign policy essays like George F. Kennan's famous "X Article" (1947).[29] Within the USSR, foreign policy specialists were putting the country on a similar ideological footing, thus resulting in a tit-for-tat escalation of ideological hostilities that would continue until the détente of the Nixon–Brezhnev era in the early 1970s.

Cultural producers of the popular persuasion responded with aplomb. In the wake of World War II, the visual arts, film, drama, literature, radio, and television all became sites of cultural production that shaped American and Western European geographical perceptions of the socialist states of the Eastern Bloc in general and the Soviet Union in particular. Infused with ideological vigor, the USSR and its allies (and, to a lesser extent, neutral-but-socialist Yugoslavia) came to be portrayed as a collective, quintessential Other through popular media such as *Reader's Digest* and other "news-lite" sources (see Sharp 2000). Comic books such as the Catholic Guild's *Treasure Chest* got in on the act as well, with the special anti-Soviet series "This Godless Communism" (1961–1962), drawn by Reed Crandall. The comic presents a slanted and vituperative reading of the rise of Bolshevism and Stalin's path to power before going on to imagine an America after a Soviet takeover, replete with religious persecution, mass arrests, gulags, and, most horrifying to a 1960s-era audience, the loss of personal wealth and property. Anti-Soviet imagery flowed through Western culture, tapping into long-seated prejudices associated with "ancient and immutable Rus" as well as new ones generated by the fear of a worldwide Communist victory. Hoary tropes about the Russian "peasant" were transformed into region-wide stereotypes about the Eastern Bloc "communist" (the ability to suffer, lack of individuality, misplaced messianism, etc.) (see Brandt 2003). More subtly, science fiction and horror films like *Invasion of the Body Snatchers* (1956) and *The Return of Dracula* (1958) drummed up Cold War anxieties of infiltration and domination by foreign powers, namely the Communists (see Hendershot 2001). As the Cold War entered a new period of intensity through the Thatcher-Reagan Anglophone geopolitical condominium, political thrillers and action films provided a host of geographical palimpsests upon which the conflict could be fought, from the northern Pacific in *The Sandbaggers* (1978–1980) to the icy streets of Moscow as evoked in Martin Cruz Smith's *Gorky Park* (1981) to the arid landscapes of Afghanistan in *Rambo III* (1988). The popular geographies of the Cold War were steadily hitched to the geopolitical aims of politico-economic elites in the West.

While many Cold War-era imaginaries of the "Soviet menace" were rooted in fear of revolution at home or military defeat abroad, others tilted towards the laughable characteristics of those on the other side of the Iron Curtain. Cartoonists were particularly fond of developing instantly recognizable effigies of Communist states with fanciful and fictitious names. From Disney studio animator Carl Barks' "Brutopia," the hostile country in the Donald Duck and Scrooge McDuck comics, to "Pottsylvania," the home of the bumbling (Soviet) spies Boris Badenov and Natasha Fatale in *The Adventures of Rocky and Bullwinkle and Friends* (1959–1964), to Hergé's totalitarian "Borduria" depicted in *The Adventures of Tintin* (1929–1976), sequential artists proved they possessed a particularly profound knack for debasing the Communist East through humor. The playful popular geopolitics of the Cold War, however, certainly extended beyond comic strips. Professional wrestling provided a fecund environment for sculpting the global conflict for the masses, pitting the Soviet "heels" against flag-waving patriot-warriors. Defeating the Soviets on the mat became a national ritual that was performed nightly in small towns like Dothan, AL, Jackson, TN, and Pine Bluff, AR—a ritual that inevitably provoked howls of triumphalist laughter from the jingoist spectators who reveled in the ludic thwarting of the "Commie" threat. As the USSR entered into its precipitous economic decline under Mikhail Gorbachev, Madison Avenue even got in on the joke. Most famously, the American fast-food chain Wendy's aired a series of "choice-based" advertisements lampooning the

Soviet Union. Subtly likening its competitors—McDonalds and Burger King—to the Soviets, Wendy's hawked its burgers as the freedom-loving (read American) alternative to "no choice," epitomized by the collapsing, anti-consumerist system that Americans had come to know (and laugh at) by the end of the 1980s.

Yet, throughout the Cold War, the USSR worked to combat negative images associated with the country and its political system in the Anglo-American world. Various efforts at cultural diplomacy were explored during these decades, particularly during times of détente. Nikita Khrushchev's visit to the United States in 1959 resulted in a measurable improvement in the overall national image of the USSR among Americans; however, with the Cuban Missile Crisis a few years later, fears of the Soviets returned with a vengeance. Regardless, the USSR was always sensitive to its national image in the Anglophone West, as well as other parts of the Western bloc and the neutral capitalist countries of Europe (Sweden, Switzerland, and Ireland). With enormous resources at the disposal of the state, the USSR strived to represent the country as a force for world peace, scientific innovation, and responsible social policies. As one group of scholars has argued:

> Given the closeness of the Soviet information space, the level of interest in the Soviet social experiment was determined not so much by the country's real accomplishments as the way they were reflected in information provided by information channels specially created for the purpose (the Cominform, Sovinformbureau, TASS and others). (Semeneko, Lapkin, and Pantin 2007, 77)

While the end of the Soviet system signaled the loss of many mechanisms of control over Russia's national image, certain elements of the totalitarian era remain wholly intact or little changed. Moscow's control of the country's news agencies represents a good example, as TASS and other media organs serve to promote the "correct" image of Russia in the post-Cold War world (Rantanen 2001). These tsarist- and Soviet-era outlets been buttressed by the new globally oriented Russia Today (RT) television network, which broadcasts in multiple languages and streams its content over the World Wide Web, as well as other forms of informational outreach from *Sputnik* (http://sputniknews.com/) to Twitter feeds of the Russian presidency (https://twitter.com/kremlinrussia_e). A variety of other media-centric efforts have also been undertaken by the state, from the Russkiy Mir project to sport promotion to advertising in high-profile international publications like *The New York Times* (see Chapter 8). However, as is the case with any branding strategy, real-world events often overshadow packaged promotional efforts about a given country, as will be explored in depth in the coming pages.

Given the power of national image combined with the growing relevance of *Fremdbebilderung* (or our being engulfed by foreign images), it is important to understand the dynamics of image flows in international relations. While there are still many in the academic community who doubt the importance, even existence, of the popular culture–world politics continuum (Grayson, Davies, and Philpott 2009), when contextualized in the realm of national identity, these intellectual dismissals naturally fade. In the coming chapter, the focus will turn to the form of postmodern statecraft deemed nation branding. Rather than suggesting that this is a completely new phenomenon, I will argue that it has been around in some shape or form since the beginning of diplomatic relations between states. In doing so, the aim is demonstrate how the "packaging" of national images can be used as a tool for advancing the national interest.

Notes

1 According to Robert J. Smith, this is partially due to social scientists' traditional hesitation to ascribe essential traits to any particular group; however, "the person-on-the-street has fewer qualms about applying national character suppositions that commonly devolve into stereotypes as explanation for behavior" (2008, 465–466). But, as we will see, not all thinkers are averse to associating particular traits with specific nations.

2 This outcome is the result of a long process of interaction between elites and masses, who over time ultimately come to agree on the perceptual bonds that unite them. Kenneth Boulding distinguishes between elites, "the small group of people who make the actual decisions which lead to war or peace, the making or breaking of treaties, the invasions or withdrawals, alliances, and enmities which make up the major events of international relations," and the masses, "ordinary people who are deeply affected by these decisions but who take little or no direct part in making them" (1959, 121).

3 A brief word on terminology is in order: I use "state" to refer to the politico-governmental authority structure that maintains a monopoly on the use of legitimate physical force within a given territory; a state need not be internationally recognized as independent and sovereign to achieve this status (e.g., Iraqi Kurdistan or Transnistria). A "nation" is a group of people who believe they possess a shared identity exclusive of all other communities based on the confluence of two or more of the following traits: language, lineage, culture, common legal rights, a common economy, and/or historical association with a particular territory. Nations need not possess a state to enjoy such a status (e.g., the Sami or Assyrians). A "nation-state" combines these two entities, creating a single state that operates in the name and interests of a particular nation; Denmark and South Korea represent paragons of the nation-state, while countries where immigration was common (e.g., Canada and Uruguay) are on the opposite end of the spectrum.

4 Further complicating this structure, shame can also serve as determiner of national image, as was the case for the Germans after the Holocaust. Recognition of Nazi crimes against European Jews formed a key component of post-WWII national identity in the German Federal Republic (see Fulbrook 1999).

5 However, it should be noted that nations are increasingly being de-tethered from territories as cyberspace and other forms of networked community enable national projects that are deterritorialized (see Appadurai 1996 and Saunders 2010).

6 As Fiebig-von Hase (1997) points out, the German nation is a paragon of the former while the American nation is the most emblematic of the latter. Much has been written about the "dividing line" between these two forms of nationalism, with some theorists suggesting that ethno-nationalism is prevalent east of the Rhine, while civic nationalism dominates to its west (see Kohn 1944). However, such a schema is far too simplistic; suffice it to say, the former dominates in the Europe-Africa-Asia supercontinent (i.e., the "Old World"), whereas the latter is the norm in the immigrant societies of the Americas and Antipodes (i.e., the "New World").

7 Not surprisingly for a text that focuses on popular culture, political marketing, and propaganda, this definition of the nation is solidly situated in the so-called "constructivist" school of thought, but is strongly influenced by the "functional" school (see Lawrence 2005).

8 Often referred to as the "X Article," the essay began as a private report, but was cleared for publication under a pseudonym to distance Kennan, the former Deputy Chief of Mission of the United States to the USSR (1944-46), and the US government from its content.

9 Following Beller and Leerssen (2007), the terms "spectant" (the viewing nation) and "spected" (the nation being viewed) will also be used in this analysis of *Fremdbild*.

10 Originally, the term derives from the printing process, referring to the metal duplicate of a relief printing surface.

11 Of special note is the special issue of *European Journal of English Studies* (December 2009), dedicated to the ways in which the British national identity has been influenced by stereotypical depictions of internal and external Others and the "rhetoric of national character from *Alfred the Great* to *Fawlty Towers*" (Hoenselaars and Leerssen 2009, 251).

12 Anglo-American propaganda expropriated the medieval trope of the "Hun" and applied it to Germany during this period, playing on deeply resonant fears associated with the "monstrous races" of the eastern frontier (see Hoppenbrouwers 2007).

13 The converse may also be true. The Russian blogger Sergey Armeyskov (2015) has provided substantive analysis of the concept of the "xenopatriot," i.e., an individual who views all markers of their own identity as negative (political leadership, religion, society, military, homeland, etc.), whereas those associated with foreigners are always seen as positive. His critique of Russophobia among Russians reflects the realization that mediatization of Russianness has become so widespread that it disrupts national identity production and maintenance within the Russian Federation.

14 While contemporary studies demonstrated a measurable distinction between the images of the *state* and images of the *people* (see, for instance, Willis 1968), national image tends to be a conflation of the two (as we will see from the Cold War tendency to refer to all Soviet citizens as "Russians," whether they were Latvians, Tajiks, or Udmurts).

15 The role of public diplomacy in national image construction, as well as the importance of "soft power," will be explored in greater depth in Chapter 2.

16 Even within authoritarian societies, there are gray areas here, with certain films which have been labeled as "propaganda" also receiving recognition as pillars of historical cinema, e.g., Leni Riefenstahl's *Olympia* (1938) and Sergei Eisenstein's masterpiece *Battleship Potemkin* (1925).

17 However, a variety of other media also served to bolster various nations' images abroad, including comic books and popular novels (see Dunnett 2009). Consumer items were important as well, from automobiles to Coca-Cola (see, for instance, Wagnleitner 1994 and Ross 1996).

18 Tota's aforementioned book is titled *The Seduction of Brazil: The Americanization of Brazil during World War II*, a reprise of the title of Richard Kuisel's (1993) earlier work *Seducing the French: The Dilemma of Americanization*.

19 This process also had important implications for subaltern populations within the country, as "fictive acts" in popular media "facilitated the domination of the internal U.S. 'Other'" (Crum 2015, 42).

20 Today, the American popular culture machine is facing competition on a variety of fronts, with polycentrism in global media becoming a reality, including massive (and highly profitable) film industries in India and Nigeria, as well as from other hubs which take advantage of digitally networked systems and the post-broadcast media revolution to target specific geolinguistic regions to reach consumers in Hispanophone, Arabaphone, Francophone, and Russophone realms (see Sinclair 2004; Nadeau and Barlow 2007; Gorham 2011; and Lahlali 2011).

21 However, it should be noted that by the 1960s, a plethora of cultural products shifted in their "representations" of both the US and UK, showing a much more nuanced view of these societies. Seminal and controversial films such as *Guess Who's Coming to Dinner* (1967), counter-cultural comic book heroes like *The Incredible Hulk* (1962–), and international sports figure Cassius Clay's (Muhammad Ali) very public conversion

to Islam (1964) and refusal to be inducted into the US Army (1967) disrupted the carefully constructed, monolithic visage of a tolerant and halcyon "America," both at home and abroad.

22 The most influential organizations to come out of these talks included the International Bank for Reconstruction and Development (1944), International Monetary Fund (1945), and General Agreement on Tariffs and Trade (1948). Subsequently, other international governmental organizations (IGOs) evolved as well, including but not limited to the European Economic Community (1957), Association of Southeast Asian Nations (1967), and East African Community (1967).

23 That is to say "European," given that the members of the Romanov family hailed from a variety of different royal lineages from across the continent.

24 Immediately thereafter, Burnaby writes: "*Grattez le Russe et vous trouverez le Tartare, ça c'est une insulte aux Tartares.* [Scratch a Russian and you will find a Tatar, but that is an insult to the Tatars.] This is a hackneyed expression; however, it is a true one. It requires but little rubbing to disclose the Tartar blood so freely circulating in Muscovite veins" (1876, 82).

25 The Bolsheviks relocated the capital to Moscow following their consolidation of power.

26 Ironically, Ioseb Besarionis dze Jughashvili, a.k.a. Joseph Stalin, was not Russian at all but Georgian.

27 This series included the following films: *Prelude to War* (1942); *The Nazis Strike* (1943); *Divide and Conquer* (1943); *The Battle of Britain* (1943); *The Battle of Russia* (1943); *The Battle of China* (1944); and *War Comes to America* (1945).

28 Not coincidentally, the wartime Anglophone nomenclature of "Russia" reverted to the "Soviet Union" as tensions increased.

29 Kennan's article, also known as "the Long Telegram" or "The Sources behind Soviet Conduct," is recognized as a major plank in the ideological fundament that would come to gird American Sovietology. In his essay, he laid out the following "truths" about the USSR: (1) Moscow saw itself in a perpetual conflict with the capitalist states of the world; (2) the Soviet Union viewed non-aligned socialists in other countries as enemies which were as dangerous as any capitalist opponent; (3) the Kremlin's ideology was influenced by deeply entrenched xenophobia, insecurity-based neurosis, and Russian nationalism; and (4) the structure of the Communist Party of the Soviet Union was impervious to rational thought and prevented accurate depictions of either internal or external realities that would otherwise shape and constrain any system of governance (see Miscamble 1992).

References

Ahonen, Sirkka. 2001. "Stereotypes of Peoples and Politics in Estonian and Finnish History." *Euroclio Bulletin* no. 14 (1):25–28.

Alami, Ahmed Idrissi. 2013. *Mutual Othering: Islam, Modernity, and the Politics of Cross-Cultural Encounters in Pre-Colonial Moroccan and European Travel Writing.* Albany, NY: SUNY Press.

Allworth, Edward. 1990. *The Modern Uzbeks: From the Fourteenth Century to the Present, a Cultural History.* Stanford, CA: Hoover Press.

Amienyi, Osabuohien P. 2005. *Communicating National Integration: Empowering Development in African Countries.* Farnham: Ashgate.

Anderson, Benedict. 1991. *Imagined Communities: Reflections on the Origin and Spread of Nationalism.* London: Verso.

Anholt, Simon. 2002. "Foreword." *Brand Management* no. 9 (4–5):229–239.

Appadurai, Arjun. 1996. *Modernity at Large: Cultural Dimensions of Globalization.* Minneapolis: University of Minnesota Press.

Armeyskov, Sergey. 2015. "Worldview of a Xenopatriot." *Russian Universe*, available at http:// russianuniverse.org/2015/03/03/xenopatriot-worldview/ [last accessed 14 December 2015].

Arnold, Samantha. 2012. "Constructing an Indigenous Nordicity: The 'New Partnership' and Canada's Northern Agenda." *International Studies Perspectives* no. 13:105–120.

Beller, Manfred. 2007a. "Perception, Image, Imagology." In *Imagology: The Cultural Construction of National Characters – A Critical Survey*, edited by Joep Leerssen and Manfred Beller, 3–16. Amsterdam and New York: Rodopi.

——. 2007b. "Stereotype." In *Imagology: The Cultural Construction of National Characters—A Critical Survey*, edited by Joep Leerssen and Manfred Beller, 429–434. Amsterdam and New York: Rodopi.

Beller, Manfred, and Joep Leerssen. 2007. *The Construction and Literary Representation of National Characters: A Critical Survey*. Amsterdam and New York: Rodopi.

Bernal, Victoria. 2006. "Diaspora, Cyberspace and Political Imagination: The Eritrean Diaspora Online." *Global Networks* no. 6 (2):161–179.

Bieber, Florian. 2000. "Cyberwar or Sideshow? The Internet and the Balkan Wars." *Current History* no. 99 (635):125–129.

Boulding, Kenneth E. 1959. "National Images and International Systems." *Journal of Conflict Resolution* no. 3 (2):120–131.

Brandt, Peter. 2003. "German Perceptions of Russia and the Russians in Modern History." *Debatte* no. 11 (1):39–59.

Breuilly, John. 1994. *Nationalism and the State*, Second Ed. Chicago: University of Chicago Press.

Brinkerhoff, Jennifer M. 2009. *Digital Diasporas: Identity and Transnational Engagement*. Cambridge, UK: Cambridge University Press.

Buell, Frederick. 1994. *National Culture and the New Global System*. Baltimore, MD: Johns Hopkins University Press.

Bulag, Uradyn E. 2010. *Collaborative Nationalism: The Politics of Friendship on China's Mongolian Frontier*. Lanham, MD: Rowman & Littlefield Publishers.

Burnaby, Frederick. 1876. *A Ride to Khiva: Travels and Adventures in Central Asia*. London, Paris, and New York: Cassell, Petter & Galpin.

Burton, Orville Vernon. 2013. "The South as 'Other,' the Southerner as 'Stranger'." *Journal of Southern History* no. 79 (1):7–50.

Calhoun, Craig. 2008. "Cosmopolitanism and Nationalism." *Nations and Nationalism* no. 14 (3):427–448.

Cameron, Keith. 1999. *National Identity*. Exeter, UK: Intellect Books.

Castañeda, Jorge G. 2012. *Manana Forever? Mexico and the Mexicans*. New York: Knopf Doubleday Publishing Group.

Chatterjee, Aneek. 2010. *International Relations Today: Concepts and Applications*. New Delhi: Dorling Kindersley.

Cheauré, Elisabeth. 2010. "Infinite Mirrorings: Russian and Eastern Europe as the West's Other." In *Facing the East in the West: Images of Eastern Europe in British Literature, Film and Culture*, edited by Barbara Korte, Eva Ulrike Pirker, and Sissy Helff, 25–41. Amsterdam: Rodopi.

Chew, William L. 2006. "What's in a National Stereotype? An Introduction to Imagology at the Threshold of the 21st Century." *Language and Intercultural Communication* no. 6 (3/4):179–187.

Coxe, William. 1784. *Travels into Poland, Russia, Sweden and Denmark*. London: J. Nichols.

Croucher, Sheila L. 2004. *Globalization and Belonging: The Politics of Identity in a Changing World*. Lanham, MD: Rowman & Littlefield Publishers, Inc.

Crum, Jason. 2015. "'Out of the Glamorous, Mystic East': Techno-Orientalism in Early Twentieth-Century U.S. Radio Broadcasting." In *Techno-Orientalism: Imagining Asia in Speculative Fiction, History, and Media*, edited by David S. Roh, Betsy Huang, and Greta A. Niu, 40–51. New Brunswick, NJ: Rutgers University Press.

David-Fox, Michael. 2011. *Showcasing the Great Experiment: Cultural Diplomacy and Western Visitors to the Soviet Union, 1921–1941*. Oxford: Oxford University Press.

de Gobineau, Arthur. 1915 [1853–1855]. *The Inequality of the Human Races*. Translated by Adrian Collins. New York: G. P. Putnam's Sons.

Deutsch, Karl. 1953. *Nationalism and Social Communication: An Inquiry into the Foundations of Nationality*. Cambridge, MA: The Technology Press of the Massachusetts Institute of Technology.

Duncan, Randy, and Matthew J. Smith. 2009. *The Power of Comics: History, Form & Culture*. New York: Continuum.

Dunnett, Oliver. 2009. "Identity and Geopolitics in Hergé's *Adventures of Tintin*." *Social & Cultural Geography* no. 10 (5):583–599.

Edensor, Tim. 2002. *National Identity, Popular Culture and Everyday Life*. Oxford: Berg.

Elsaesser, Thomas. 2005. "German Cinema Face to Face with Hollywood: Looking into a Two-Way Mirror." In *Americanization and Anti-Americanism: The German Encounter with American Culture after 1945*, edited by Alexander Stephan, 166–185. New York and London: Berghahn Books.

Eriksen, Thomas Hylland. 2007. "Nationalism and the Internet." *Nations & Nationalism* no. 13 (1):1–17.

Falk, Andrew. 2011. *Upstaging the Cold War: American Dissent and Cultural Diplomacy 1940–1960*. Amherst: University of Massachusetts Press.

Fiebig-von Hase, Ragnhild. 1997. "Introduction." In *Enemy Images in American History*, edited by Ragnhild Fiebig-von Hase and Ursula Lehmkuhl, 1–40. Providence, RI and Oxford: Berghahn Books.

Fischer, John. 1947. *Why They Behave Like Russians*. New York and London: Harper & Brothers.

Freud, Sigmund. 1915 [1991]. *Civilization, Society and Religion: Group Psychology, Civilization and Its Discontents and Other Works*. New York: Penguin.

Fulbrook, Mary. 1999. *German National Identity after the Holocaust*. Hoboken, NJ: Polity.

Gaufman, Elizaveta, and Katarzyna Wałasek. 2014. "The New Cold War on the Football Field: .ru vs .pl." *Digital Icons: Studies in Russian, Eurasian and Central European New Media* (12):55–75.

Gellner, Ernest. 1983. *Nations and Nationalism*. Ithaca, NY: Cornell University Press.

Gemünden, Gerd. 1998. *Framed Visions: Popular Culture, Americanization, and the Contemporary German and Austrian Imagination*. Ann Arbor: University of Michigan Press.

Gienow-Hecht, Jessica C. E. 1997. "Friends, Foes, or Reeducators? *Feindbilder* and Anti-Communism in the U.S. Military Government in Germany, 1946–1953." In *Enemy Images in American History*, edited by Ragnhild Fiebig-von Hase and Ursula Lehmkuhl, 281–299. Providence, RI and Oxford: Berghahn Books.

Giffard, C. Anthony, and Nancy K. Rivenburgh. 2000. "News Agencies, National Images, and Global Media Events." *J&MC Quarterly* no. 77 (1):8–21.

Ginzburg, Lyubov A. 2014. "Review of *Understanding Russia in the United States: Images and Myths, 1881–1914*." *Journal of American History* no. 101 (2):606–607.

Gorham, Michael. 2011. "Virtual Rusophonia: Language Policy as 'Soft Power' in the New Media Age." *Digital Icons: Studies in Russian, Eurasian and Central European New Media* no. 5:23–48.

Gottlieb, Evan. 2007. *Feeling British: Sympathy and National Identity in Scottish and English Writing, 1707–1832*. Lewisburg, PA: Bucknell University Press.

Grayson, Kyle, Matt Davies, and Simon Philpott. 2009. "Pop Goes IR? Researching the Popular Culture–World Politics Continuum." *Politics* no. 29 (3):155–163.

Guibernau, Montserrat. 1996. *Nationalisms: The Nation-State and Nationalism in the Twentieth Century*. Cambridge, UK: Polity Press.

Halperin, Charles J. 2009. *The Tatar Yoke: The Image of the Mongols in Medieval Russia*. Bloomington, IN: Slavica.

Hendershot, Cynthia. 2001. *I Was a Cold War Monster: Horror Films, Eroticism, and the Cold War Imagination*. Madison, WI: Popular Press.

Hoenselaars, Ton, and Joep Leerssen. 2009. "The Rhetoric of National Character: Introduction." *European Journal of English Studies* no. 13 (3):251–255.

Holsti, Ole R. 1962. "The Belief System and National Images: A Case Study." *Journal of Conflict Resolution* no. 5 (3):244–252.

Hoppenbrouwers, Peter. 2007. "Medieval Peoples Imagined." In *Imagology: The Cultural Construction of National Characters—A Critical Survey*, edited by Joep Leerssen and Manfred Beller, 45–62. Amsterdam and New York: Rodopi.

Hunt, G. W. 1877. "Macdermott's War Song." London: Hopwood & Crew.

Hutchinson, John, and Anthony Smith. 1994. *Nationalism*. Oxford: Oxford University Press.

Inkeles, Alex. 1997. *National Character: A Psycho-Social Perspective*. New Brunswick, NJ: Transaction Publishers.

Jervis, Robert. 1970. *The Logic of Images in International Relations*. Princeton, NJ: Princeton University Press.

Joireman, Sandra. 2003. *Nationalism and Political Identity*. London: A&C Black.

Jourdan, Annie. 2007. "Symbol." In *Imagology: The Cultural Construction of National Characters—A Critical Survey*, edited by Joep Leerssen and Manfred Beller, 434–437. Amsterdam and New York: Rodopi.

Jowett, Garth S., and Victoria O'Donnell. 2006. *Propaganda and Persuasion*. Thousand Oaks, CA: SAGE.

Kennan, George F. 1947. "The Sources of Soviet Conduct." *Foreign Affairs* no. 25 (4):566–582.

Kibria, Nazli 2011. *Muslims in Motion: Islam and National Identity in the Bangladeshi Diaspora*. New Brunswick, NJ: Rutgers University Press.

Kinsey, Dennis F., and Myojung Chung. 2013. "National Image of South Korea: Implications for Public Diplomacy." *Exchange: The Journal of Diplomacy* no. 4 (1):5–16.

Kligman, Gail. 2001. "On the Social Construction of 'Otherness': Identifying 'the Roma' in Post-Socialist Communities." *Review of Sociology* no. 7 (2):61–78.

Kohn, Hans. 1944. *The Idea of Nationalism: A Study in its Origins and Background*. New York: Macmillan.

Koshkin, Pavel. 2013. "How Myths about Russia Embrace US Identity." *Russia Direct*, available at www.russia-direct.org/qa/how-myths-about-russia-embrace-us-identity# [last accessed 14 July 2015].

Koski, Anne. 2011. "Organising Political Consensus: The Visual Management of Diplomatic Negotiations and Community Relations in the Finnish Accession to the EU." In *Images in Use: Towards the Critical Analysis of Visual Communication*, edited by Matteo Stocchetti and Karin Kukkonen, 91–112. Amsterdam and Philadelphia, PA: John Benjamin Publishing Co.

Kramer, Lloyd S. 2011. *Nationalism in Europe and America: Politics, Cultures, and Identities since 1775*. Raleigh, NC: UNC Press Books.

Kroes, Rob. 1996. *If You've Seen One, You've Seen the Mall*. Urbana and Chicago: University of Illinois Press.

Kuisel, Richard F. 1993. *Seducing the French: The Dilemma of Americanization*. Berkeley and Los Angeles: University of California Press.

Kunczik, Michael. 1997. *Images of Nations in International Public Relations*. Mahwah, NJ: Lawrence Erlbaum Associates.

Lahlali, El Mustapha. 2011. *Contemporary Arab Broadcast Media*. Edinburgh: Edinburgh University Press.

Langhorne, Richard. 2001. *The Coming of Globalization Its Evolution and Contemporary Consequences*. Houndmills, UK: Palgrave.

Lawrence, Paul. 2005. *Nationalism: History and Theory*. Harlow, UK: Pearson Education Ltd.

Leerssen, Joep. 2000. "The Rhetoric of National Character: A Programmatic Survey." *Poetics Today* no. 21 (2):267–292.

———. 2007. "Imagology: History and Method." In *Imagology: The Cultural Construction of National Characters—A Critical Survey*, edited by Joep Leerssen and Manfred Beller, 17–32. Amsterdam and New York: Rodopi.

———. 2014. "National Identity and National Stereotype." *Imagologica*, available at: www.imagologica.eu/leerssen [last accessed 29 July 2014].

Leontief Alpers, Benjamin. 2003. *Dictators, Democracy, and American Public Culture: Envisioning the Totalitarian Enemy, 1920s–1950s*. Chapel Hill: University of North Carolina Press.

Li, Xiufang, and Naren Chitty. 2009. "Reframing National Image: A Methodological Framework." *Conflict & Communication Online* no. 8 (2):1–11.

Lippmann, Walter. 1922. *Public Opinion*. New York: Harcourt, Brace and Company.

Lüthi, Barbara. 2013. "Germs of Anarchy, Crime, Disease, and Degeneracy: Jewish Migration to the United States and the Medicalization of European Borders around 1900." In *Points of Passage: Jewish Migrants from Eastern Europe in Scandinavia, Germany, and Britain 1880–1914*, edited by Tobias Brinkmann, 27–45. London and New York: Berghahn Books.

Mandler, Peter. 2006. *The English National Character: The History of an Idea from Edmund Burke to Tony Blair*. New Haven, CT: Yale University Press.

Mankekar, Purnima. 2015. *Unsettling India: Affect, Temporality, Transnationality*. Durham, NC: Duke University Press.

Marlin, Randal 2002. *Propaganda and the Ethics of Persuasion*. Peterborough, ON: Broadview Press.

McAlister, Melani. 2005. *Epic Encounters: Culture, Media, and U.S. Interest in the Middle East since 1945*. Berkeley: University of California Press.

McCrae, Robert R. 2013. "The Inaccuracy of National Character Stereotypes." *Journal of Research in Personality* no. 47 (6):831–842.

McDermott, Rose. 2004. *Political Psychology in International Relations*. Ann Arbor: University of Michigan Press.

Metzeltin, Michel. 1998. "L'imaginaire roumain de l'Occident: Questions de méthode et essais d'application." In *Imaginer l'Europe*, edited by Danièle Chauvin, 173–179. Grenoble: Iris.

Mežs, Ilmārs. 2010. *The People of Latvia*. Riga: The Latvian Institute.

Miscamble, Wilson D. 1992. *George F. Kennan and the Making of American Foreign Policy, 1947–1950*. Princeton, NJ: Princeton University Press.

Mitchell, W. J. T. 1986. *Iconology: Image, Text, Ideology*. Chicago: University of Chicago Press.

Morgenthau, Hans J. 1985 [1948]. *Politics among Nations: The Struggle for Power and Peace*. Columbus, OH: McGraw-Hill Companies.

Motyl, Alexander J. 1992. "The Modernity of Nationalism: Nations, State and Nation-States in the Contemporary World." *Journal of International Affairs* no. 45 (2):311–323.

Nadeau, Jean-Benoît, and Julie Barlow. 2007. *The Story of French: From Charlemagne to Cirque du Soleil*. Toronto: Random House of Canada.

Neumann, Birgit. 2009. "Towards a Cultural and Historical Imagology: The Rhetoric of National Character in 18th-century British Literature." *European Journal of English Studies* no. 13 (3):275–291.

Neumann, Iver B. 1999. *Uses of the Other. The "East" in European Identity Formation*. Minneapolis: University of Minnesota Press.

O'Dell, Tom. 2011. "*Raggare* and the Panic of Mobility: Modernity and Hybridity in Sweden." In *Car Cultures*, edited by Daniel Miller, 105–132. Oxford: Berg.

O'Shaughnessy, Nicholas J. 2003. *The Marketing Power of Emotion*. Oxford: Oxford University Press.

Ong, Aihwa. 1999. *Flexible Citizenship: The Cultural Logistics of Transnationality*. Durham, NC: Duke University Press.

Pile, Steve. 2010. "Emotions and Affect in Recent Human Geography." *Transactions of the Institute of British Geographers* no. 35:5–20.

Prieto, Eric. 2011. "Geocriticism, Geopoetics, Geophilosophy, and Beyond." In *Geocritical Explorations: Space, Place, and Mapping in Literary and Cultural Studies*, edited by Robert T. Tally Jr., 13–27. Basingstoke, UK: Palgrave Macmillan.

Rantanen, Terhi. 2001. "The Old and the New: Communications Technology and Globalization in Russia." *New Media & Society* no. 3 (1):85–105.

Rash, Felicity. 2012. *German Images of the Self and the Other: Nationalist, Colonialist and Anti-Semitic Discourse 1871–1918*. Houndsmills, UK: Palgrave Macmillan.

Ross, Kristin. 1996. *Fast Cars, Clean Bodies: Decolonization and the Reordering of French Culture*. Cambridge, MA: MIT Press.

Rusciano, Frank L. 1998. *World Opinion and the Emerging International Order*. Westport, CT: Praeger.

Rusciano, Frank L., and Bosah Ebo. 1998. "National Consciousness, International Image, and the Construction of Identity." In *World Opinion and the Emerging International Order*, edited by Frank L. Rusciano, 59–88. Westport, CT: Praeger.

Sari Karademir, Burcu. 2012. "Turkey as a 'Willing Receiver' of American Soft Power: Hollywood Movies in Turkey during the Cold War." *Turkish Studies* no. 13 (4):633–645.

Saunders, Robert A. 2010. *Ethnopolitics in Cyberspace: The Internet, Minority Nationalism, and the Web of Identity*. Lanham, MD: Lexington Books.

Semeneko, Irina, Vladimir Lapkin, and Vladimir Pantin. 2007. "Russia's Image in the West (Formulation of the Problem)." *Social Sciences* no. 38 (3):79–92.

Sharp, Joanne P. 2000. *Condensing the Cold War: Reader's Digest and American Identity*. Minneapolis: University of Minnesota Press.

Shlapentokh, Dmitry. 2007. *Russia Between East and West: Scholarly Debates on Eurasianism*. Leiden, Netherlands: Brill.

Simons, Greg. 2013. "Nation Branding and Russian Foreign Policy." *Swedish Institute of International Affairs Occasional Papers* no. 21:1–19.

Sinclair, John. 2004. "Geolinguistic Region as Global Space: The Case of Latin America." In *The Television Studies Reader*, edited by Robert C. Allen and Annette Hill, 130–138. London: Routledge.

Smith, Anthony. 1988. *The Ethnic Origin of Nations*. Oxford: Blackwell.

———. 1991. *National Identity*. Reno: University of Nevada Press.

Smith, Robert J. 2008. "In Defense of National Character." *Theory & Psychology* no. 18 (4):465–482.

Tenorio-Trillo, Mauricio. 1996. *Mexico at the World's Fairs: Crafting a Modern Nation*. Berkeley: University of California Press.

Thien, Deborah. 2005. "After or Beyond Feeling? A Consideration of Affect and Emotion in Geography." *Area* no. 37 (4):450–454.

Thrift, Nigel. 2008. *Non-Representational Theory: Space/Politics/Affect*. London and New York: Routledge.

Tota, Antonio Pedro. 2010. *The Seduction of Brazil: The Americanization of Brazil during World War II*. Austin: University of Texas Press.

Treisman, Daniel. 2011. *The Return: Russia's Journey from Gorbachev to Medvedev*. New York: Simon & Schuster.

Wagnleitner, Reinhold. 1994. *Coca-Colonization and the Cold War: The Cultural Mission of the United States in Austria After the Second World War*. Translated by Diana Wolf. Raleigh: University of North Carolina Press.

Wang, Jian. 2008. "The Power and Limits of Branding in National Image Communication in Global Society." *International Political Communication* no. 14 (2):9–24.

Westphal, Bertrand. 2011. *Geocriticism: Real and Fictional Spaces*. New York: Palgrave Macmillan.

Willis, Richard H. 1968. "Ethnic and National Images: Peoples vs. Nations." *Public Opinion Quarterly* no. 32 (2):186–201.

Wingfield, Nancy Meriwether. 2003. *Creating the Other: Ethnic Conflict and Nationalism in Habsburg Central Europe*. Oxford and New York: Berghahn Books.

Zhuravleva, Victoria I. 2012. *Ponimanie Rossii v SShA: obrazy i mify, 1881–1914 [Understanding Russia in the United States: Images and Myths, 1881–1914]*. Moscow: Russian State University for the Humanities.

2 The supermarket of nations

Competitive identity and the brand state

A nation's reputation has long been recognized for its effect on relations between countries and as a tool of foreign policy. Joseph S. Nye Jr. (2004) suggests that a positive national image provides the ability to entice and attract other countries, which in turn leads to acquiescence or imitation. More than simply a tool of statecraft, world opinion of a country's reputation is integral to that nation's negotiation of its identity (Rusciano 2003, 361). Consequently, national image and country reputation are major components of alliance building and figure prominently in international conflict (see Keohane 1984), while simultaneously serving to bind together the domestic political community. While it has always been true that a country's image "results from its geography, history, proclamations, art and music, famous citizens and other features" (Kotler and Gertner 2002, 251), there are many other inputs into this image-centric calculus. Taken together, the sum of all "descriptive, inferential and informational beliefs one has about a particular country" (Kleppe 2002, 62) is an increasingly important aspect of modern diplomacy and world politics. This state of affairs has resulted in a frothy discussion of what scholars call the "nation brand."

Borrowing from the field of business management, international relations scholars have begun to explore the management of national image through strategic policy initiatives, targeted public diplomacy, "managed" cultural production, and specialized advertising and marketing programs intended to alter global perception on the elite and mass levels. This new field of inquiry investigates nation or state branding, a phenomenon that is distinct from national image (as discussed in Chapter 1) due to the active rather than passive nature of policy elites and other agents of the nation in crafting, maintaining, and changing their country's image. Practitioners and scholars alike view nation branding as an attempt to "commodify the physical and symbolic dimensions of place and space, as well as the ideas of collective identity and solidarity associated with nationhood" (Kaneva 2011, 131). As an integral "component of national policy" (Anholt 2008, 22), positive national image and its maintenance are increasingly being weaved into the larger field of international relations.[1] This stems from the neoliberal perception that a nation brand is an asset, and that like all "profitable assets" it needs to be managed (Aronczyk 2013, 3). Nation branding includes not only the activities of a given state's governmental actors but also its celebrities, multinational corporations, and non-governmental organizations. Positive associations with various products are also vital to national brands. In the current era of global commerce, corporate brands such as KFC, Nissan, Real Madrid, and Virgin have a significant role to play, as do BBC, Bollywood, Al Jazeera, and other international media purveyors and platforms.

While older countries enjoy well-established national images at home and abroad, the past century of decolonization has seen the emergence of roughly one hundred new nations that face a double challenge. These "young" nations are charged first with crystallizing a coherent country image in the domestic realm, i.e., determining, defining, and/or delimitating a national identity which, theoretically, applies to *and* is accepted by (as close to as possible in some cases) 100 percent of its citizenry. Second, these states bear the burden of transmitting a viable country image to the world community. Such an image must be clear and intelligible ("brand definition"), it must be distinguishable and mutually exclusive from the images of all other nations ("brand differentiation"), and it must be seen as positive ("brand value"). Assuming all these factors are working in harmony, a state may be said to have achieved brand equity, or to have reached a point where the country provides a unique cachet to those communities with which it interacts (that is, alliance and trade partners, investors, international organizations, non-governmental organizations, tourists, etc.) (see Anholt 2007). As discussed earlier, the current global information age, and specifically its digital challenges to the nation, greatly heightens the importance of managing national image and country reputation to achieve these ends. The spread of new media technologies, increased mobility of tourists and immigrants, deepening of complex economic interdependence, and globalized consumerism have made governments ever more attentive to their national images through nation branding.

As a result, nation branding has become *de rigueur* among newer states and a necessary undertaking for older ones, particularly in those zones of the world where dis/association with existing states, ideological systems, and geopolitical structures is paramount (e.g., Latin America, post-socialist Europe/Eurasia, sub-Saharan Africa, etc.). Importantly, the practice functions as a highly visible manifestation of the postmodern realties of contemporary geopolitics, though this has only surfaced quite recently in the literature. Influenced by theoreticians-cum-consultants such as Simon Anholt, Keith Dinnie, and Wally Olins and aided by a host of marketing firms from Interbrand to Ogilvy & Mather, countries as disparate as Ireland and the United Arab Emirates have embraced brand management strategies for their respective national images. Recognizing that public diplomacy is no longer simply an adjunct to traditional forms of diplomacy, contemporary strategies are attitudinal in nature (Szondi 2009). Rather than just trying to motivate an overseas population to do (or not do) something (e.g., influence their own governments in negotiations with external powers, resist a push to war, etc.), state branders want various foreign polities to *feel* something about their country or, more specifically, its government, people, products, and/or geography. Nation branding also serves as a container of "distinct identities and loyalties," thus serving as a tool for a variety of ancillary projects including expanding sovereignty (Aronczyk 2013, 5). Consequently, state branding thus aims at establishing a "competitive identity"

(Anholt 2007) for a given country, positively distinguishing it through a series (and more importantly, a confluence) of key measurements which are discussed later in this chapter. Recognizing that a positive and resonant brand is necessary for any country seeking foreign investment, influence in international affairs, and a host of other functions of contemporary statehood, most states now engage in some form of structured national-image promotion on the global stage. However, as a number of critical theorists have found, nation branding often produces wholly unpredictable results (see, for instance, Unwin 1999; Curry Jansen 2008; Kaneva 2012; Bardan and Imre 2012), sometimes disrupting national identity and even provoking internal strife within countries. Furthermore, as more and more nations gravitate towards this globalized neoliberal practice, there is a palpable devaluing or at least diminution of what it actually means to be a nation, or, put another way, a sacrifice of the national essence on the altar of marketization. Nonetheless, it seems that as the Cold War gave way to the "End of History" new world order (Fukuyama 1992), selling nations has become the way of the world.

The nation as brand: The prehistory of and precursors to a concept

Determining the origins of nation branding is a difficult undertaking, as evidenced by the common refrain espoused by key scholars in the field that there is "nothing new" about nation branding in the literature (see, for instance, Olins 2006; Halsall 2008; Teslik 2009).[2] Certainly, the oldest precursors of nation branding are to be found in the gifts given by rulers in initial negotiations or as mementos of good relations. In ancient times (and today), monarchs, heads of state, and other dignitaries exchanged tokens of (potential) friendship. Typically, the items were semiotic markers of the society or place from which they originated (or a symbol of the territorial reach or trading capacities of the state). In antiquity, the Byzantine Empire made presents of silk to its Western counterparts, while the Tang empress Wu Zetian (625–705) sent a pair of pandas to the Japanese emperor as a sign of China's uniqueness. Famous modern examples of the practice include the Statue of Liberty, gifted by France to the US (1886), and Japan's gift of cherry tree plantings (1912), which today define the green areas of Washington, D.C.

Some scholars see the genesis of nation branding in the rise of the post-Westphalian state system. In his widely cited article "Branding the Nation—The Historical Context" (2002), Wally Olins argues that the reign of the "Sun King" Louis XIV (1654–1715) is an appropriate starting point. His use of spectacle to advance the image of France certainly shares some of the hallmarks of contemporary efforts at image management by policymakers and nation branders alike. In Olin's estimation, Louis XIV's activities ignited a spark which his revolutionary (and Napoleonic) successors continued and expanded, marketing the essence of the French *état* far and wide. Olins goes on to suggest that numerous other nations—great and small—have engaged in strategic and tactical refashioning of their international images over the past few centuries, using a variety of tools and techniques that would be familiar to today's nation branders. He points to

Bismarck's Prussia and the use of folklore, language, commerce, and industry to establish the new "Germany" brand in the second half of the nineteenth century. Whereas the Second Reich's proto-branding was linked to the formation of an empire, more often than not these national "makeovers" were outgrowths of decolonization, including the creation of a Kemalist Turkey, a post-Dutch Indonesia, and a Rhodes-free Zimbabwe. While Olins admits his "rebranding" of history as a history of "branding" might be a bit controversial, he counters: "[If] I had used words like identity, national image, national identity and so on, no well-educated person with any historical knowledge would have raised an eyebrow" (2002, 245).

Melissa Aronczyk chooses to situate the "prehistory" of nation branding in the establishment of state-affiliated merchant enterprises with a global reach, including the East India Company and the Russia Company; however, she points out that participation in international events like Hyde Park's Great Exhibition in the Crystal Palace (1851) undoubtedly heralded the beginning of modern state branding (see Figure 2.1). Such forums allowed for the "staging of national culture" through a wide variety of "material and symbolic representations" from architecture to folk costumes (Aronczyk 2013, 4). Katherine Smits and Alix Jansen echo this notion: "By their very nature extraordinary and 'artificial' events, international exhibitions have been, from their origins, sites at which national identity was constructed and displayed, in an explicitly and normatively global context" (2012, 173). Then as today, the relationship between enterprise and the nation was tightly bound, creating what some have deemed "commercial nationalism" (Volčič and Andrejevic 2015).

Figure 2.1 The Great Exhibition (Library of Congress)

According to Emma Björner, "At the earliest expositions national business corporations operated as signifiers in the process of displaying the nation, its culture, history and future prospects, and the interests of the nation and the corporations went hand in hand" (2010, 8). As a by-product of the process of industrialization, international expositions often informed or reinforced associations of particular states (or world regions) with modernity and imperial power (Moore 2013), while simultaneously mooring more traditional societies in geopolitical imaginaries of savagery and primordialism through "ethnic shows," "exotic performance," and "spectacles of difference" (Blanchard, Boetsch, and Snoep 2011; Musée du quai Branly 2012).

While international expositions marketed the nation to foreign populaces, the national museum—an institution which came into existence at roughly the same time—served to *brand* (or at least narrate) the nation to itself as well as "publicize" and "authenticate" this brand abroad (Cooke 2014, 79). As a repository of knowledge and objects of value, a museum articulates social ideas, enshrines memory, emphasizes community values, and promotes civic cohesion (Karp 1992). Viewed through the lens of national identity, museums are thus "culture machines" that establish spaces of transformative ritual akin to premodern shrines, obelisks, and other monumental loci of worship, but rather than gods it is the state (or in some cases, corporations associated with the state) which deserves reverence (Greenblatt 1991). According to Theodore Low, "No one can deny that museums have powers which are of the utmost in any war of ideologies. They have the power to make people see the truth" (1942 [2004], 30). Consequently, the national museum, with its focus on "imagined communities" (Anderson 1991), is a key plank in the nation-building process, which, through the hegemonic power of persuasion, makes the differences between one nation and another (despite any number of important similarities) appear both natural *and* constitutive. As the ultimate forum of representation and a site where the nation is both "produced and consumed" (Jackson 2008, 125), the national museum possesses the "unassailable" power of the curatorial voice (Walsh 1997). Inside its walls, taxonomies are reified and hierarchies promoted through the structuring of "ways of seeing" (Jay 1988) and how to remember the past; such practices ultimately inform the core values of the nation, thus elevating certain ideals, symbols, and narratives, while diminishing, omitting, or silencing others. Through the use of the "disciplining knowledges" of exhibition and display (Crang 2003), such museums are charged with "making" history and "manifesting" heritage, a process which is intimately tied to narrow political pedagogical agendas (Karp 1992). The national museum thus packages the nation into a consumable form, a visual-informational product that could be processed and internalized in a matter of hours, but which would—in theory—produce a lifetime of (geo) political dividends.

More than words: Distinguishing nation branding from other oft-conflated concepts

Given that some scholars of political communication contend that the distinction between public relations and propaganda at the international level is semantic (see Kunczik 1997), it is possible to employ the rise of targeted messaging that crossed state borders as the beginning of the phenomenon now labeled nation branding. As explored in the previous chapter, the dawn of the twentieth century ushered in the era of mass-mediated propaganda. With the introduction of broadcast media, novel forms of political communication were made available to political elites, which quickly tethered these revolutionary tools to projects of national identity production (at home) and projection (overseas). Radio broadcasts, motion pictures, and eventually television signals all delivered images and textual content that invariably served to shape national image. These new platforms added to the influence of the newspapers, posters, and other forms of print media that had been the mainstay of the previous century. The apex of broadcast media and the modern state allowed for expanded capacities for influencing the image of the nation. Such manipulation of national image by Soviet Russia, Nazi Germany, Imperial Japan, the United States, and Western European governments alike remains an issue that colors the perception of nation branding campaigns to this day, in that the term "propaganda" is frequently linked to state branding efforts (see, for instance, Kahn 2006; Curry Jansen 2008; Marat 2009). However, as Nadia Kaneva (2011) rightly points out, nation branding is not synonymous with propaganda, nor is nation branding simply "advertising" the nation. While a film like *Triumph of the Will* (1935) might be seen as both propaganda and (fascist) nation branding, the vast majority of twentieth-century psychological operations (psy-ops) activity falls outside the scope of nation branding. Even the para-propaganda conducted by the US in Latin America discussed in the previous chapter falls far from the nation branding tree, as it was largely intended to deliberately manipulate political attitudes (at home) rather create the sort of "emotional attachment" to a country (abroad), that end to which all nation branders strive.

Conceptually more closely aligned to nation branding is the field of public diplomacy, which is broadly defined as "a nation's attempt to shape its image and influence public opinion in other nations" (Curry Jansen 2008, 123). Such efforts focus on cultural exchange, thus resulting in the linked concept of "cultural diplomacy" (see Gienow-Hecht and Donfried 2013). Purportedly coined in 1965 by Edmund Gullion, Dean of the Fletcher School (though in use as early as the late 1800s), public diplomacy became a catchphrase for US information campaigns during the Cold War (Cull 2006).[3] Unlike traditional diplomacy, activities are not exclusive to a professional class and tend to take place over a longer time horizon. Focused on reaching the masses (rather than just decision-makers), public diplomacy needed to embrace new methods of persuasion.

> Diplomats are, or should be, fully prepared to *change their minds* about any country at any point. Publics, on the other hand, have neither the expertise, the experience, the habit nor the desire to consider the actions of foreign governments so carefully and in so even-handed a manner, and their responses to those governments' policies are likely to be directly and substantially conditioned by their perceptions of the country as a whole. (Anholt 2006, 273)

Differing from both propaganda and nation branding, public diplomacy (in its ideal form) is about establishing pathways of communication that go beyond traditional diplomacy, and are not about simply broadcasting (dis)information. While often conflated in academic literature, there are important conceptual differences that distinguish public diplomacy from nation branding. György Szondi (2008) points out that public diplomacy is a peculiarly American concept, and a practice which is more associated with great powers with regional, supra-regional, and even global foreign policy objectives. Nation branding, on the other hand, is practiced by all sorts of states, and has proved especially popular with smaller countries hoping to profit from the openness of the neoliberal system. Other scholars have cheekily delineated the distinction between the two by noting that public diplomacy targets "hearts and minds" (see Lord 2006; Gilboa 2008), while nation branding is more concerned with investment portfolios, bank accounts, pocketbooks, and wallets. Whereas public diplomacy attempts to move people's ideological opinions through factual information or symbolic interactions (Zhang 2006), nation branding is often more focused on affectual content and other sorts of non-tangibles, a recognition of postmodern sensibilities which privilege the sublime over the rational, affect over effect, and feeling over knowing (see Hoffmann 2005). A public diplomacy campaign might be solely composed of verifiable information about a particular topic (for instance, freedom of religion in the US) or a single act (e.g., tsunami relief in Indonesia); however, any nation branding campaign worth its salt must appeal to a minimum of six dimensions: emotional, physical, financial, leadership, cultural, and social (Kinsey and Chung 2013, 7). While public diplomacy and nation branding are certainly intertwined (particularly in internationally influential countries like the US, UK, China, and Germany) and sub-national actors have important roles to play in both areas, it is possible to have one without the other, as small countries like Slovenia (Damjan 2005) and Liechtenstein (Passow, Fehlmann, and Grahlow 2005) have demonstrated.

Last among the concepts which are frequently conflated with nation branding is that of "soft power." Originally articulated by Harvard professor Joseph S. Nye in a number of books and articles including *Bound to Lead* (1990), *The Paradox of American Power* (2002), and *Soft Power* (2004), a country's soft power rests on co-opting rather coercing other states. Extrapolating from the work of Antonio Gramsci (1929–1935 [1999]), Nye argues that agenda-setting and determining the discursive framework of the debate are vital precursors for obtaining and wielding soft power. According the Nye, "Soft power is . . . more than persuasion or the ability to move people by argument. It is the ability to entice and attract. And attraction often leads to acquiescence or imitation" (2002, 9). A country may achieve the outcomes it desires in world politics because other countries want to follow it or have agreed to a system that produces the same effect. The soft power of a country rests primarily on three resources: its culture (in places where it is attractive to others), its political values (when it lives up to them at home and abroad), and its foreign policy (when such policy is seen as legitimate and having moral authority) (Nye 2006). Summarized by Nye (2004) as a directing, attracting, and imitating force, soft power is derived mainly from intangible resources like national cohesion, culture, ideology, and influence on international institutions. While soft power is often associated with the US and the country's Cold War-era success in public diplomacy

(Wang 2006), other countries possess substantive resources of soft power, including China (Lueck, Pipps, and Lin 2014), Germany (Pamment 2013), Russia (Bogomolov and Lytvynenko 2012), Brazil (Budaev 2014), South Africa (Sidiropoulos 2014), South Korea (Kinsey and Chung 2013), and the supranational European Union (Khanna 2004). It seems, in fact, that to be without soft power in the twenty-first century is to be powerless; however, soft power is *not* nation branding. Like public diplomacy, soft power is a preferred tool of those seeking to motivate other states (usually in a clearly delineated "sphere of influence") to do something rather than buy, feel, or visit something. While a country with vast reserves of soft power is likely to have a marketable nation brand, the reverse—i.e., a country with a world-class brand possesses substantial soft power—is not necessarily true (a case in point: Switzerland). As such, it becomes clear that public diplomacy, soft power, and nation branding—while potentially linked—are separate concepts with differing tools, techniques, and outcomes. Nation branding, while often yoked to a state's relative power in the international system, is something altogether more nuanced than is often portrayed. A nation brand is a halo that positively distorts the image in front of the viewer, winnowing away warts and wrinkles and forcing the spectant to raise their gaze to the heavens, invariably projecting the spected against that ethereal realm.

Unlocking the matrix: Developing and measuring competitive identity

Why does a country seek to brand itself? The answers to the question are nearly infinite, but will likely revolve around a neoliberal desire to assist in one or more of the following areas: (1) to solidify answers to questions surrounding national identity; (2) to create a climate where innovation flourishes; (3) to increase chances of winning investment or attracting international events; (4) to promote tourism; (5) to improve foreign media representation; (6) to boost exports; (7) to assist efforts to integrate into regional trade and/or geopolitical organizations (Anholt 2007). If you are a country looking for branding services, there is no shortage of organizations, think tanks, and firms that are willing to assist (for a price). As the maven of all things nation branding, Simon Anholt (www. simonanholt.com) is probably the most recognized player in the field, especially given that he is credited with introducing the term "nation branding" into the vernacular of academia. As an independent policy advisor, Anholt's corporate web site states that he assists national, regional, and city governments in their efforts to "develop and implement strategies for enhanced economic, political and cultural engagement with other countries." At the national level, his client list includes Chile, Botswana, Korea, Jamaica, and Bhutan, as well as nearly 50 other countries. He maintains a strong influence over the field through his collaboration with German market research firm GfK, and the Sarasota, Florida-based Roper Center for Public Opinion Research, to produce the Anholt-GfK Roper Nation Brands Index (2009–present). He is also the founding editor of the academic journal *Place Branding and Public Diplomacy* and a member of the UK Foreign Office's Public Diplomacy Board.

Besides Anholt, there are number of other high-profile consultants who specialize in state branding. Robert Govers is the managing research partner of the "Good Country Index" (www.goodcountry.org) and the co-editor of the *International Place Branding Yearbook* series (2010–). Keith Dinnie, who literally wrote the (text)book on the subject—*Nation Branding: Concepts, Issues, Practices* (2008)—heads Brand Horizons (www.brandhorizons.com), an independent consultancy focused on nation branding and public diplomacy. Until his death in 2014, Wallace "Wally" Olins, former chairman of Wolff Olins and later Saffron Brand Consultants (http://saffron-consultants.com), was also an important force, with a list of clients that included Mauritius, Northern Ireland, Poland, Portugal, and Lithuania. In addition to these specialists, large marketing, corporate communications, and public relations firms also have practices specializing in the management of state brands, including Bell Pottinger, Ketchum, Hill & Knowlton, and LEVICK. There are also a good number of boutique firms, such as Anholt's former company Placebrands (www.placebrands.net) and Bloom Consulting (www.bloom-consulting.com). These consultancies provide services that go far beyond increasing tourism, a typical misconception about the focus of nation or place branding. Importantly, some states have decided to invest in developing their own branding expertise, including the International Marketing Council (www.brandsouthafrica.com), the "official custodians of Brand South Africa," and Iceland's public–private partnership "Promote Iceland." Other countries have significantly expanded their tourism boards and/or investment promotion agencies to include activities that go far beyond the scope of attracting visitors.

A variety of agencies track nation brands using a wide variety of measurements, including formal surveys (see Table 2.1). The aforementioned Anholt-GfK Roper Nation Brands Index is the most cited; however, there are other influential polls as well. Bloom Consulting produces the Country Brand Ranking, while the consultancy FutureBrand (www.futurebrand.com) produces its own Country Brand Index.[4] Brand Finance (www.brandfinance.com) has also introduced an Annual Report on Nation Brands. The global consulting firm Reputation Institute (www.reputationinstitute.com) produces an annual Country Reputation report which measures the top 50 countries in terms of trust, admiration, respect, and affinity. Individual elements of a nation's brand are also regularly ranked by a host of other organizations, intergovernmental bodies, and private firms.

The following areas are most often referenced by international brand consultants as key sources for promoting (or hindering) a given country's image abroad: (1) good governance; (2) an optimistic and friendly people; (3) suitability for foreign direct investment; (4) personal safety and security of property; (5) reliability and desirability of export products; (6) a unique cultural heritage; (7) attractive geography; (8) significant national treasures and tourism sites; (9) environmental and natural resource protection; and (10) adaptability to globalization (see van Ham 2001; Kotler and Gertner 2002; Anholt 2007; Aronczyk 2013). As Table 2.2 demonstrates, "First World" or developed countries enjoy marked advantages over "Third World" (developing) countries, as do older states versus newer ones. As the subsequent chapters will elucidate, the countries of the

Table 2.1 Sampling of nation brand report rankings, 2014–2015 (top 10)

Anholt-GfK Roper (2014)	FutureBrands (2014)	Brand Finance (2014)	Reputation Institute (2014)	Bloom Consulting— Tourism (2014–2015)	Bloom Consulting— Trade (2014–2015)
Germany	Japan	United States	Switzerland	United States	United States
United States	Switzerland	China	Canada	Spain	China
United Kingdom	Germany	Germany	Sweden	Germany	United Kingdom
France	Sweden	United Kingdom	Finland	Hong Kong	Hong Kong
Canada	Canada	Japan	Australia	France	Singapore
Japan	Norway	Canada	Norway	Thailand	France
Italy	United States	France	Denmark	Australia	Brazil
Switzerland	Australia	India	New Zealand	Macao	Australia
Australia	Denmark	Australia	Netherlands	China	India
Sweden	Austria	Brazil	Germany	Italy	Canada

Sources: Anholt-GfK Roper, FutureBrands, Brand Finance, Reputation Institute, and Bloom Consulting

post-Second World do not form a coherent group vis-à-vis nation branding. Certain countries such as Slovenia and Estonia score quite well across most of these categories, whereas other countries such as Turkmenistan and Tajikistan rank towards the bottom in more than a few (if not most) relevant areas.

Large, well-developed countries are most likely to command the top spots in nation brand reports. Not coincidentally, English-speaking nations (and countries with high propensity for the use of English like Germany, Switzerland, and Sweden[5]) account for more than half of the top 10 performers (see Figure 2.2).[6] The reasons for the skewered performance of highly developed countries stems from three factors: (1) embeddedness of national image; (2) respondent pools; and (3) use of radar-style data measurements. As discussed in Chapter 1, Western European countries possess the most deeply embedded national images, both inside and outside of Europe (a by-product of the dominance of Western civilization in the modern era combined with the legacy of imperial domination of much of the globe from 1500 to 1945). Beyond Western Europe, the countries of the United States, Japan, Russia, China, and India possess the most resonant national images, though the latter few of these suffer from Orientalist distortion/Occidentalist self-adulation. Reflecting post-imperial power structures embedded in the current world system, Canada, Australia, and New Zealand, while not European, are part of the so-called "White Dominions" (More 2014) of the British Empire, and are thus positively distinguished in ways that result in excellent nation

Table 2.2 Branding categories, factors, and high/low-scoring countries

Category	Factors	Representative high scorer	Representative low scorer
Governance	Rule of law, government effectiveness, control of corruption, political stability, accountability	Finland	Somalia
People	Openness to foreign visitors, gender/racial equality, happiness and optimism indexes	New Zealand	Bolivia
FDI	Advanced economy, regulatory regime, ease of doing business, workforce, return on investment (ROI), transparency	Singapore	North Korea
Security	Crime rates, prevalence of violence/conflict, government respect for private property	Hong Kong (China)	Zimbabwe
Exports	Attractive products, perceived quality, consumer awareness, authenticity, differentiation	United States	South Sudan
Culture	Historical patrimony, distinct character, cultural production, promotion of national values	France	Turkmenistan
Geography	Physical landscapes, flora/fauna, differentiation, climate, accessibility/transportation networks	Switzerland	Central African Republic
Tourist sites	Internationally recognized/historically important sites, ease of access, preservation efforts	Italy	Tajikistan
Environment	Protection of biodiversity, water/air/soil quality, regulation of pollution, health impacts	Australia	Haiti
Globalization	Global connectivity/integration, immigration, openness to change, does "good'" in the world	Belgium	Laos

Sources: United Nations, UNESCO, World Bank, World Economic Forum, Gallup, Legatum Institute, GfK, FutureBrand, Reputation Institute, Environmental Performance Index, A. T. Kearney/*Foreign Policy*

brands.[7] Countries from other geopolitical regions of the world (Eastern Europe, Latin America, Middle East, Central and Southern Asia, Southeast Asia, Oceania, and Sub-Saharan Africa) tend to score poorly by comparison.[8] Undoubtedly, there is a material economy of states that exists in the current structure of international relations. Whether one refers to the aforementioned "global supermarket" of nation brands (Anholt 2007), the "competition state" (Cerny 1997), or the natural outcome of the transition to "capitalist sovereignty" (Woodley 2015), states are cognizant of the rules of the neoliberal world system or the so-called "new economy," which originated in the post-World War II economic reforms instituted by the United States through the Bretton Woods system, was refined through the transition from heavy industry to an information-based economy, and was ultimately made global with the end of the Cold War in 1989 (see Saunders forthcoming). Consequently, states—whether they be ancient empires or novel postcolonial constructs—ineluctably started the process of self-commoditization.

The criteria for selecting respondents only magnifies these issues. The Reputation Institute, for example, polled 27,000 consumers in the G8 countries[9] in order to prepare its 2013 report (Adams 2013). The Anholt-GfK Roper Nation Brands Index survey approximately 20,000 consumers in 20 developed countries (Casanova 2013). Both reports limit their analyses to 50 pre-selected countries, predominately from the Global North, a fact which Anholt begrudgingly admits is "one of the great inequalities of the world" (2007, 126). Those reports that eschew "popularity contest"-style consumer voting, in which name recognition is paramount (e.g., Brand Finance), tend result in a more globally equitable distribution in their rankings (e.g., Malaysia scored in the top 10 in 2013). Lastly, the various methodologies of the above-mentioned reports all issue overall scores based on a multifaceted matrix of factors. The most visible manifestation of this approach is Anholt's methodology, which employs the Nation Brand Hexagon, a registered

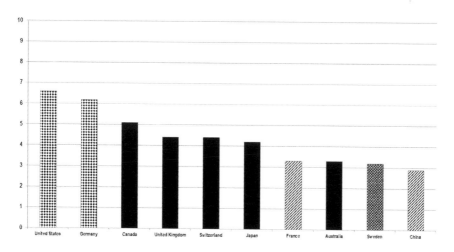

Figure 2.2 Composite nation brand scores: Top performers, 2014–2015

trademark. The Anholt system measures six categories: tourism, people, exports, culture/heritage, investment/(in-bound) immigration, and governance. In order to achieve the highest brand rankings, countries must receive stellar ratings in all categories (just a few will not do the job). While certain nations may score very well on tourism and geography (e.g., Thailand or Costa Rica), they may score poorly on other metrics such as governance or attracting inward investment. Other nations perform very well on investment or export metrics (e.g., United Arab Emirates and South Korea, respectively), but do not rank highly in fields such as cultural patrimony or the friendliness of the population towards foreign visitors.[10] Perhaps it goes without saying that with such a complicated array of metrics, the winners and losers of nation branding seem to be almost totally predetermined (at least for the foreseeable future).

Tools of the trade, or how to become a brand state

During the middle of the Cold War, IR scholar Robert Jervis declared, "Many factors about a state that contribute heavily to its image are permanent or semipermanent and thus beyond the control of its decision-makers" (1970, 13). However, in the era of postmodern diplomacy, this is not as true as it once was. Nations are now able to engage in "geopolitical staging" (Gagnon 2007) on an international level, a sort of latter-day, digitized model of the aforementioned "Potemkin Village," though one which—if sustained—can yield long-term results abroad as well at home. Using the tools and techniques of nation branding, states have enormous latitude to manipulate perceptions that, in a previous era, might have been seen as fixed and immutable. While a nation's brand must avoid overt fabrication and identifiable falsehoods, perceptions of authenticity can be achieved through hard work and a sustained commitment to "living the brand" (Aronczyk 2008). While each case of state branding is a unique undertaking (paralleling the haecceity of each nation), there is an established set of mechanisms that gird any campaign.

Whether a campaign is conducted internally or with the help of highly paid international consultants, a team of knowledgeable and dedicated individuals who are willing to oversee the cost-intensive project for years (perhaps decades) is an absolute necessity. Stakeholders in a variety of fields must be engaged and supportive of the undertaking; at a minimum, this includes key actors in government (the head of state and/or government needs to be "on board" and "on message"),[11] business, finance, industry, media, academia, civil society, and the culture industries. Best practices dictate that the tourism board and/or investment promotion agencies are invaluable members of the team, but the two should be balanced against one another to allow for other voices to be heard. Too much power in the hands of either will skew the brand, and prevent overall performance against the measurement matrixes.

Any good campaign should begin with a SWOT analysis, identifying strengths, weaknesses, opportunities, and threats to the brand. An open and honest debate about the issues at stake, as well as the content of the country's national image, must occur. Potential liabilities should be identified and addressed (if they cannot fixed, they should at least be understood in the larger context). These might include corruption, environmental degradation, human rights issues, societal/ethnic strife, and gender/racial inequality; notably, all these are Western neoliberal value sets, thus predetermining that postcolonial polities will have an uphill battle to positively brand themselves. The key to success lies in the realistic appraisal of how a given country is seen by the following: (1) its own citizens (including ethnic, linguistic, and religious minorities, as well as country's most disadvantaged socio-economic groups); (2) its immediate neighbors; (3) other states in the region but which are not contiguous; (4) great powers farther afield (US, China, Russia, and the EU); (5) intergovernmental organizations (UN, WTO, etc.); and (6) major transnational corporations. When and where possible, political, corporate, and other stakeholders should seek to alleviate grossly negative impacts on the brand, e.g., improve human rights, address environmental issues, root out corruption, etc.

The first major step in a brand management program is to identify core values for a country, which in turn will help define a coherent message for delivery to external audiences (see Passow, Fehlmann, and Grahlow 2005). Identifying the unique qualities of the nation is often one of the most delicate parts of this process. Elucidating the quiddity of a nation is often most difficult for its denizens, as they inherently understand what separates them from all other nations, but when they try to translate this to an outsider, the message often falls flat. The problem is that once you get beyond the immediate region, whether it is the Balkans, Southeast Asia, or West Africa, smaller countries tend to blend together in the minds of the larger international community (assuming a state does not possess nuclear weapons). Working through this blind spot can be difficult; however, as Anholt counsels, "Most countries, if they look hard enough, will find something that is theirs, and inherently competitive" (2007, 73). Opportunities for future differentiation should also be identified, allowing for growth over time (e.g., a small country may engage in substantial foreign aid and/or conflict resolution; by recognizing this as a brand asset, subsequent action in this area can be maximized and redound to the brand).

According to the literature, a vision for the nation is typically the first major deliverable, providing a clear and concise set of guidelines for promoting the desired image of the country. This is typically a mission statement that defines the most important "talking points" and how to engage outsiders.[12] Second, the scope of the campaign must be established, including regional targeting and what industries or sectors to pursue (e.g., exports, tourism, investment, education, immigration, etc.). Target audiences should also be mapped, with specific countries, regions, and multinational corporations being identified for the campaign. Third, discrete aims should be stipulated; according to Anholt (2007), the possible goals include all of the following: targeting, correcting, expanding, enhancing, revitalizing, improving, refuting, surprising, contextualizing, and deemphasizing. Next, slogans, signs, and symbols need to be identified for use in representing the nation. These will likely include: a country slogan, motto, or "strapline" (Olins 1999); a logo or stylized version of the national flag; a color palette for promotional material (not necessarily replicating the colors of the flag, which may be historic but not necessarily aesthetically pleasing); and stock photos representing the idealized "ethnos" (e.g., gingers for Ireland or *pardos* for Brazil) and/or an idyllic representation of space and place (geysers for Iceland or beaches for the Bahamas). Geographical and architectural icons should also be selected, chosen for their resonance as well as appeal, and properly situated in the large schema, as should key aspects of the country's cultural heritage (books, works of art, and historical artefacts). Intangible cultural heritage (language, song, music, drama, skills, lifeways, cuisine, and craft) should likewise be integrated. Important myths, historical figures, and celebrities should also be factored into the strategy, paying attention to how these might be adjusted or reoriented for foreign consumption. Products, corporations, and sports teams should also be considered for inclusion in the campaign.

Once ready for deployment, a branding program should be delivered across multiple platforms (print, television, Internet, etc.). However, the messaging should be generally consistent across all venues. Print campaigns are often used to target foreign investors and the diplomatic elite. A tried-and-true method, though one which is arguably a bit overdone, is the targeted "special advertising supplement." Both *The New York Times* and *The Telegraph* have made ample space for inserts by states including the Russian Federation and the People's Republic of China, among others. The glossy international relations magazine *Foreign Policy* is the favored venue for these branding efforts, as the publication allows countries to insert an 8–16-page spread with vital statistics about the country, color graphics, advertising space for important local industries, and interviews with key policy-makers and business executives. With a circulation of more than 200,000 diplomatic elites, almost 18,000 of whom are in Washington, D.C. alone, *FP* markets its readership as the "Who's Who of Influentials." However, the obvious problem with this form of media-based branding is that everyone else is doing it too. Angola, Cape Verde, São Tomé and Principe, Mozambique, Ghana, and Myanmar have all opted for the *FP* special advertising supplement in recent years, each with a comparatively similar set of things to say about economic growth, a vibrant population, increasing stability, reduced levels of corruption, and so on. More specific targeting is often required for FDI, including industry periodicals and various forms of direct and indirect marketing, including brochures, billboards, etc.

Television advertising and "infomercials" are also a commonly used mechanism for developing the nation brand, particularly in the area of tourism. The global satellite channels CNN International, Euronews, and Eurosport are the most popular venues for such endeavors, though other media networks carry similar advertisements. Typically delivered in English, such ads are used to attract visitors to the country, while also serving to redefine (or, in some cases, define) the national image of the country. Just as *Foreign Policy* special advertising supplements target a niche group, so do these TV spots. By reaching out to a jet-setting global elite, such commercials and informational programs influence opinion-makers, who in turn have a powerful capacity to become follow-on brand agents after returning from their travels to new countries. With household penetrations of 60% or greater in the UK, Ireland, France, Portugal, the Benelux countries, Germany, Austria, Switzerland, and Poland, Euronews has been able to attract long-term contracts from the tourism boards of Russia, Azerbaijan, and the Philippines, among other countries. CNN has similar deals with Cambodia, Malawi, and Croatia. Besides commercial spots, developing close relationships with content providers allows for more in-depth and "unbiased" reporting on historical sites, tourism attractions, and natural landscapes. Allowing overseas-based reality TV shows such as *Survivor* (2000–), *Man vs. Wild* (2006–2011), and *The Amazing Race* (2001–) has also been a boon to many small countries seeking eco-tourism, including Nicaragua, Vanuatu, and Palau. "Edutainment" venues such as the *Worlds Apart* (2003–) television program on National Geographic Channel have also been used for nation branding efforts (see Sinha Roy 2007). Likewise, the international platform of Eurovision (discussed in Chapter 8) provides smaller states with a venue for making their identities known and valued (at least within a European context).

The use of the Internet and various new media tools have now become vital to any successful nation branding strategy. The projection of a positive image online, and calculated social-networking site messaging associated with a state's key "brand assets," is now as important as offline national image management. Strategies to project a state's brand in cyberspace should always begin with developing an "attractive, easy-to-navigate, and content-rich web site or portal for the national government (president, parliament, ministry of tourism, etc.), though any digital branding program should also include substantial 'push' content alongside 'pull' content like web sites" (Saunders 2014, 148). As the Internet replicates and reinvents all previous platforms of media transmission (text, radio, film/TV, etc.), while also generating new structures of communication (instant messaging, blogs, podcasts, social networking sites, etc.), any branding campaign must be multifaceted. Nation branding via new media should include web pages, interactive content, Twitter feeds, a Facebook page, and online videos delivered via a dedicated YouTube channel, as well as other content aggregators. Additionally, a wellspring of downloadable content that positively represents the brand (screensavers, fact sheets, recipe books, folk songs, etc.) should be made available. Management of a country's web presence is also important to ensure that when one searches for Chile, one ends up on the front page of the national portal and not on a web site for a barbecue restaurant in northern Texas.

Lastly, overseas "brand agents" should be tapped and leveraged as the campaign is rolled out. These resources might include a diaspora, foreign companies with a market presence in the country, domestic firms with an international presence, etc. Certain countries have influential cultural diplomacy resources that can be utilized, e.g., the British Council (UK), Goethe-Institut (Germany), or the Confucius Institutes (China); however, most states lack such resources. Instead, small-scale facilities to inform and educate, as well as promote the brand, may be established. Once the brand campaign is in place, stakeholders and brand managers must constantly revise and revitalize the effort, paying special attention to domestic and foreign activities that might interrupt or disrupt the campaign. As the subsequent chapters detail, the most pernicious influences come in the form of popular-cultural production, especially Hollywood (though the actions of domestic actors can also be a problematic issue). Other deleterious forces include war, government malfeasance, ethnic conflict, environmental disasters, and exposure of corruption. Like any large corporation, a nation should be on guard and prepared for damage control to deal with such challenges to the nation brand.

The brand state: An overview of three successful branding campaigns

There have been many attempts at (re)branding states over the past few decades, with differing levels of success. Some efforts have been lambasted as half-hearted or counter-productive, most notably the United States' abortive attempt to use branding techniques to rehabilitate its image in the wake of 9/11 and the US-led invasion of Afghanistan and Iraq. Headed by former ad executive turned undersecretary of state for public diplomacy and public affairs Charlotte Beers, this attempt at blunting anti-Americanism ended in ridicule, with the *Guardian* newspaper running headlines like "America is Not a Hamburger" (Klein 2002) and quips about "Uncle Sam" being switched for "Uncle Ben's," a popular instant rice formerly marketed by Beers (Carlson 2001). While the "selling of America" (Pappas 2001) via a Madison Avenue-style image makeover clearly failed, other efforts have proved much more successful, particularly among small states with underdeveloped national images or in instances where the state has undergone intense political and socio-economic change. New Zealand and Qatar are representative examples of the former, while Spain serves as a paragon of the latter.

New Zealand: Distant, pure, and Tolkienesque

Island states are choice candidates for brand campaigns. As a consequence of their geography, these nations, generally speaking, tend to possess significant physical distance from their neighbors, easily defensible borders, and attractive landscapes. In turn, these attributes endow such countries with easily distinguishable national images that are most often associated with serenity and pleasure. These generalizations do not apply to all insular states (certainly Great Britain and Japan, as powerful—and historically aggressive—imperial powers, defy this abstraction); however, island nations score well on people, tourism, and geography indexes. Yet, economics is typically a weak area as the cost of transportation and logistical challenges hamper development. Additionally, many Caribbean and Oceanic nations have suffered from political instability following decolonization, leading to problems with governance and personal security metrics. However, New Zealand is an archipelago nation which performs well across all categories, typically ranking in the top 20 of all nation brands.

A former British colony in the southwestern Pacific Ocean, New Zealand has a population of 4.5 million and a GDP per capita equivalent to that of Italy. Relations between the indigenous Māori population and descendants of British settlers (Pākehā) are comparatively amiable,[13] and the country is renowned for its unique and striking landscapes. Once described as "little green garden at the bottom of the world" (Gold and Kelly 2006, 233), the country has employed Tourism New Zealand (TNZ) to leverage its picturesque geography, temperate climate, English-speaking population, and Anglo-Saxon sensibilities to create a powerful brand state for the twenty-first century. While stepping out of the shadow of its larger neighbor Australia has proved difficult, New Zealand has become adept at managing its image, especially through the use of its verdant locales, iconically showcased in native-born Peter Jackson's *Lord of the Rings* (2001–2003) and *Hobbit* trilogies (2012–2014).[14] The images of unspoiled nature effortlessly reinforce the country's brand identity based on its "pure" natural environment and ecologically sound produce (BFJ 2012).

Launched in late 1999 and initially targeting the US, UK, Japan, Germany, and Singapore, the "100% Pure" brand is the basis of the country's brand architecture, which is managed by the government to generate tourism, attract investment, and promote exports (Morgan, Pritchard, and Morgan 2002). Once a victim of its antipodean geography, New Zealand has mobilized the perceptual "shrinking of the world" via globalization (especially cheaper flights and cargo shipping); this has redounded to the nation's brand, which is now happily based on *distance from* rather than *proximity to* the rest of the world. Over its first decade, the campaign produced measurable dividends, including a 60% increase in international visitors (Hickton 2010), with 6% of these visitors claiming *LoTR* as the primary reason for their trip (TNZ n.d.). As the national economy is dependent on developing its exports, the campaign had "clear economic goals from the outset" (Dinnie 2008, 28) and worked to combat the fact that New Zealand had been hitherto "invisible on the world business stage" (Brodie and Sharma 2011, 7); specific target products include sports teams, salmon, wine, and hops. In 2007, TNZ launched its "The Youngest Country" campaign to market the country as a workplace destination for skilled immigrants and returning Kiwis.

Qatar: Old, new, and wealthy

Qatar, another small country with a history of British rule, has also done well in burnishing its national image in recent decades, recently being labeled the world's "fast growing" (Trade Arabia 2014) and "best performing" nation brand (BF 2014), worth some $256 billion. As a region, the Middle East evokes a variety of associations that can hamper an individual nation's brand: (political) instability, (religious) intolerance, (geographical/ethnic) sameness, etc. However, the smaller Arab nations of the Persian Gulf have proven adept at distinguishing their country brands in recent decades, using techniques such as cutting-edge architecture, lavish national museums, and heritage investment to project an image of "tribal modern" via the spectacularization of progress (see Cooke 2014). With small (indigenous) populations and enormous hydrocarbon wealth, the Gulf States have transformed their countries from sleepy maritime outposts into sites for international investment, global logistics, world-class universities, and tourism. With its terraforming projects (Palm Island), the world's tallest building (Burj Khalifa), and the first seven-star hotel (Burj al-Arab), the UAE typically receives the most international attention; however, Qatar ranks as one of the world's great branding stories.

Sharing its only land border with Saudi Arabia, Qatar is a small peninsular state that juts into Persian Gulf. A former Ottoman province and British protectorate, this Muslim country is now ruled by the Al Thani family. With a population of about 2 million, Qatar is the richest country in the world and possesses vast reserves of natural gas and oil. Suffering from an twentieth-century image founded on "every Middle Eastern cliché in the book" (Mattern 2008, 479), post-millennial Qatar is a known and positive quotient due to the "determined image building and reputation management strategy" implemented by Hamad bin Khalifa, father of the current emir (North 2012). Aiming to diversify its economy and raise its international profile, Qatar has arrayed a unique suite of tools to position its brand as "business-orientated, modern, savvy," as well as one which is "positive, populist and enlightened" (Roberts 2012, 236). Most notably, these include the international news broadcaster Al Jazeera, world-class sporting events like the 2022 FIFA World Cup,[15] the Qatar Science and Technology Park, and the I. M. Pei-designed Museum of Islamic Art. With money being "no object," Qatar has emerged as the premier "laboratory" of place branding (Wardrop 2012). Qatar's reputation has also benefited from its overseas activities, including investment in foreign sports teams, erecting huge skyscrapers like London's "Shard,"[16] and an activist foreign policy in the Arab world, "wedding political and economic power to cultural power" (Cooke 2014, 170). Attention to detail has characterized this micro-state's approach, including everything from an Arabic logo that suggests "global symbolic exchange" (Mattern 2008) to leveraging Qatar Airways to raise the country's international profile (Peterson 2006).

Spain: Sunny, savvy, and democratic

Nation branding campaigns present old nations with a double-edged sword; every country can benefit from sprucing up its national image in the current era of global commerce, but being seen as inauthentic can do real damage (as was the case with the aforementioned Bush-era "rebranding" of America), while tinkering too much with established brands can prompt confusion, ridicule, and even hostility (as some scholars have claimed occurred with the "Cool Britannia" campaign of the late 1990s) (see Werther 2011). As previously argued, the countries of Western Europe enjoy the most deeply ingrained nation images (for good or ill), so any government looking to retrofit its identity faces innumerable challenges. However, in the case of 1980s Spain, a country emerging from decades of authoritarian rule by the dictator Francisco Franco, action was clearly necessary if the population hoped to convince the outside world that change was really afoot.

Dominating the Iberian Peninsula between the Atlantic Ocean and the Mediterranean, Spain is a geophysically diverse country of highland plateaus, mountains, coastal lowlands, and Mediterranean islands. Its temperate climate and sunny beaches are mainstays of its tourism industry, alongside historically important cities like Madrid, Barcelona, Seville, and Santiago de Compostela. Its population of 48 million is divided between the numerically dominant Castilians, along with minority populations of Catalans, Gallegos, Basques, and other groups. A major player in European and international politics for centuries, Spain's history is robust, as are its cultural contributions to world literature, cuisine, art, and music. Following Franco's death in 1975, Spain began democratic reforms and the process of reintegration into the (Western) European family of nations. The most tangible manifestations of this shift in political alignments were NATO membership (1982) and admission to the European Economic Community (1986).

Given a success which transformed associations of Spain with poverty and despotism into feelings of warmth and good cheer engendered by the now-famous "Sol de Miró" logo, many scholars of nation branding see Spain's revitalization of its international image as proof that a country can be "rebranded" (see Gilmore 2002; Aronczyk 2007; Carmichael 2008). In fact, some have gone even further, suggesting that the campaign actually changed the composition of the Spanish state:

> It would be hard to find someone who would argue that Spain's branding efforts were not the key to its modern transformation. Yes, it had a long and rich history and fabulous tourist destinations, but without "collecting" them under a unifying and heavily promoted brand, the development of the tourism industry, and with it the nation as a whole, could not have been achieved. (Cromwell and Kyriacou 2014)

Certainly, the most important branding success came with the winning and ultimate hosting of the 1992 Olympiad in Barcelona. However, there were numerous other elements of Spain's repositioning in the global supermarket of nations along the way, including cultural production (especially the films of Pedro Almodóvar), architectural innovation (e.g., the Guggenheim Museum in Bilbao and the work of Santiago Calatrava), economic integration with Europe, and the development of "mass tourism" (see Martínez Expósito 2011). With the dawn of the new millennium, Spain doubled down on the active management of its national image, launching the Marca España, an overarching branding architecture intended to "improve the image of [the] country both domestically and beyond [its] borders for the common good." The new program focused on Spain as an avant-garde, tech-centric, and fully modern European country (Marca España 2014). Mobilizing stakeholders from sport, commerce, and finance (including international heavyweights such as Zara, Real Madrid, F.C. Barcelona, and Banco Santander), Spain produced a successful follow-up to its branding success of the late twentieth century (see Rius Ulldemolins and Martín Zamorano 2014). While the country's brand has faced challenges since the 2008–2009 global economic recession (most notably, its inclusion among the so-called "PIGS" of Europe, alongside Portugal, Italy, and Greece), Spain remains a testament to the benefits of a long-term commitment to the ethos of being a brand state.

While these three brand states represent only small fraction of the myriad activities currently going on in the protean realm of twenty-first-century geopolitics, they do suggest how markedly different international relations has become in the new millennium. Whereas states once relied on diplomats, treaties, and armies to make their presence known and to gain respect, the focus is increasingly shifting into the realm of popular culture, product placement, and "feel-good" promotion. This sea change in global affairs percolates through many strata of power, affecting identity and loyalties along the way. While it is common to witness hyperbole when it comes to the importance of brand in the international community, it is foolhardy to assume that it is a trivial or even marginal aspect of contemporary statecraft. From Kinshasa to Kyoto, from Cagliari to Caracas, from Seoul to San Francisco, agents of the nation are seeking to distinguish their countries on the global stage, using any and all tools at their disposal. Consequently, when cultural producers in another country turn their gaze on a particular state (especially when that eye is a satirical, jaundiced, or overtly antagonistic), problems are bound to ensue. Recent furores over films such as *The Interview* (2014), *Borat: Cultural Learnings of America for Make Benefit Glorious Nation of Kazakhstan* (2006), or *Hostel* (2005) just scratch the surface. In the next chapter, we will explore the role played by mediated representations in sculpting international views of various nations, and the larger implications of the relationship between popular culture and world politics.

Notes

1 It can be argued that a given state's nation brand is that which makes a country special in relation to all others and is always represented in positive terms, whereas national identity is a neutral concept, reflecting the good and bad of any given people's overall identity. Nationalism can be distinguished from the previous two fields of identity in that it is an ideology based on the supremacy of one's own nation over all others without the need for factual information to support the supposition (or alternatively, the ideology of another nation's supremacy which *pro forma* is flawed as it does not recognize the self-endowed supremacy of the *viewing* nation). National image is perhaps the most difficult of all to pin down on a conceptual level as it cuts both ways, reflecting the negatives and positives of *Selbstbild* as well as *Fremdbild* depending on the vantage of the spected/spectant.

2 In my own writing, I have identified the origins of nation branding in the Cold War, arguing that the competition for "hearts and minds" that characterized the propaganda battles between Moscow and Washington steadily morphed into a struggle for mind-space market share based on geo-capitalist imagineering of "dreams" of other places and spaces and an international acceptance of the universality of "trademarks" in the neoliberal global marketplace, phenomena that reached full fruition with the end of command-and-control economies in the Eastern Bloc and (to all intents and purposes) the world's largest market, China, in 1991 (see Saunders forthcoming).

3 Historically, the relationship between mass media and public diplomacy in the United States is a close one, with magazines, radio programming, film, and television playing vital roles. The fact that the current undersecretary of state for public diplomacy and Public Affairs, Richard Stengel, was the managing editor of *Time* magazine (2006–2014) before assuming his position only serves to underscores this.

4 The firm also produces a "Made In" report focusing on export products and the "country of origin effect" (see Dinnie 2004); in the 2013 report, the United States scored the highest, followed by France and Germany.

5 In Education First's English Proficiency Index (2013), Sweden scored the highest among measured non-English-speaking countries, while Germany (14th) and Switzerland (17th) both ranked in the "High Proficiency" categories.

6 Regarding the methodology, each country was given a score based on placement in each of the reports listed in Table 2.1 (for Bloom Consulting, the "Trade" and "Tourism" scores were combined). A first-place ranking earned 10 points, second-place 9 points, and so on. These scores were then averaged across all reports. The highest-ranked country, the United States, scored 6.6 out of a possible 10. Countries with a solid background are from Western Europe, whereas North American countries are checkered, Asian countries have diagonal lines, and the nations of Oceania are shown with bubbles.

7 In keeping with the notion of Gramscian hegemony (Gramsci 1929–1935[1999]), such perceptions of "excellence" derive from invisible yet nonetheless predetermined expectations of what is valued in a world where capitalism is the only possible system.

8 There are a few notable exceptions to this rule, especially among the so-called "Asian Tigers" (South Korea, Taiwan, Singapore, and Hong Kong). Brazil and South Africa have also enjoyed levels of success that distinguish them from their regional counterparts. "Boutique nations" (see Klieger 2012) also do well, e.g., in FutureBrand's 2012–2013 report, the island nations of Mauritius and the Maldives both scored in the top 20.

9 This report was compiled before Russia's suspension from the group of leading industrialized economies in 2014, following the annexation of Crimea.

10 Dinnie (2011) identifies South Koreans' sense of ethnic superiority, rejection of "multicultural families," and "poor treatment" of foreigners as the country's greatest challenges in developing its nation brand.

11 In recent years, there have been a number of nation branding campaigns that were smothered in the cradle due to presidential interference in or alienation from the project. In other cases, executive actions have undermined campaigns by engaging in highly visible actions which short-circuit the brand, e.g., engaging in human rights violations or supporting anti-homosexuality legislation as was the case in Uganda (see Amalu 2013). By contrast, best-case scenarios involve executive-level implementation of a "brand-informed policy" (see Anholt 2007, 31–2).

12 For a representative example, see "Brand Estonia—Welcome to Estonia," available at http://photos.visitestonia.com/eng/videos/oid-1577/?#id=1577.

13 Unlike in Australia, where intense racism, forced resettlement, and cultural genocide were the norm, Māori–Pākehā interactions have tended to be fairly positive, with the exception of the latter third of the 19th century when land disputes soured relations.

14 It is difficult to quantify the value of this exposure given that hundreds of millions have viewed these films worldwide, but it is regularly referenced in relation to expanding the New Zealand brand as the result of Jackson's assertion that his homeland is the "perfect setting" for realizing J. R. R. Tolkien's Middle-earth (TNZ n.d.).

15 Due to a number of scandals within the FIFA, Qatar's winning of the 2022 World Cup has cut both ways, with the emirate being seen as a willing participant in the organization's widespread corruption.

16 In October 2014, Al Jazeera relocated its European operations to the sixteenth floor of the new building, further linking the image of Qatar with this imposing newcomer to the London skyline.

References

Adams, Susan. 2013. "The World's Most Reputable Countries, 2013." *Forbes*, available at www.forbes.com/sites/susanadams/2013/06/27/the-worlds-most-reputable-countries-2013/ [last accessed 9 September 2014].

Amalu, Ngozika. 2013. "A Brand Apart: Nation Branding in a More Competitive Africa." *Brenthurst Foundation*, available at www.thebrenthurstfoundation.org/a_sndmsg/news_view.asp?I=134877&PG=288 [last accessed 12 September 2014].

Anderson, Benedict. 1991. *Imagined Communities: Reflections on the Origin and Spread of Nationalism*. London: Verso.

Anholt, Simon. 2006. "Public Diplomacy and Place Branding: Where's the Link?" *Place Branding* no. 2:271–275.

——. 2007. *Competitive Identity: The New Brand Management for Nations, Cities and Regions*. Houndsmill, UK: Palgrave Macmillan.

——. 2008. "From Nation Branding to Competitive Identity – The Role of Brand Management as a Component of National Policy." In *Nation Branding: Concepts, Issues, Practice*, edited by Keith Dinnie, 22–23. Oxford: Butterworth-Heinemann.

Aronczyk, Melissa. 2007. "New and Improved Nations: Branding National Identity." In *Practicing Culture*, edited by Craig Calhoun and Richard Sennett, 105–128. New York: Routledge.

——. 2008. "'Living the Brand': Nationality, Globality and the Identity Strategies of Nation Branding Consultants." *International Journal of Communication* no. 2:41–65.

——. 2013. *Branding the Nation: The Global Business of National Identity*. Oxford and New York: Oxford University Press.

Bardan, Alice, and Anikó Imre. 2012. "Vampire Branding: Romania's Dark Destinations." In *Branding Post-Communist Nations: Marketizing National Identities in the "New" Europe*, edited by Nadia Kaneva, 168–192. New York and London: Routledge.

BF. 2014. "Annual Report on Nation Brands." *Brand Finance*, available at www.brandfinance.com/knowledge_centre/reports/brand-finance-nation-brands-2014 [last accessed 27 July 2014].

BFJ. 2012. "The "Pure" New Zealand Brand." *Brand Finance Journal*, available at www.brandfinance.com/knowledge_centre/stories/the-pure-new-zealand-brand [lastaccessed 15 September 2014].

Björner, Emma. 2010. *Nation Branding at World Expositions: Sweden's Brand Architecture at Expo 2010*. Stockholm: Stockholm University.

Blanchard, Pascal, Gilles Boetsch, and Nanette Jacomijn Snoep. 2011. *Human Zoos: The Invention of the Savage*. New York: Distributed Art Publications.

Bogomolov, Alexander, and Oleksandr Lytvynenko. 2012. "A Ghost in the Mirror: Russian Soft Power in Ukraine." *The Aims and Means of Russian Influence Abroad Series*. London: Chatham House. Available at www.chathamhouse.org/publications/papers/view/181667 [last accessed 23 March 2016].

Brodie, Roderick J., and Rahul Sharma. 2011. "Nation Branding for New Zealand Exports: Developing Distinctive Meaning." *University of Auckland Review* no. 14 (1):7–17.

Budaev, A. 2014. "Brazil: Soft Power in Foreign Policy." *International Affairs: A Russian Journal of World Politics, Diplomacy & International Relations* no. 60 (4):63–70.

Carlson, Margaret. 2001. "Can Charlotte Beers Sell Uncle Sam?" *Time*, available at http://content.time.com/time/nation/article/0,8599,184536,00.html [last accessed 14 February 2009].

Carmichael, Ben. 2008. "From Despotism to Destination." *Print* no. 62 (1):72–75.

Casanova, Marco. 2013. "Rated Ranking: Anholt-GfK Roper Nation Brands Index 2013." *Branding-Institute CMR*, available at www.branding-institute.com/rated-rankings/anholt-gfk-roper-nation-brands-index [last accessed 9 September 2014].

Cerny, Philip G. 1997. "Paradoxes of the Competition State: The Dynamics of Political Globalisation." *Government and Opposition* no. 49 (4):595–625.

Cooke, Miriam. 2014. *Tribal Modern: Branding New Nations in the Arab Gulf*. Berkeley and Los Angeles: University of California Press.

Crang, Mike. 2003. "On Display: The Poetics, Politics and Interpretation of Exhibitions." In *Cultural Geography in Practice*, edited by Alison Blunt, Pyrs Gruffudd, Jon May, Miles Ogborn, and Pinder Pinder, 255–268. Oxford and New York: Oxford University Press.

Cromwell, Thomas, and Savas Kyriacou. 2014. "Nation Branding: The Concept and Benefits of Nation Branding." *Diplomatic Traffic*, available at www.diplomatictraffic. com/opinions_archives.asp?ID=75 [last accessed 16 September 2014].

Cull, Nicholas. 2006. "'Public Diplomacy' Before Gullion: The Evolution of a Phrase." *USC Public Diplomacy Blog*, available at http://uscpublicdiplomacy.org/blog/060418_public_diplomacy_before_gullion_the_evolution_of_a_phrase [last accessed 11 September 2014].

Curry Jansen, Sue 2008. "Designer Nations: Neo-liberal Nation Branding—Brand Estonia." *Social Identities* no. 14 (1):121–142.

Damjan, Janez. 2005. "Development of Slovenian Brands: Oldest Are the Best." *Place Branding* no. 1:363–372.

Dinnie, Keith. 2004. "Country-of-Origin 1965–2004: A Literature Review." *Journal of Customer Behaviour* no. 2 (1):165–213.

——. 2008. *Nation Branding: Concepts, Issues, Practice*. Oxford: Butterworth-Heinemann.

——. 2011. "Nation Branding, Cultural Diplomacy and Economic Objectives: Recent Practice in South Korea, Japan, China and France." In *Berlin International Economics Congress*, edited by Institute for Cultural Diplomacy. Berlin.

Education First. 2013. "English Proficiency Index." *Education First*, available at www. ef.edu/epi/ [last accessed 9 September 2014].

Fukuyama, Francis. 1992. *The End of History and the Last Man*. New York: Free Press.

Gagnon, Serge. 2007. "L'intervention de l'État québécois dans le tourisme entre 1920 et 1940." *Hérodote* no. 4 (127):151–166.

Gienow-Hecht, Jessica C. E., and Mark C. Donfried. 2013. *Searching for a Cultural Diplomacy*. London and New York: Berghahn Books.

Gilboa, Eytan. 2008. "Searching for a Theory of Public Diplomacy." *Annals of the American Academy of Political and Social Science* no. 616 (1):55–77.

Gilmore, Fiona. 2002. "A Country—Can It Be Repositioned? Spain—The Success Story of Country Branding." *Brand Management* no. 9 (4–5):281–293.

Gold, Sarah, and Shannon Kelly. 2006. *New Zealand 2007*. New York: Fodor's Travel Publications.

Gramsci, Antonio. 1929–1935 [1999]. *Selections from the Prison Notebooks*. New York: International Publishers.

Greenblatt, Stephen. 1991. "Resonance and Wonder." In *Exhibiting Cultures: The Poetics and Politics of Museum Display*, edited by Ivan Karp and Steven D. Lavine, 42–56. Washington, DC: Smithsonian Institution Press.

Halsall, Robert. 2008. "From 'Business Culture' to 'Brand State': Conceptions of Nation and Culture in Business Literature on Cultural Difference." *Culture & Organization* no. 14 (1):15–30.

Hickton, George. 2010. *Celebrating 10 Years of 100% Pure New Zealand*. Wellington: Tourism New Zealand.

Hoffmann, Gerhard. 2005. *From Modernism to Postmodernism: Concepts and Strategies of Postmodern American Fiction*. Amsterdam: Rodopi.

Jackson, Daniel M. 2008. "Sonic Branding—Capturing the Essence of a Nation's Identity." In *Nation Branding: Concepts, Issues, Practice*, edited by Keith Dinnie, 124–127. Oxford: Butterworth-Heinemann.

Jay, Martin. 1988. "Scopic Regimes of Modernity." In *Vision and Visuality*, edited by Hal Foster, 3–23. Seattle, WA: Bay Press.

Jervis, Robert. 1970. *The Logic of Images in International Relations*. Princeton, NJ: Princeton University Press.

Kahn, Jeremy. 2006. A Brand-New Approach. *Foreign Policy*, November–December, 92.

Kaneva, Nadia. 2011. "Nation Branding: Toward an Agenda for Critical Research." *International Journal of Communication* no. 5:117–141.

——. 2012. "Who Can Play This Game? The Rise of Nation Branding in Bulgaria, 2001–2005." In *Branding Post-Communist Nations: Marketizing National Identities in the "New" Europe*, edited by Nadia Kaneva, 99–123. New York and London: Routledge.

Karp, Ivan. 1992. "Introduction." In *Museums and Communities: The Politics of Public Culture*, edited by Ivan Karp, Christine Mullen Kreamer and Steven Lavine, 1–17. Washinton, DC: Smithsonian Institution Press.

Keohane, Robert. 1984. *After Hegemony: Cooperation and Discord in the World Political Economy*. Princeton, NJ: Princeton University Press.

Khanna, Parag. 2004. "The Metrosexual Superpower." *Foreign Policy* no. 143:66–68.

Kinsey, Dennis F., and Myojung Chung. 2013. "National Image of South Korea: Implications for Public Diplomacy." *Exhange: The Journal of Diplomacy* no. 4 (1):5–16.

Klein, Naomi. 2002. "America is Not a Hamburger: President Bush's Attempts to Rebrand the United States are Doomed." *The Guardian*, available at www.theguardian.com/media/2002/mar/14/marketingandpr.comment [last accessed 14 February 2009].

Kleppe, Ingeborg Astrid. 2002. "Country Images in Marketing Strategies: Conceptual Issues and an Empirical Asian Illustration." *Journal of Brand Management* no. 10 (1):62.

Klieger, P. Christiaan. 2012. *The Microstates of Europe: Designer Nations in a Post-Modern World*. Lanham, MD: Lexington Books.

Kotler, Philip, and David Gertner. 2002. "Country as Brand, Product, and Beyond: A Place Marketing and Brand Management Perspective." *Journal of Brand Management* no. 9 (4–5):249–261.

Kunczik, Michael. 1997. *Images of Nations in International Public Relations*. Mahwah, NJ: Lawrence Erlbaum Associates.

Lord, Carnes. 2006. *Losing Hearts and Minds? Public Diplomacy and Strategic Influence in the Age of Terror*. Westport, CT: Greenwood Publishing Group.

Low, Theodore. 1942 [2004]. "What is a Museum?" In *Reinventing the Museum: Historical and Contemporary Perspectives on the Paradigm Shift*, edited by Gail Anderson, 30–43. Lanham, MD: AltaMira Press.

Lueck, Therese L., Val S. Pipps, and Yang Lin. 2014. "China's Soft Power: A *New York Times* Introduction of the Confucius Institute." *Howard Journal of Communications* no. 25 (3):324–349.

Marat, Erica. 2009. "Nation Branding in Central Asia: A New Campaign to Present Ideas about the State and the Nation." *Europe-Asia Studies* no. 61 (7):1123–1136.

Marca España. 2014. "What is Marca España? We Are All Marca España." *Marca España*, available at http://marcaespana.es/en/quienes-somos/que-es-marca-espana.php [last accessed 16 September 2014].

Martínez Expósito, Alfredo. 2011. "Screening Nation Brands: New Zealand and Spanish Perspectives on Film and Country Image." *Journal of New Zealand Studies* no. 11:135–150.

Mattern, Shannon. 2008. "Font of a Nation: Creating a National Graphic Identity for Qatar." *Public Culture* no. 20 (3):479–496.

Moore, Sarah J. 2013. *Empire on Display: San Francisco's Panama-Pacific International Exposition of 1915*. Norman: University of Oklahoma Press.

More, Charles. 2014. *Britain in the Twentieth Century*. London and New York: Routledge.

Morgan, Nigel, Annette Pritchard, and Rachel Morgan. 2002. "New Zealand, 100% Pure. The Creation of a Powerful Niche Destination Brand." *Brand Management* no. 9 (4/5):335–354.

Musée du quai Branly. 2012. "Human Zoos: The Invention of the Savage." *Musée du quai Branly*, available at www.quaibranly.fr/en/programmation/exhibitions/last-exhibitions/human-zoos.html [last accessed 10 September 2014].

North, Samantha. 2012. "Qatar: All That Glitters . . ." *Places: A Critical Geographical Blog*, available at http://blog.inpolis.com/2012/11/22/qatar-all-that-glitters/ [last accessed 16 September 2014].

Nye, Joseph S., Jr. 1990. *Bound to Lead: The Changing Nature of American Power*. New York: Basic Books.

———. 2002. *The Paradox of American Power: Why the World's Only Superpower Can't It Alone*. New York: Oxford University Press.

———. 2004. *Soft Power: The Means to Success in World Politics*. New York: Public Affairs.

———. 2006. "Think Again: Soft Power." *Foreign Policy*, available at www.foreignpolicy.com/articles/2006/02/22/think_again_soft_power [last accessed 9 December 2008].

Olins, Wally. 1999. *Trading Identities: What Countries and Companies Can Learn from Each Other's Roles*. London: Foreign Policy Centre.

———. 2002. "Branding the Nation—The Historical Context." *Brand Management* no. 9 (4/5):241–248.

———. 2006. "Making a Nation Brand." In *The New Public Diplomacy: Soft Power in International Relations*, edited by Jan Melissen. London: Palgrave Macmillan.

Pamment, James. 2013. "Time, Space and German Soft Power: Toward a Spatio-Temporal Turn in Diplomatic Studies?" *Perspectives: Central European Review of International Affairs* no. 21 (2):5–25.

Pappas, Charles. 2001. "The Selling of America." *Advertising Age* no. 51:1–23.

Passow, Tanja, Rolf Fehlmann, and Heike Grahlow. 2005. "Country Reputation—From Measurement to Management: The Case of Liechtenstein." *Corporate Reputation Review* no. 7 (4):309–326.

Peterson, J. E. 2006. "Qatar and the World: Branding for a Micro-State." *Middle East Journal* no. 60 (4):732–748.

Rius Ulldemolins, Joaquim, and Mariano Martín Zamorano. 2014. "Spain's Nation Branding Project *Marca España* and Its Cultural Policy." *International Journal of Cultural Policy* doi 10.1080/10286632.2013.877456:1–21.

Roberts, David B. 2012. "Understanding Qatar's Foreign Policy Objectives." *Mediterranean Politics* no. 17 (2):233–239.

Rusciano, Frank L. 2003. "The Construction of National Identity—A 23-Nation Study." *Political Research Quarterly* no. 56 (3):361–366.

Saunders, Robert A. 2014. "Mediating New Europe-Asia: Branding the Post-Socialist World via the Internet." In *New Media in New Europe-Asia*, edited by Jeremy Morris, Natalya Rulyova, and Vlad Strukov, 143–166. London: Routledge.

———. Forthcoming. "'Brand' New States: Post-Socialism, the Global Economy of Symbols, and the Challenges of National Differentiation." In *The Future of Post-Socialism*, edited by Danijela Lugarić and Dijana Jelača. Albany, NY: SUNY Press.

Sidiropoulos, Elizabeth. 2014. "South Africa's Emerging Soft Power." *Current History* no. 113 (763):197–202.

Sinha Roy, Ishita. 2007. "Worlds Apart: Nation-branding on the National Geographic Channel." *Media, Culture & Society* no. 29 (4):569–592.

Smits, Katherine, and Alix Jansen. 2012. "Staging the Nation at Expos and World's Fairs." *National Identities* no. 14 (2):173–188.

Szondi, György. 2008. *Public Diplomacy and Nation Branding: Conceptual Similarities and Differences*. The Hague: Netherlands Institute of International Relations "Clingendael."

——. 2009. "Central and Eastern European Public Diplomacy: A Transitional Perspective on National Reputation Management." In *Routledge Handbook of Public Diplomacy*, edited by Nancy Snow and Phillip M. Taylor, 292–313. London: Routledge.

Teslik, Lee Hudson. 2009. "Nation Branding Explained." *Council on Foreign Relations*, available at www.cfr.org/publication/14776/nation_branding_explained.html [last accessed 4 January 2009].

TNZ. n.d. "New Zealand—The Perfect Middle-earth." *100% Pure New Zealand*, available at www.newzealand.com/travel/en/media/features/film&television/film_nz-is-middle-earth_feature.cfm [last accessed 15 September 2014].

Trade Arabia. 2014. "Qatar 'World's Fastest Growing Nation Brand'." *Trade Arabia*, available at www.tradearabia.com/news/MISC_271299.html [last accessed 26 July 2015].

Unwin, Tim. 1999. "Contested Reconstruction of National Identities in Eastern Europe: Landscape Implications." *Norsk Geografisk Tidsskrift—Norwegian Journal of Geography* no. 53:113–120.

van Ham, Peter. 2001. "The Rise of the Brand State: The Postmodern Politics of Image and Reputation." *Foreign Affairs* no. 80 (5):2–7.

Volčič, Zala, and Mark Andrejevic. 2015. *Commercial Nationalism: Selling the Nation and Nationalizing the Sell*. London: Palgrave Macmillan.

Walsh, Peter. 1997. "The Web and the Unassailable Voice." *Archives and Museum Informatics* no. 11:77–85.

Wang, Jian. 2006. "Localising Public Diplomacy: The Role of Sub-National Actors in Nation Branding." *Place Branding* no. 2 (1):33.

Wardrop, Kenneth. 2012. "Qatar: As Independent As You Are." *Places: A Critical Geographic Blog*, available at http://blog.inpolis.com/2012/11/05/qatar-as-independent-as-you-are/ [last accessed 16 September 2014].

Werther, Charlotte. 2011. "Rebranding Britain: Cool Britannia, the Millennium Dome and the 2012 Olympics." *Moderna språk* no. 105 (1):1–14.

Woodley, Daniel. 2015. *Globalization and Capitalist Geopolitics: Sovereignty and State Power in a Multipolar World*. London and New York: Routledge.

Zhang, Juyan. 2006. "Public Diplomacy as Symbolic Interactions: A Case Study of Asian Tsunami Relief Campaigns." *Public Relations Review* no. 32:26–32.

3 The mind's eye

Popular culture, geographical imagination, and international relations

Nearly a century ago, Walter Lippmann remarked, "The way in which the world is imagined determines at any particular moment what men will do" (1922, 25). This assertion, while probably not specifically intended to address geography or geopolitics, nonetheless resonates within the field, particularly from the perspective of popular geopolitics. Popular geopolitics focuses on mass media and how representations of people, place, and space inform national identity, foreign policy, and international relations. Of particular interest to scholars of popular geopolitics are the process of contemporary myth-making and the outcomes of geographical imagination, particularly as these relate to a given society's images of other nations and how the world "actually" works (see, for instance, Sharp 2000a; Dodds 2008; Dittmer 2010b). Mass media, particularly popular-cultural artefacts, are continually offering up images of nations for consumption, resulting in their becoming the primary source of dissemination of information about foreign countries in the (post)modern neoliberal era. According to one commentator: "The influence of novels, films, and youth literature in forming images of foreign nations and countries should not be underestimated" (Kunczik 1997, 8). Not only does popular culture present a mirror of international relations, increasingly popular culture serves as a constitutive force in world politics, shaping how politicians engage with publics, as well as how polities understand what politicians "mean" when they speak of foreign affairs (Nexon and Neumann 2006).

As information consumption shifts inexorably towards the visual through the global rise of television consumption and spread of digitized "new media," the very notion of reality becomes ever more defined by cultural constructions of the Self/Other and influenced by the shifting contours of the scopic regime of contemporary existence; by manipulating "ways of seeing," reality is unhinged from its "real" referents, and representation becomes independent of what is being portrayed in both spatial and temporal terms (see Jay 2011). Characterized by "consumer worlds saturated by image-making" where the visual reifies "fictional stories of subject territories" and "factual control on the ground" (Daniels 2011, 183), films, television series, comic books, and other forms of visual-centric popular media situate knowledge and reinforce topographies of power associated with national identity and geopolitical prowess, persuading audiences through mimesis rather than presenting imaginaries that "actually respond to reality" (Kukkonen

2011, 56). These geographical and geopolitical imaginaries (see, for instance, Gregory 1994; Harvey 2005; Daniels 2011) are reinforced by informational (textual/audio) content drawn from newspapers, journals, magazines, radio, and other popular sources of logo-centric knowledge which, together with visual sources (maps/images/symbols), create and sustain public and popular geographical views of the "foreign" world. Such conceptual entities are strongly influenced by "mutually constitutive relations between representation and imperialism, knowledge, and power" (Anand 2007, 81), and we should remember that the "laws of logic do not apply to the world of images" (Kunczik 1997, 42) and that "every lie can be legitimated" (Kunczik 1997, 57).

"Real" and "unreal": Representation and the making of geographical imagination

Geographical imagination is the faculty wherein space and its key complements (people, flora, fauna, etc.) are *perceived* without actually having been *seen*. Like any act of imagination, this process is impacted by one's own biography, i.e., the experiences, relationships, hopes, and fears that constitute the "self." These "imaginative geographies" (Said 1994; Ó Tuathail 1996a; Gregory 2004; Al-Mahfedi 2011) are, in fact, fictional realities or performative representations of spaces which mix the *real* and *unreal* to "produce the effect that they name and describe" (Puar 2007, 39). As a result, geographical imagination is highly idiosyncratic and contested; however, societal norms and political acculturation are highly effective means of establishing commonalities within and across a given polity (see, for instance, Cosgrove and della Dora 2005). Accordingly, the spatial and the social are deeply entwined; according to Daniels (2011, 182), "geographical imagination in its various forms and meanings is a powerful ingredient of many kinds of knowledge and communication, within and beyond geography as an academic subject, as a way of envisioning the world, experiencing and reshaping it too." Like any practice of representation, the (mental) mapping of space and place is affective, situated, embodied, and partial, and thus fundamentally marked by topographies of power and knowledge (Gregory 1994). Due to geography's roughly equal reliance on visual and written representation, geographical imagining is strongly influenced by linguistic devices, including metaphor (Jarosz 1992), as well as visual symbols and synecdoche. The more culturally distant, foreign, or uncanny a place or society seems, the more likely it is to suffer from bias (Harvey 2005). Geography's close association with the sense of sight (Cosgrove 2005), or ocularcentrism, (MacDonald, Hughes, and Dodds 2010a) has a particularly acute influence on geographical knowledge, and, in turn, imagination, especially following the introduction of the photographic image (Schwartz and Ryan 2003). This has given rise to the notion of regimes based on "seeing" that inform how we "graph" the "geo" and culturally construct "both what is seen and how it is seen" (Rose 2001, 6). Keeping in mind Heidegger's politically pregnant term *Weltbild* ("world-picture"), an understanding of popular geographical imagination's influence on geopolitical attitudes and behaviors is vital.

As geographical knowledge provides the firmament upon which geopolitics operates, a brief précis of the discipline is in order. Coined as a term in 1899 by the Swedish scholar Rudolf Kjellen, "geopolitics" (or *Geopolitik*) emerged as a distinct field in the early twentieth century, offering practitioners of statecraft the "theory of the state as a geographical organism or phenomenon in space" (Kjellen 1917), and a "doctrine on the spatial determinism of all political processes" (Haushofer qtd. in Cohen 2003) based on political geography. Fluctuating between the pseudo-scientific analysis of Friedrich Ratzel, who characterized space (*Raum*) and location (*Lage*) as organic properties and envisioned frontiers as the "skins" of states, and the predominantly historical approach favored by Halford Mackinder, who developed the famous "Heartland" theory, geopolitics enjoyed a brief period of exaltation before being rejected as a "handmaiden of Nazism" and an "intellectual poison" in the wake of World War II (Dodds 2007a, 24). However, during the Cold War geopolitics witnessed a renaissance, resulting in the contemporary tripartite division of the field into formal (academic), practical (policy-making), and popular (everyday) geopolitics (see Ó Tuathail 1996a). My analysis is primarily concerned with the last of these three subsets, i.e., popular geopolitics, or the way in which quotidian understanding of how the world *really* works influences dynamics of power. The foundational trinity of popular geopolitics is narrative, myth, and stereotype. Through repetition and tradition, each plays its role in mapping beliefs about space and place, friends and foes, good and evil (Dijkink 1996). According to Dittmer (2010b), the discipline's progenitor is the postcolonial literary theorist Edward Said, whose work evinced the infrangible bonds between cultural production and imperialism, as well as the mutually constitutive nature of these two fields of power. Seconding Said's importance to our understanding of imaginative geography, Sharp (2009) contends that the various Orientalist texts often *preceded* experience, and thus projected a single, uniform culture upon the space we now call the Middle East (see Figure 3.1). In effect, we—as situated Western subjects—are prevented from "seeing" the human and physical diversity of the region because we already "know" it.

Given that most Americans will never set foot on foreign shores (less than one-third of all US citizens hold a passport) and the fact that the US commands the role of *primus inter pares* in world affairs, the role of American geopolitical imagination is not trivial. While a number of authors have linked geopolitics and imagination (see Latham 2001; Ridanpää 2009; Dittmer 2008), a clear consensus on the exact definition of the concept of "imagined geopolitics" or "geopolitical imagination" (assuming the two concepts are synonymous) has yet to emerge. Güney and Gökcan have gone the furthest in terms of fixing the meaning of geopolitical imagination, stating that it is a "constructed view of the world" (2010, 23) and a given state's place within that construct. However, the authors fail to elaborate on what exactly this means before shifting to a list of subcomponents, including geopolitical codes, geopolitical vision, etc., each of which require their own discrete definitions. For purposes of clarity, I define geopolitical imagination as the human capacity to connect fragmentary, non-present data associated with the relative power and worth of people, places, and spaces and formulate expectations

Figure 3.1 The streets of Kairouan, Tunisia, as a "typical" Middle Eastern setting
         (Library of Congress)

of the interaction of these elements based on one's experiences and informed by one's values. Like any type of imagination, geopolitical imagination can be reproductive (faithful to images associated with old experiences) or productive (creatively combining and/or rearranging old material into new forms).

As Said (1979) has demonstrated, literature, art, and mass media are ensconced in ideology, particularly as they relate to issues of Self and Other. In his words, "Representation itself has been characterized as keeping the subordinate subordinate, the inferior inferior" (Said 1994, 80). The cultural production of foreign space is characterized by "deliberately constructed and maintained" narratives that enable political elites to pursue "narrow and self-serving (even sometimes nefarious) interests" in the name of universal values, goodness, and right (Harvey 2005, 220). The West is most complicit in this process, as imaginative geographies and political imaginaries of "Otherness" are used in both explicit and subconscious ways to reify clear (and ultimately constitutive) distinctions between *us* and *them*, with the ultimate goal of locating, opposing, and casting out (see Debrix 2008). The most basic element of this imaginative alchemy is the use of stereotype. Unfortunately, in the neoliberal, globalized world in which we live, these external representations are often internalized by various polities as part of the effort to "play the game" of images, thus producing a form of commercialized nationalism as in the case of the intense corporatization of South African identity vis-à-vis the hosting of the 2010 FIFA World Cup (Volčič and Andrejevic 2015). This in turn creates a powerful feedback loop (Franklin 2006, 15) that often manifests as self-Orientalism, as evidenced by Romania's decision to embrace the notion of "dark tourism" surrounding the Dracula mythos (see Bardan and Imre 2012). Of all the peoples and places around the globe, the linked stereotypes of the "Arab" and the Middle East are arguably the most profound, pernicious, and perfidious. Said is particularly attuned to how these interweaving processes of geographical knowledge, geopolitical power, and perceptual manipulation impact the Middle East and the larger Muslim world, transforming it into a "*tableau vivant* for inspection by the distanced and detached Western audience" (qtd. in Gregory 1994, 170).

The promiscuous origins of popular geopolitics

It is important to note that popular geopolitics grew out of critical geopolitics, a subfield of political geography concerned with geopolitical thinking, the conceptualization of geopolitical space, and how structures of power advantage certain discourses over others (Ó Tuathail 2006). The nested identity of popular geopolitics within the larger field of political geography somewhat obfuscates the real utility of the budding discipline, which is actually a hybrid of cultural studies and geography. As Jason Dittmer (2010b) points out, geographers and other academics in the "classical" fields of study have, to their chagrin, been forced to accept popular culture's increasing importance in shaping reality, particularly via motion pictures, television programs, comics, serialized fiction, blogs, and user-generated content on the web. Klaus Dodds (2005b) refers to this as the "leaking" of ideas and images associated with world politics into popular consciousness, which, in turn, affects the conduct of foreign policy by elites, who must satisfy the desires and allay the concerns of their constituencies. Globalization, and particularly the transnationalization of media products, has had an acute impact on this process, as human beings—especially those who live in nodal areas of worldwide communication, migration, and commercial flows—are forced to grapple with the time/space compression that accompanies transborder flows of information, money, goods, people, and diseases. As a result of managing the demands of this new "closeness" and "newness," citizens around the world have been compelled to gain a certain fluency in what we might call *quotidian geography*. Rather than seeking some sort of geographical pansophy, people instead construct imagined geographies based on the media representations they encounter on a daily basis. Whether or not these constructs mirror reality on the ground is—to an extent—irrelevant (see Kukkonen 2011).

The mass mediation of space and place is nothing new; however, the speed at which textual and visual information travel has certainly created novel phenomena associated with our perceptions of these two fields of understanding. The confluence of the cartographic revolution and the spread of the printing press in Early Modern Europe led to a massive shift in the conceptualization of space. From the sixteenth century onwards, Europeans catalogued and codified physical spaces (and the peoples who occupied these spaces) through maps, censuses, encyclopedias, ethnographic treatises, photos, museums, and—eventually—documentaries and satellite imagery (this obsession with mapping territorial, topographical, and ethnic space came to other parts of the world somewhat later, though it was taken up with no less enthusiasm). As information about space and place expanded, so did its mediation, i.e., the dissemination of geographical knowledge (or representation of knowledge) through news content, cartography, visual art, exhibitions, etc. (see Campbell and Power 2010). Naturally, this dissemination of intelligence was accompanied by a "grassroots discussion" of what this geographical knowledge actually meant (Dittmer 2010b, 15). Our personal experiences with faraway places thus ultimately are formed mostly by "viewing" them in the mind's eye (visually and textually enabled through mass media) rather than via physically visiting these places. With the advent of the twentieth century, motion pictures played a particularly important role; according to Dodds:

> As an immensely popular form of entertainment, films are highly effective in grabbing the attention of mass audiences. The power of film lies not only in its apparent ubiquity but also in the way in which it helps to create (often dramatically) understandings of particular events, national identities and relationships to others. (2008, 1621)

In this author's own experiences teaching undergraduates, any discussion of individual countries or geographical regions in the classroom inevitably leads to students making references to popular films associated with the states in question, from *The Kingdom* (Saudi Arabia) to *Blood Diamond* (West Africa) to *Slumdog Millionaire* (India). While this may represent inherent weaknesses associated with the teaching of geography at the primary and secondary levels, it is nonetheless a state of affairs which affects interactions with the outside world.

In the Anglophone world, popular geographies contextualize space within "ideological landscapes" and "political imaginaries" (Gregory 1994) that favorably posit the agents and effects of Western civilization, the United States, and the neoliberal economic world system against the chaotic confluence of the "non-West," NATO "enemies," and alternative economic systems. As a number of scholars have argued, there is an increasingly dynamic and complex relationship between imagined popular geographies and the real-world foreign policies of states (see Bassin 2007; Muller 2008; MacDonald, Hughes, and Dodds 2010b). Given that the international system is simply a "large and complex communications network" (Kunczik 1997, 12), mass media and popular culture should not be treated in isolation from more orthodox forms of diplomacy, international relations, and foreign affairs. Consequently, visual and textual representations of foreign states and peoples—as the primary inputs into "spatial consciousness" (Harvey 2005)—are key planks of contemporary geopolitics, and phenomena which require ever closer scrutiny from scholars.

Novelists, comedians, filmmakers, cartoonists, and other geopolitical "imagineers" (Daniels 2011) bring their own cultural prisms and geopolitical agendas to the table, often interrupting the goals of those wishing to manage their national images beyond their own borders. Making ready use of metaphor and hetero-stereotypes, cultural producers are often guilty of homogenizing, simplifying, exoticizing, and even eradicating "genuine" geographical knowledge of for-eign peoples, places, and spaces, what I have labeled "anti-branding" elsewhere (Saunders 2012a). David Harvey (2005) suggests that such empty geographical "knowledge," which is deliberately formulated and maintained to further nar-row, self-serving, and sometimes abstruse interests, is a primary factor in how people view the world beyond their own territorial borders. Whereas nation brand-ers explored in this text are "missionaries" (Sussman 2012) for the nation, these alternative narrators tend to be less kind, often—to use a contemporary colloqui-alism—"hating on" others/Others in an attempt to build their own nation up. Yet both intermediaries seek to influence the "logistics of perception" (Crampton and Power 2005b) in overlapping forums, from pop songs to comic books to screen-plays. As such, a minor film like *EuroTrip* (2004), with its twisted view of what life is like in Bratislava, may end up influencing more people in the Anglophone world than all the efforts of Slovakia's diplomats, brand consultants, and tourism managers combined, given popular culture's "ability to produce and circulate feel-ings" and function as source for "political thought and action" (Street 1997, 10).

At the nexus: Situating the inter-discipline of popular geopolitics

As an evolving field with a radically interdisciplinary approach, popular geopoli-tics does not sit neatly within any one field of academic study. Popular geopolitics is certainly interested in the colonial gaze, which "distributes knowledge and power to the subject who looks, while denying or minimizing access to power for its object, the one looked at" (Rieder 2008, 7). However, the scope of analy-sis goes far beyond the colonial relations that so intrigued Said and postcolonial scholars who followed in his footsteps. Drawing on Said's work, as well as that of Antonio Gramsci (cultural hegemony), Michel Foucault (governmentality and biopolitics), Michel de Certeau (everyday life and writing as a tool of colonial-ism), and Stuart Hall (encoding/decoding identity), scholars of popular geopolitics reflect a diverse wellspring of intellectual influences. What binds these scholars is a focus on the "popular." Accordingly,

> Unlike more traditional currents of geopolitical and/or international relations research, the study of popular geopolitics is not afraid to dive into the uni-verse of 'low' and 'middle brow' culture in order to identify newly forming institutional locales where political discourses of identity, security, and often enmity are being created. (Debrix 2004, 182)

Many scholars in the field also point to the advent of social constructionism as a foundational element of the discipline. Recognizing that "meaning is shared, thereby constituting a taken-for-granted reality" (Andrews 2012), scholars of popular geopolitics study the effect and affect of cultural products on orientations, attitudes, and engagement vis-à-vis the outside world. Not surprisingly considering the discipline's intense focus on signs and symbols, the post-structuralist theories of Jacques Derrida, Jacques Lacan, Gilles Deleuze and Félix Guattari, Julia Kristeva, and Jean Baudrillard, as well as those of the Frankfurt School, are frequently cited in works of popular-geopolitical analysis.

As discussed earlier, popular geopolitics grew out of critical geopolitics and was strong influenced by both feminist geopolitics, and political and cultural geography. As a result of its focus of analysis on outcomes that impact relations between states (via the influence of popular culture on national image), popular geopolitics is linked to international relations theory as well—particularly critical IR theory, a sub-discipline that has evolved in tandem with popular geopolitics. In reality, many of the pivotal works in the burgeoning field of popular geopolitics have been produced by scholars in the field of IR, most notably Jutta Weldes, Cynthia Weber, Michael Shapiro, Daniel Drezner, and Iver Neumann. While none of these scholars would likely consider themselves to be geographers, they can certainly be labeled as major contributors to the field of popular geopolitics. Theories and approaches drawn from sociology, anthropology, and history also influence popular geopolitical analyses. However, with its focus on texts (films, comic books, novels, etc.) and discourse, popular geopolitics shares much with disciplines outside the social sciences, including comparative literature, literary theory, linguistics, and especially media studies and cultural studies. Consequently, the methodologies and primary sources of popular geopolitics have parallels in newer disciplines including postcolonial studies, globalization/globality studies, and imagology, as well as the smallish fields of museum, fan, and performance studies.

Given its wide range of theoretical and methodological approaches and diverse suite of academic influences, it is then appropriate to think of popular geopolitics as an inter-discipline or a post-discipline. In terms of methodological approaches to popular geopolitics, there are a variety of methods that engage and/interrogate the subject matter. Drawing inspiration from cultural studies, many scholars focus on the institution and/or processes associated with the production of (pop) cultural artefacts. Other scholars are more interested in decoding the ideological orientations of the narrative, characters, and imagery of the text with an eye towards connecting the work to larger geopolitical questions. A third avenue explores audience reception and interpolates various "meanings" and how these are processed, utilized, and/or transformed by the viewer, reader, or listener (see Weldes 1999; Dittmer 2010b; Dittmer and Gray 2010). In all these modalities, there is a stated aim of connecting the political to the popular in an effort to better understand the continuum and interplay between the two.

In the first instance, the focus is typically on the cultural producers, as well as the financial and market structures that influence content, tone, visuality, etc. In some cases, this also involves cooperation or cooptation by political elites, as demonstrated by Matthew Alford's *Reel Power: Hollywood Cinema and American Supremacy* (2010), which explores the robust relationship between "American" cinema[1] and the Pentagon in sculpting acceptable narratives regarding the role of the US military in global affairs. Matthew J. Costello's *Secret Identity Crisis: Comic Books and the Unmasking of Cold War America* (2009) is paragon of the second modality, engaging in a historiographic analysis of American superhero narratives in conjunction with the changing nature of US national identity, Cold War geopolitics, and various dynamics of diversity (gender, race, socio-economics, etc.) therein mapping how popular culture serves as both as a forge and a mirror of politics (see Nexon and Neumann 2006). In terms of studies associated with reception, the analytical structure tends to focus on how viewers/readers/listeners process popular culture and then articulate geopolitical visions associated with media. As one scholar of media representations of Arabs in post-9/11 popular culture states, "[E]lisions between televised fiction and historical reality are common across fan forums" (Alsultany 2012, 35). An excellent representative example of this type of analysis is Jason Dittmer's article "The Geographical Pivot of (the End of) History" (2008), which analyzes fans' geopolitical imaginations associated with Christian eschatological pulp series *Left Behind*.[2]

Medium–message relations: Platform, genre, subject

Despite the comparative newness of the discipline, scholars of popular geopolitics have engaged in analysis of a wide variety of media platforms and genres over the past two decades. If one dates the emergence of the field to Joanne Sharp's (1993) groundbreaking work on *Reader's Digest*, the first medium to be explored through the lens of popular geopolitics was that of tabloid journalism, a form of textual mass media which specifically catered to literate but less-educated members of society. However, a host of other scholars (mostly, but not exclusively, from the field of critical geography) were quick to follow in her footsteps, producing studies on the framing of geopolitical content of cinema, popular literature, comic books, television series, videogames, popular music, radio programming, cartoons, stand-up/sketch comedy, toys/games/fads, and advertisements, as well as the performing and visual arts (see Table 3.1). As the field develops, new media platforms (blogs, social networking sites, Internet memes, etc.), sport as entertainment (media coverage of international events, celebrity sport culture, and professional wrestling), and media-centric practices such as international activism by celebrities are also being interrogated, as is the "pop-culture turn" in foreign policy/IR scholarship (see van Munster and Sylvest 2015). While analysis of the travelogue is still considered the bailiwick of scholars operating in the humanities (particularly those engaged in image studies, as discussed in Chapter 1), this particular form of creative expression also fits well within the spectrum of primary sources to be considered by using the tools and techniques of popular-geopolitical analysis.

Certainly, film—a media platform rich in semiotic, visual, audio, and textual content and one particularly well-suited to academic analysis—has provided the most fecund field of investigation. There are myriad genres within cinema, each endowed with a specific set of allusions, scenarios, and settings that influence geographical imagination, invoke cultural meaning, and influence ways of seeing (Hopkins 1994); however, they all share an unmatched capacity for creating

Table 3.1 Previously examined and potential sources of popular-geopolitical analysis

Media	Prior to 1946	Cold War era	1991–present
Literature/ novels	*Dracula* (↕), *Heart of Darkness* (↕), Fu Manchu (↕), Tarzan (↕), *Jungle Book* (↕), *The Saint* (↕)	*The Quiet American* (↕), James Bond (↕), Jason Bourne (↕), *The Hunt for Red October* (↕)	*Kite Runner*, *World War Z* (↕), *Left Behind* (↕), *Whiteman, Body of Lies* (↕)
Film	*Duck Soup, The Sheik, White Zombie, The Great Dictator, Casablanca*, The Road films→	*The Third Man, Dr. Strangelove*, Rambo→(↕), *The Killing Fields*, Indiana Jones→(↕), *Salvador*	*Patriot Games, True Lies, Syriana, Blood Diamond, Slumdog Millionaire, Babel*
Television series	–	*The Avengers* (↕), *Mission: Impossible* (↕), *Star Trek*→(↕), *The Prisoner, Airwolf*	*The West Wing*, 24 (↕), *Homeland, Strike Back, Spooks, The Amazing Race*
Comic books/ graphic novels	Tintin→(↕), Superman→(↕), Captain America→(↕), Blackhawk→	*This Godless Communism*, Iron Man→(↕), Captain Britain→, *The 'Nam, Maus*	300 (↕), *Palestine, Persepolis, Palomar, Ex Machina, Holy Terror, Marzi*
Cartoons/ comic strips	Thomas Nast, Theodor "Dr. Seuss" Geisel, David Low→, *Punch*→, *Puck, New Yorker*→	Johnny Hazard, Buzz Sawyer, Spy vs. Spy→(↕), Doonesbury→(↕), Steve Bell→	*Jyllands Posten*'s "Muhammad" cartoons, Daryl Cagle, *Weltschmerz*
Board games/ videogames	–	*Risk*→, *Missile Command, Twilight 2000, Balance of Power, Castle Wolfenstein*→	*Street Fighter II*→(↕), *Call of Duty, Resident Evil* (↕), *Second Life, America's Army*
Popular music	"By Jingo," "I Didn't Raise My Boy to Be a Soldier," "Hunting the Hun"	"Back in the USSR" (The Beatles), "Biko" (Peter Gabriel), "Rock the Casbah" (The Clash)	"Holiday" (Green Day), "Taliban Song" (Toby Keith), "Crumbs from Your Table" (U2)
Radio programs	*Chandu the Magician*→, *The Man Called X*→, *It's That Man Again*→	*Dick Barton—Special Agent, Douglas of the World, Bob Barclay*	–
Comedy/ variety shows	Vaudeville	Late-night talk shows/ *SNL*→, Yakov Smirnoff→	Achmed the Dead Terrorist, parodic YouTube videos

Toys, fads, ads, and memes	Dinky military vehicles, Army Men→, "Kilroy Was Here"→	Air raid promos, G.I. Joe→(↕), *Battleship*→(↕), fallout shelters, Wendy's "choice" ads	Terrorist "hunting license," Guy Fawkes/ anonymous "face"
Travelogues	*The Innocents Abroad, The Road to Oxiana, Arabian Sands*	*Iron and Silk, Stranger in the Forest, A Barbarian in Asia*	*Dark Star Safari, In a Sunburned Country*
Tabloid journalism	Newsreels→, *Reader's Digest*→, *Life*→, *National Geographic*→	"I Was a Communist for the FBI" (↕), *Encounter, Globe*→	*O'Reilly Factor, Slate, Daily Mail, Huffington Post*

→ Continues into subsequent historical periods; (↕) Migration to other media platforms

imaginary realms and worlds of meaning. Anticipating the emergence of the field of popular geopolitics, Stuart Aitken and Leo E. Zonn argue that there is "little difference between our political culture and our celluloid culture, between *real*-life and *reel*-life . . . cinematic representations needs to be a part of geographic investigation" (1994, 5). In the words of another scholar:

> Visual images can move us in this way not because they harbor a mysterious power over us, but because, through carrying and condensing meanings in forms that involve us emotionally, they mobilize a power that is already ours. They set us into motion, "taking us places" . . . Cinematic moving images, through their melding of temporally sequenced visual display and sound, move us all the more forcefully. They take us on journeys (of a sort), and through the movement they exhibit and elicit, they give shape to imagined or perceived worlds. Cinema, in this sense, is a form of world-production. (Ivakhiv 2011, 191–192)

Undoubtedly, action-thrillers, including war, espionage, disaster, and crime films, have proved the most tantalizing to scholars in the field due to their exotic locales, geopolitical intrigues, diverse international characters, use of spectacle, employment of cutting-edge technologies, and fast-moving narratives. Supernatural cinema is also a common source of popular-geopolitical analysis, with horror, science fiction, and fantasy films providing a wealth of allegorical content about controversial issues of "contemporary concern" (Franklin 2006), which can be read through the lens of critical geopolitics (Saunders 2015). Historical epics and period dramas are also pregnant with contemporary geopolitical meaning and serve as powerful reminders of "national values" (see Higson 2002; Williams 2002), thus explaining why a number of researchers have focused their efforts on these films; however, as with sci-fi and fantasy, metaphorical readings are often employed (e.g., pagan Romans as Nazis/Communists in "sword-and-sandal" movies or medieval Saracens as modern-day Islamist fighters). Lastly, the comedy genre, including parodies, spoofs, and satire, provides quite a different set of optics through which to view foreign peoples, places, and space, intended to provoke laughter at the Other (rather than fear or loathing).[3]

Perhaps the best example the early evolution of the field comes from Klaus Dodds (2003, 2005b), who interrogated the *James Bond* film series (1962–) and contextualized its themes and content in Britain's post-imperial role in international affairs. He has also explored how Hollywood scrambled to give new meaning to the geopolitics of the Greater Middle East in the wake of 9/11 (Dodds 2008) and the centrality of strategic cinema in shaping perceptions of surveillance and international threats (Dodds 2011; MacDonald, Hughes, and Dodds 2010b). However, other scholars have also examined the role of Hollywood's "fixed ideological parameters" (Alford 2010, viii) in shaping workaday understandings of geography and foreign affairs, particularly in the wake of the 9/11 terrorist attacks on the World Trade Center and the Pentagon (see Martin and Petro 2006; Carter and McCormack 2006; Howie 2011). These include studies of how quasi-historical films as *Saving Private Ryan* (1998), *Gladiator* (2000), *The Kingdom Of Heaven* (2005), and *300* (2006) (see Crampton and Power 2005a; Dalby 2008; Es 2011; Dittmer 2011), as well as the current spate of comic-book superhero, sci-fi, fantasy, and zombie films, influence understandings of contemporary issues in global politics (see Whitehall 2003; Nexon and Neumann 2006; Blanton 2013). Recent scholarship has also focused on lighter fare, including how satirical depictions of foreigners color geopolitical attitudes in the era of globalization (see Wallace 2008; Saunders 2008a; Kassabova 2010; Saunders 2012a). While some cinema scholars have begun to look at earlier films through the lens of popular geopolitics (see Lipschutz 2001; Hendershot 2001; Shapiro 2008), there is much work still to be done, particularly as it relates to the popular geopolitics of late Cold War cinema, including the ideologically rich content of the Reagan-Thatcher era (e.g., the *Rambo* series [1982–2008], *Red Dawn* [1984], *Top Gun* [1986], and *Russkies* [1987]), as well as geographically informed and politically framed films from earlier decades (in addition to other formats, including videogames, toys, comics, etc.).

While the novel lacks the sensory of power of film, it can make up for it with a surfeit of space for extended textual analysis of its subject matter and ideological matrixes, including nations and countries, lifestyles and landscapes, political attitudes and identities—all of which are of interest to the scholar of popular geopolitics (see Ridanpää 2010). Literature often provides the original source material for many of the motion pictures analyzed by scholars of critical geopolitics, most notably novels of international espionage. Ian Fleming's James Bond, Tom Clancy's Jack Ryan, and Robert Ludlum's Jason Bourne all began their lives as characters in novels before moving on to the silver screen.[4] Interestingly, all three of these characters are products of the Cold War, yet have seen their identities transformed for "duty" in a post-Cold War era (see Sharp 1998). However, there is a long list of fictional personages whose globe-trotting adventures and derring-do (or perfidy) have evinced geographical imaginaries with strong political connotations, including fin-de-siècle and early-twentieth-century figures such as Allan Quatermain (1885), Sherlock Holmes (1887), Dracula (1897), Tarzan (1912), Hercule Poirot (1920), Simon Templar (1928), and Fu Manchu (1933). Each of these characters enjoyed significant runs in popular novel series, as well as a successful migration to other media, including film, radio, comic books, and television. With politically topical content and "exotic settings" (African savannahs, Amazonian jungles, Balkan capitals, etc.), such serialized novels present scholars of popular geopolitics with a treasure trove of potential source material.

During the Cold War, strategic fiction played an important part in the social construction of reality vis-à-vis distant threats and complex political machinations. Throughout the period, a steady raft of post-nuclear, apocalyptic novels, including *The Chrysalids* (1955), *On the Beach* (1959), *Canticle for Liebowitz* (1960), *Z for Zachariah* (1974), *Warday* (1984), and *Swan Song* (1987), sculpted a dire future that served to reinforce loyalty at home and suspicion of the outside world.[5] However, the geopolitical content of such literature paled in comparison to the overt framing of classic works such as George Orwell's *Animal Farm* (1945), Paul Bowles' *The Sheltering Sky* (1949), Graham Greene's *The Quiet American* (1955), James Clavell's *Tai-pan* (1966), and Thomas Harris's *Black Sunday* (1977), which explored a variety of themes, from decolonization to international political economy to terrorism. In the wake of 1989, the geopolitical novel became much more cosmopolitan in nature; while continuing to situate its narrative arc in specific spaces and places, the old stereotypes of "us" and "them" became blurred in a neoliberal world of transnational corporations, the rapid movement of people and products, and global connectivity (see Irr 2013). However, the shocking events of 11 September 2011 reset the paradigm, partially re-Orientalizing popular literature and creating an environment where the geopolitics of fear reigned supreme (see Pain and Smith 2008; Pain 2009; Saunders 2012b). While often overlooked in the field, many works of science fiction, fantasy, and horror fiction written since 2001 have manifested palpable geopolitical frames in their narratives (see Scott 2014).

There is perhaps no medium is better suited to popular-geopolitical analysis that that of the comic book. Originally little more than collected volumes of comic strips, the medium—an outgrowth of the larger medium of sequential art—came into its own in the 1930s. During the so-called Golden Age of Comic Books (late 1930s–early 1950s), a unique visual-narrative structure with its own artistic conventions and structural vocabulary evolved alongside popular "American" superheroes such as Superman, Batman, Wonder Woman, and Captain Marvel; the threat of and eventual war with the Axis prompted story lines in which these characters took on foreign and fifth-column threats, as new titles aimed at increasing youthful patriotism, including *Blackhawk*, *Shield*, and *Young Allies* (see Murray 2000; Smith 2001; Emad 2006; Scott 2007, 2011; Dipaolo 2011; Brooker 2013). During the same period, the boyish Belgian reporter Tintin achieved stardom in Francophone Europe, later being (re-)published in English versions (see Dunnett 2009). While the medium has always been characterized by a diversity of genres (war, teen romance, horror, crime, comedy, etc.), it is the superhero comic that continues to dominate after nearly a century, particularly as Marvel Comics unleashed its "Cold-War creations" (Costello 2009)—*Fantastic Four*, *The Hulk*,

Iron Man, and *Uncanny X-Men*—on the world in the early 1960s (all of which moved onto the big screen in the new millennium).

The don of the popular geopolitics of comics is incontestably Jason Dittmer, whose work on Captain America (as well as Captain Britain and the Canadian superhero Captain Canuck) has established many of the parameters of the discipline of popular geopolitics (see Dittmer 2005, 2007a, 2007b; Dittmer and Larsen 2007; Dittmer 2010a, 2012). As a "closed ideological text, imposing on the reader preferred meanings" (McAllister, Sewell, and Gordon 2001, 3), comic books—particularly those with a patriotic bent—create a conceptual environment that reifies and reinforces geopolitical values, although comics may occasionally be employed as critiques of power, e.g., Judge Dredd in *2000AD* (1977–2000) or Alan Moore's *V for Vendetta* (1982–1985) and *Watchmen* (1986–1987) (see How 2000; Paik 2010; Dipaolo 2011), or function as an "anti-geopolitical eye" (Ó Tuathail 1996b) as is the case with the works cartoonist-historian Joe Sacco and the graphic novels *Maus* and *Persepolis* (see Holland 2012). Through the use of vantage, positionality, colors, and facial features, the medium is particularly disposed towards "Othering" of the foreigner. American comic books, in particular, have a long history of demonizing foreign enemies (Germans, Japanese, Arabs, Russians, Vietnamese, etc.), while also casting aspersions on potential enemies on the home front (African Americans, immigrants, communists, "hippies," etc.) (see Shaheen 1994; Singer 2002; Wright 2003; DuBose 2007; Hogan 2009; Duncan and Smith 2009; Madison 2013).[6]

Sharing much with the mediums of film and comic books, but with its roots in "radio theatre," is the narrative television series (see Butler 2012). Developed in the 1920s and emerging as the dominant broadcast platform in the 1950s, serial television became a powerful tool for inculcating national values and shaping political culture. During the Cold War, there was no shortage of content that engaged geopolitical themes; however, show creators (and their sponsors) often trod lightly, using allegory and fantasy to deliver subtle rather overt messages about enemies abroad. The late 1960s represents a high-water mark of such strategic television; classic examples from the period include *The Avengers* (1961–1969), *Mission: Impossible* (1966–1967), *Star Trek* (1966–1969), and *The Prisoner* (1967–1968). With the advent the Second Cold War (1979–1985), geopolitical television programming saw a resurgence with the series *The Sandbaggers* (1978–1980), *The A-Team* (1983–1987), *G.I. Joe: A Real American Hero* (1983–1986), *Airwolf* (1984–1987), and *MacGyver* (1985–1992) (see Fox 2014). The television miniseries *The Day After* (1983) and *Amerika* (1987) dealt with the elevated Cold War paranoia of the period head-on, depicting a nuclear war and a Soviet takeover of the United States, respectively (see Shaw 2007). After 9/11, television became a major medium for Americans and Britons to cope with the challenges of a new threat paradigm. In some cases, this was done through allegory, as in *Lost* (2004–2010), *Battlestar Galactica* (2004–2009), *The Walking Dead* (2010–), and *Game of Thrones* (2011–) (see Porter, Lavery, and Robson 2008; Haugevik 2008; Kaye 2010; Neumann 2011; Drezner 2011; Morrissette 2014), while other programs manifested a clearly stipulated geopolitical agenda through the use of "real" enemies, places, and spaces, e.g., *24* (2001–2010), *NCIS* (2003–), *Sleeper Cell* (2005–2006), *Homeland* (2011–), *The Americans* (2013–), and *Tyrant* (2014–) (see Orr 2006; Ingram and Dodds 2011; Morey and Yaqin 2011; Johnson 2011; Alsultany 2012; Hall 2015).[7] Given that former US vice president Al Gore has commented on the geopolitical power of the "boob tube," stating, "[T]he visual imagery on television can activate parts of the brain involved in emotions in a way that reading about the same event cannot" (qtd. in Debrix and Lacy 2009, 2), it is not surprising that scholars of popular geopolitics have increasingly trained their critical eyes towards the medium since 9/11.

Like comic books, videogames are a medium that has proved particularly fertile for pop-culture geopolitical analysis. While early videogames such as *PONG* (1972) and *Tetris* (1984) were characterized by simple graphics, advances in personal computing technology and arcade game consoles soon allowed for the evolution of increasingly robust cyberworlds; yet, even in the first decade of computer games, geopolitics and the Cold War permeated the gameworlds of *Missile Command* (1980), *Castle Wolfenstein* (1981), and *Carmen Sandiego* (1983). The popular home and arcade game *Street Fighter II: The World Warrior* (1991) represents an important milestone in the geopolitical evolution of the medium, with its array of national stereotypes (see Table 3.2).[8] As game consoles evolved, the first-person shooter style rose to preeminence, with software companies scrambling to create new and ever more threatening environments for players, often with "highly racialized (and often racist) virtual topographies" (Shaw and Warf 2009, 1337), including *Call of Duty*, *Halo*, and the James Bond-inspired *GoldenEye 007*. In turn, this has resulted in the evolution of what Rachel Hughes (2010) calls "gameworld geopolitics."

The close ties of the gaming industry to the military industrial complex (Robinson 2012) has led to the use of real battlefields like Fallujah, Iraq (Power 2007), as well as the creation of mythical sites of danger (typically in the Greater Middle East, but also Eastern Europe, Latin America, and Africa) as spaces of "future-war storytelling" (see Davis 2006). Representative examples of these imaginary locales include: the Central Asian republic of "Adjikistan" (*SOCOM: US Navy SEALs Combined Assault*); the Mediterranean island-state of "Altis" (*ARMA 3*); and the Latin American nation of San Esperito (*Just Cause 2*). However, the most (in)famous of all geopolitical videogames is *America's Army*, a free online game used by the US military to assist in recruitment (see Dittmer 2010b; Salter 2011). The role of immersive videogame environments (exemplified by the highly successful series *Grand Theft Auto*, *World of Warcraft*, and *Second Life*) as a "digital hearth," where young people acquaint themselves with the outside world (Ash and Gallacher 2011), "identify" its dangers (Šisler 2008), and "code" their responses to it (Shaw and Sharp 2013), reflects how important such new media technologies are in influencing geographical imagination (see Bos 2015).

Another area of mass media attracting attention from the field is popular music and radio programming. Lyrics, as well as imagery (album covers, music videos, etc.), can convey a variety geopolitical messaging, often in ways that other media cannot. While some work has been done on the use of radio broadcasting as a tool

Table 3.2 Character options for *Street Fighter II*

Character	Nationality/ description	Special move	National characteristics
Ryu	A Japanese martial artist	"Tsunami Kick"	*Karategi* and *hachimaki* (bandana); belt with the *Fūrinkazan* (battle standard of the Sengoku period)
Guile	A U.S. Air Force major on a mission to avenge his best friend, who was killed in action	"Sonic Boom"	Blond hair and blue eyes; flat-top haircut; military fatigues and flag tattoo
Chun-Li	A female Chinese martial artist and Interpol officer	"Hundred Rending Legs"	Mandarin gown (*qipao*) and "ox horns" hairstyle; name translates as "Spring Beauty"
Dhalsim	An emaciated yogi and magician from India	"Yoga Inferno"	Thuggee skull necklace, shaved head, Sikh bracelets
Ken Masters	Ryu's Japanese-trained American sparring partner	"Rising Dragon Fist"	Blond hair; friendliness combined with arrogance
Zangief	A professional wrestler from the USSR, a.k.a. "Red Cyclone"	"Atomic Suplex"	Bear-wresting scars, Cossack beard, vodka-drinking
E. Honda	A sumo wrestler from Japan	"Hundred Hand Slap"	*Chonmage* hairstyle; sumo *mawashi* (loincloth)
Blanka	A animal-human hybrid from the jungles of Brazil	"Electric Thunder"	Mutated by Amazonian electric eels; piranha-like teeth and green skin

Source: *Street Fighter II* and "Street Fighter Wikia" (http://streetfighter.wikia.com/wiki/Street_Fighter_Wiki)

of foreign propaganda (see, for instance, Rawnsley 1996; Horten 2002; Cummings 2009; Somerville 2012), Alasdair Pinkerton and Klaus Dodds have identified the fact that there is a major lacuna in popular geopolitics associated with radio and popular music that "deserves to be remedied" (2009, 14).[9] That being stated, some intrepid scholars—many taking direction from Lily Kong's trenchant essay "Popular Music in Geographical Analyses" (1995)—have addressed these issues, focusing on popular music's impact on indigenous–settler politics in Australia (Dunbar-Hall and Gibson 2000), how music contributes to the exoticization of foreign cultures (Jazeel 2005), American country music's identification of external threats after 9/11 (Boulton 2008), and the use of music to politicize human rights issues on a global scale (Davies and Franklin 2015). However, little work has been done on the role of pop music's lyrical pedagogy vis-à-vis distant places and spaces. The late Cold War was a particularly provocative period for geopolitical content, with the release of the Sex Pistols' "Anarchy in the UK" (1976), The Clash's protest album *Sandinista!* (1980) and single "Rock the Casbah" (1982), Peter Gabriel's anthem "Biko" (1980) and Paul Simon's partnering with Ladysmith Black Mambazo as critiques of *apartheid* in South Africa, U2's interrogation of "The Troubles" in Northern Ireland in "Sunday Bloody Sunday" (1983), and Billy Bragg's critiques of American imperialism and the Falklands War (see Mitchell 2000; Barnes 2003; Dodds 2005a; Dunn 2011; Jackiewicz and Craine 2012). This period revived a tradition of the late 1960s, when songs such as The Beatles' "Back in the USSR" (1968), the Rolling Stones' "Street Fighting Man" (1968), Creedence Clearwater Revival's "Fortunate Son" (1969), and the Plastic Ono Band's "Give Peace a Chance" (1969) directed North American, British, and Australian minds to the outside world and its problems (see Kruse 2012).

The ludic realm is luxuriant with geopolitical meaning. In recent years, researchers have undertaken detailed studies of how various types of "play" and the instruments of such play frame the world beyond our shores (Ó Tuathail 2000). War toys, including Dinky military vehicles and rockets (MacDonald 2008), G.I. Joe/Action Man action figures (Woodyer 2013), and Desert Storm trading cards represent key links in the so-called "military-industrial-entertainment complex" (see Der Derian 2009; Höglund 2014). Like videogames, cartography-centric board and role-playing games like Risk "play active and creative roles" in everyday spatial understanding of the outside world, while also perpetuating threat perceptions (Cosgrove 2008, 8). The world of humor also contributes to popular conceptualizations of foreign places and spaces. Stand-up comedy, a form of humor that frequently employs stereotyped representations of the Other (Schutz 1989) yet which possesses "tremendous potential for subversion" (Alford 2010, 61), has recently come under the scrutiny of scholars in the field, notably with Darren Purcell's examination of the comedian-ventriloquist Jeff Dunham's "Achmed the Dead Terrorist" skit and anti-Arab jokes since 9/11 (Purcell, Scott Brown, and Gokmen 2010). Sketch comedy, late-night talk shows, "fake news" programs, and variety television programming have also helped to form deep (though often disjointed and distorted) impressions of foreigners and foreign affairs (see Baym 2005; Saunders 2008b; Colletta 2009; Grayson, Davies, and Philpott 2009; Day and Thompson 2012). With their simplistic and condescending representations of foreigners, children's cartoons, including *Looney Tunes* (1939–1969), *The Rocky & Bullwinkle Show* (1959–1964), *Johnny Quest* (1964–1965), and *The Simpsons* (1989–) also convey the "logic and meanings of contemporary geopolitics" as they apply to "common citizens" (see Ridanpää 2009). Political cartoons, while rarely "funny" in the strictest sense of the word, also script associations and ideologies with the "foreign," from Dr. Seuss's anti-Axis jingoism (Minear 2012) to the stereotypical renderings of the bearded Arab/Muslims (Culcasi and Gokmen 2011) to the Sovietesque depiction of current Russian presidents (Foglesong 2007); however, more often than not, political cartoons can function as transgressive forms of popular geopolitics lampooning those in power (see Dodds 2007b, 2010; Hammett 2014). Regardless, the ready use of stereotypes by political cartoonists is one of the most visible forms of enemy-Othering in popular culture and, with little need for reflection or critical assessment, can be seen as the most populist form of geopolitical imagination.

One area that has received little attention from the field but which possesses great promise is the analysis of ads and fads. If the purpose of popular geopolitics is to explore everyday practice and performance (Dittmer and Gray 2010) and to pay attention to the "little things" (Thrift 2000) around us, then analyses of the anodyne and banal—from TV commercials and magazine advertisements to bumper stickers and tattoos—is certainly in order. Another greenfield opportunity for popular geopolitics is new media (beyond previously mentioned studies of video/online gaming). Microblogging, YouTube videos, personal web pages, Internet memes, and social networking sites all contribute to geopolitical (in)competence and (mis)interpretation of international affairs, yet there is a dearth of scholarship on this subject—though Saara Särmä's (2015) recent work on collage and laughter suggests a variety of new directions in research. The popular geopolitics of sport is an additional sub-field that shows enormous promise. While the literatures linking (elite) geopolitics and international political economy to sport are robust (see, for instance, Beacom 2000; Nauright 2004; Kuper 2006; Foer 2009), only a few scholars of popular geopolitics have turned their gaze to the relationship between media coverage and popular conceptualizations of global politics and geographical space (see Inthorn 2010; Manzo 2012; Merkel 2013).[10]

As mentioned earlier, the tabloid magazine *Reader's Digest* (1922–) is ground-zero in the field, given its role in Joanne P. Sharp's (1993, 1996, 2000a, 2000b) foundational studies of popular geopolitics. Tabloid journalism, with its sensationalism, patriotic leanings, and fluid approach to "facts," has undeniably shaped public attitudes to the outside world. The nineteenth-century tabloids *Punch* (1841–1992) and *Puck* (1871–1918) engaged in satirical, often jingoistic representations of British and American allies and rivals, while *Life* (1883–2000), *National Geographic* (1888–), and other photo-centric magazines introduced many Americans to the diverse world beyond their shores (often framing political attitudes along the way). In their study of the "scopic regime" of Africa, David Campbell and Marcus Power (2010) demonstrated how *Time*'s cover stories on

Africa were limited to two sets of images: wild animals, and starving women and children. Other scholars of critical geography have been increasingly focused on how the juxtaposition of images and text, framed through the lens of the Other, has influenced public perception of conflict, poverty, and corruption in "faraway" and "primitive" places (see Jarosz 1992; Myers, Klak, and Koehl 1996). With the advent of cable news networks in the 1980s, tabloid journalism migrated to the medium of television, opening the door to "news" programs such as *The O'Reilly Factor* (1996–) and *Fareed Zakaria GPS* (2011–). A mixture of talk, conjecture, colorful graphics, and political spin (supported by a modicum of factual reporting), these programs represent shift towards populist geopolitics, or what François Debrix (2004, 2008) calls "tabloid realism" (see Chapter 6). He points to a sea change in the post-9/11 relationship between geopolitics, national security, and public opinion, triggered in part by the public intellectuals (Zbigniew Brzezinki, Samuel Huntington, Robert Kaplan, et al.) who have sought to rally a once uninterested populace in policy-making by proxy.[11]

The Internet as a site of unregulated information exchange (generally beyond the control of the state) has fueled this trend (see Dodds 2007a). While tabloid newspapers focus mostly on domestic issues, globalization has created an environment where international events are becoming the bread-and-butter of the online arms of gossip rags like the *Daily Mail*, now the world's most popular source for online information (Oremus 2012). Similarly, online media like *Slate* and the *Huffington Post* are increasingly engaged with foreign affairs, furthering the shift towards a tabloidization of geopolitics in cyberspace. Within the tabloid realm, celebrity activism is another burgeoning area of investigation; as Hollywood "A-List" actors like Angelina Jolie, George Clooney, and Ben Affleck and pop musicians like Bono, Madonna, and Tom Morello use entertainment tabloid television coverage to highlight humanitarian and environmental crises in places like Darfur, Malawi, and Central America, scholars are taking increased interest in how this impacts quotidian views on foreign spaces and peoples. Celebrity support for the Free Tibet Movement is an important case in point, especially given that such activism often pits high-profile actors and popular musicians against their own governments, as the latter seek to maintain amiable relations with Beijing, which holds that Tibet is part of China (Anand 2007; George Daccache and Valeriano 2012).

As the three previous chapters have demonstrated, popular visions of space and place are deeply imbricated in geopolitics and international relations. National image, while malleable, is surprisingly stable and changes little over time, despite the best efforts of brand managers to enhance their countries' respective reputations through various forms of geopolitical staging. Popular culture, on the other hand, has the capacity to influence geographical imagination, and frequently does so, even when there is an acute danger of distorting reality. In the coming chapters, the analysis turns to a series of examples of such alternative narration of the nation, focusing on representation of post-Soviet peoples as dangerous agents in global affairs or simply as rubes hoping to get a piece of the Western-consumerist dream, before turning to aesthetic treatments of Eurasia that present a vast, inscrutable realm where evil lurks. The ensuing analysis attempts to unpack an array of imagined truths, many of which lack any geopolitical grounding whatsoever, while simultaneously seeking to explain how popular-culture phantasms create real-world outcomes. In excavating the hidden messages behind Anglophone popular culture's representations of post-Soviet Eurasia, the following chapters are intended to ask and answer critical questions about the role that mass media play in international relations in the neoliberal world order.

Notes

1 I place the descriptor "American" in quotes due to the highly transnational nature of the motion picture industry which gathers under the synecdochic term "Hollywood" (see Stenger 2004).
2 For a more in-depth analysis of the genealogy of popular geopolitics, see Saunders and Strukov (forthcoming).
3 Certainly, other genres lend themselves to similar analysis, e.g., Westerns can be read as allegories for imperialism, civilizational conflict, etc., while romance may serve up content that addresses international issues, e.g., the propagandistic romance *Casablanca* (1942) or Costa-Gavras' geopolitical melodrama *Missing* (1982).
4 These characters first appeared in *Casino Royale* (1953), *The Hunt for Red October* (1984), and *The Bourne Identity* (1980), respectively.
5 These were complemented by non-nuclear apocalyptic literature, including works such as *Earth Abides* (1949), *The Drowned World* (1962), and *Lucifer's Hammer* (1977) wherein disease, climate change, and a massive comet strike stand in for a full-scale nuclear exchange, respectively.
6 Anti-Semitism in comic books remained rather tepid during the twentieth century, owing to a strong representation of Jews in the industry (most famously the creators of *Superman*, writer Jerry Siegel and artist Joe Shuster); however, it was not totally absent (see Weinstein 2006; Buhle 2008; and Brod 2012).
7 Specific to the "War on Terror," Alsultany points out that television executives often construct imaginary countries to allow for more "salacious story lines" than might be possible if US allies (or even enemies) were identified; however, she notes that viewers clearly understand that these are just stand-ins for Afghanistan, Saudi Arabia, Qatar, etc. (2012, 26).
8 While *Street Fighter II* was originally a Japanese product, it was extremely influential in the United States and elsewhere, and a variety of changes were made to the US release to adhere to Western tastes and sensibilities, while maintaining the integrity of the "Japan as Number One" image (see Vogel 1979).
9 While more like a television program in its narrative structure, radio theatre (now almost extinct) also provides scholars with a concentration of artefacts from an earlier era, including 1930s-era Orientalist action-adventure programs such as *Chandu the Magician* (1931–1936) and *Terry and the Pirates* (1937–1948) (see Crum 2015), wartime propaganda shows like *It's That Man Again* (1939–1949), and early Cold War-era anti-Communist spy-thrillers, e.g., *Top Secret* (1950), *The Silent Men* (1951–1952) and *Douglas of the World* (1952–1953).
10 While more pantomime than sport, professional wrestling is also a cultural practice/ product that merits scrutiny from the field. Though not an academic work, David Shoemaker's (2013) book on the history of professional wrestling includes a section on geopolitics that highlights the use of foreign "heels" as villains in times of overseas conflict/rivalry, including World War II (Germans and Japanese), the Cold War (Russians), and beyond (Iranians after 1979 and, more recently, Arabs).

11 To this list, we might now add Richard Haas, Ayaan Hirsi Ali, Thomas Friedman, and Anne Applebaum, among others.

References

Aitken, Stuart, and Leo E. Zonn. 1994. "Re-Presenting the Place Pastiche." In *Place, Power, Situation, and Spectacle: A Geography of Film*, edited by Stuart Aitken and Leo E. Zonn, 3–25. Lanham, MD: Rowman & Littlefield.

Al-Mahfedi, Mohamed Hamoud Kassim. 2011. "Edward Said's 'Imaginative Geography' and Geopolitical Mapping: Knowledge/Power Constellation and Landscaping Palestine." *Criterion* no. 2 (3):115–140.

Alford, Matthew. 2010. *Reel Power: Hollywood Cinema and American Supremacy*. New York: Pluto Press.

Alsultany, Evelyn. 2012. *Arabs and Muslims in the Media: Race and Representation after 9/11*. New York: NYU Press.

Anand, Dibyesh. 2007. *Geopolitical Exotica: Tibet in Western Imagination*. Minneapolis: University of Minnesota Press.

Andrews, Tom. 2012. "What is Social Constructionism?" *Grounded Theory Review* no. 11 (1):online. Available at http://groundedtheoryreview.com/2012/06/01/what-is-social-constructionism/ [last accessed 24 March 2016].

Ash, James, and Lesley Anne Gallacher. 2011. "Cultural Geography and Videogames." *Geography Compass* no. 5 (6):351–368.

Bardan, Alice, and Anikó Imre. 2012. "Vampire Branding: Romania's Dark Destinations." In *Branding Post-Communist Nations: Marketizing National Identities in the "New" Europe*, edited by Nadia Kaneva, 168–192. New York and London: Routledge.

Barnes, Trevor J. 2003. "Introduction: 'Never Mind the Economy, Here's Culture'." In *Handbook of Cultural Geography*, edited by Kay Anderson, Mona Domosh, Steve Pile, and Nigel Thrift, 89–97. London and Thousand Oaks, CA: SAGE.

Bassin, Mark. 2007. "Civilisations and Their Discontents: Political Geography and Geopolitics in the Huntington Thesis." *Geopolitics* no. 12:351–374.

Baym, Geoffrey. 2005. "The Daily Show: Discursive Integration and the Reinvention of Political Journalism." *Political Communication* no. 22 (3):259–276.

Beacom, Aaron. 2000. "Sport in International Relations: A Case for Cross-Disciplinary Investigation." *The Sports Historian* no. 20 (2):1–23.

Blanton, Robert G. 2013. "Zombies and International Relations: A Simple Guide for Bringing the Undead into Your Classroom." *International Studies Perspectives* no. 14 (3):1–13.

Bos, Daniel. 2015. "Military Videogames, Geopolitics and Methods." In *Popular Culture and World Politics: Theories, Methods, Pedagogies*, edited by Federica Caso and Caitlin Hamilton, 101–109. Bristol, UK: E-International Relations.

Boulton, Andrew 2008. "The Popular Geopolitical Wor(l)ds of: Post-9/11 Country Music." *Popular Music and Society* no. 31 (3):373–387.

Brod, Henry. 2012. *Superman Is Jewish? How Comic Book Superheroes Came to Serve Truth, Justice, and the Jewish-American Way*. New York: Simon and Schuster.

Brooker, Will. 2013. *Batman Unmasked: Analyzing a Cultural Icon*. New York and London: Bloomsbury Publishing.

Buhle, Paul. 2008. *Jews and American Comics: An Illustrated History of an American Art Form*. New York: New Press.

Butler, Jeremy G. 2012. *Television: Critical Methods and Applications*. London and New York: Routledge.

Campbell, David, and Marcus Power. 2010. "The Scopic Regime of Africa." In *Observant States: Geopolitics and Visual Culture*, edited by Fraser MacDonald, Rachel Hughes, and Klaus Dodds, 167–197. London and New York: I. B. Tauris.

Carter, Sean, and Derek P. McCormack. 2006. "Film, Geopolitics and the Affective Logics of Intervention." *Political Geography* no. 25:228–245.

Cohen, Saul Bernard. 2003. *Geopolitics of the World System*. Lanham, MD: Rowman & Littlefield.

Colletta, Lisa. 2009. "Political Satire and Postmodern Irony in the Age of Stephen Colbert and Jon Stewart." *Journal of Popular Culture* no. 42 (5):856–874.

Cosgrove, Denis E. 2005. "Apollo's Eye: A Cultural Geography of the Globe." In *Geographical Imagination and the Authority of Images*, edited by Denis E. Cosgrove, 7–28. Munich: Franz Steiner Verlag.

——. 2008. *Geography and Vision: Seeing, Imagining and Representing the World*. London: I. B. Tauris.

Cosgrove, Denis E., and Veronica della Dora. 2005. "Mapping Global War: Los Angeles, the Pacific, and Charles Owens's Pictorial Cartography." *Annals of the Association of American Geographers* no. 95 (2):373–390.

Costello, Matthew J. 2009. *Secret Identity Crisis: Comic Books and the Unmasking of Cold War America*. London: Bloomsbury.

Crampton, Andrew, and Marcus Power. 2005a. "Frames of Reference on the Geopolitical Stage: *Saving Private Ryan* and the Second World War/Second Gulf War Intertext." *Geopolitics* no. 10 (2):244–265.

——. 2005b. "Reel Geopolitics: Cinemato-graphing Political Space." *Geopolitics* no. 10:193–203.

Crum, Jason. 2015. "'Out of the Glamorous, Mystic East': Techno-Orientalism in Early Twentieth-Century U.S. Radio Broadcasting." In *Techno-Orientalism: Imagining Asia in Speculative Fiction, History, and Media*, edited by David S. Roh, Betsy Huang, and Greta A. Niu, 40–51. New Brunswick, NJ: Rutgers.

Culcasi, Karen, and Mahmut Gokmen. 2011. "The Face of Danger: Beards in the U.S. Media's Representation of Arabs, Muslims, and Middle Easterners." *Aether: The Journal of Media Geography* no. 8:82–96.

Cummings, Richard H. 2009. *Cold War Radio: The Dangerous History of American Broadcasting in Europe, 1950–1989*. Jefferson, NC: McFarland.

Dalby, Simon. 2008. "Warrior Geopolitics: *Gladiator, Black Hawk Down* and *The Kingdom Of Heaven*." *Political Geography* no. 27:439–455.

Daniels, Stephen. 2011. "Geographical Imagination." *Transactions of the Institute of British Geographers* no. 36:182–187.

Davies, Matt, and M. I. Franklin. 2015. "What Does (the Study of) World Politics Sound Like?" In *Popular Culture and World Politics: Theories, Methods, Pedagogies*, edited by Federica Caso and Caitlin Hamilton, 120–147. Bristol, UK: E-International Relations.

Davis, Doug. 2006. "Future-War Storytelling: National Security and Popular Film." In *Rethinking Global Secuirty: Media, Popular Culture, and the War on Terror*, edited by Andrew Martin and Patrice Petro, 13–44. New Brunswick, NJ: Rutgers University Press.

Day, Amber, and Ethan Thompson. 2012. "Live From New York, It's the Fake News! *Saturday Night Live* and the (Non)Politics of Parody." *Popular Communication* no. 10 (1/2):170–182.

Debrix, François. 2004. "Tabloid Realism and the Revival of American Security Culture." In *11 September and Its Aftermath: The Geopolitics of Terror*, edited by Stanley D. Brunn, 151–190. London and Portland, OR: Frank Cass.

——. 2008. *Tabloid Terror: War, Culture, and Geopolitics*. London and New York: Routledge.

Debrix, François, and Mark Lacy. 2009. *The Geopolitics of American Insecurity: Terror, Power and Foreign Policy*. London and New York: Routledge.

Der Derian, James. 2009. *Virtuous War: Mapping the Military-Industrial-Media-Entertainment Network*. London: Routledge.

Dijkink, Gertjan. 1996. *National Identity and Geopolitical Visions: Maps of Pride and Pain*. New York and London: Routledge.

Dipaolo, Marc. 2011. *War, Politics and Superheroes: Ethics and Propaganda in Comics and Film*. Jefferson, NC: McFarland.

Dittmer, Jason. 2005. "Captain America's Empire: Reflections on Identity, Popular Culture, and Post-9/11 Geopolitics." *Annals of the Association of American Geographers* no. 95 (3):626–643.

——. 2007a. "Captain Britain and the Narration of Nation." *The Geographical Review* no. 101 (1):71–87.

——. 2007b. "The Tyranny of the Serial: Popular Geopolitics, the Nation, and Comic Book Discourse." *Antipode* no. 39:247–268.

——. 2008. "The Geographical Pivot of (the End of) History: Evangelical Geopolitical Imaginations and Audience Interpretation of *Left Behind*." *Political Geography* no. 27 (3):280–300.

——. 2010a. "Comic Book Visualities: A Methodological Manifesto on Geography, Montage and Narration." *Transactions of the Institute of British Geographers* no. 35:222–236.

——. 2010b. *Popular Culture, Geopolitics, and Identity*. Lanham, MD: Rowman & Littlefield.

——. 2011. "American Exceptionalism, Visual Effects, and the Post-9/11 Cinematic Superhero Boom." *Environment and Planning D: Society and Space* no. 29 (1): 114–130.

——. 2012. *Captain America and the Nationalist Superhero: Metaphors, Narratives, and Geopolitics*. Philadelphia, PA: Temple University Press.

Dittmer, Jason, and Nicholas Gray. 2010. "Popular Geopolitics 2.0: Towards Methodologies of the Everyday." *Geography Compass* no. 4 (11):1664–1677.

Dittmer, Jason, and Soren Larsen. 2007. "*Captain Canuck*, Audience Response, and the Project of Canadian Nationalism." *Social & Cultural Geography* no. 8 (5):735–753.

Dodds, Klaus. 2003. "License to Stereotype: Popular Geopolitics, James Bond and the Spectre of Balkanism." *Geopolitics* no. 8 (2):125–156.

——. 2005a. *Global Geopolitics: A Critical Introduction*. Harlow, UK: Prentice Hall.

——. 2005b. "Screening Geopolitics: James Bond and the Early Cold War Films (1962–1967)." *Geopolitics* no. 10: 266–289.

——. 2007a. *Geopolitics: A Very Short Introduction*. Oxford and New York: Oxford University Press.

——. 2007b. "Steve Bell's Eye: Cartoons, Geopolitics and the Visualization of the 'War on Terror'." *Security Dialogue* no. 38:157–177.

——. 2008. "Hollywood and the Popular Geopolitics of the War on Terror." *Third World Quarterly* no. 29 (8):1621–1637.

——. 2010. "Popular Geopolitics and Cartoons: Representing Power Relations, Repitition and Resistance." *Critical African Studies* no. 2 (4):113–131.

——. 2011. "Gender, Geopolitics, and Surveillance in *The Bourne Ultimatum*." *Geographical Review* no. 101 (1):88–105.

Drezner, Daniel W. 2011. "What Can Game of Thrones Tell Us about Our World's Politics?" *Foreign Policy*, available at www.foreignpolicy.com/posts/2011/06/23/what_can_game_of_thrones_tell_us_about_our_worlds_politics [last accessed 2 January 2012].

DuBose, Mike S. 2007. "Holding Out for a Hero: Reaganism, Comic Book Vigilantes, and Captain America." *Journal of Popular Culture* no. 40 (6):915–935.

Dunbar-Hall, Peter, and Chris Gibson. 2000. "Singing about Nations within Nations: Geopolitics and Identity in Australian Indigenous Rock Music." *Popular Music and Society* no. 24 (2):45–74.

Duncan, Randy, and Matthew J. Smith. 2009. *The Power of Comics: History, Form and Culture*. New York: Continuum.

Dunn, Kevin C. 2011. "'Know Your Rights': Punk Rock, Globalization, and Human Rights." In *Popular Music and Human Rights: British and American Music (Vol. 1)*, edited by Ian Peddie, 27–37. Farnham, UK and Burlington, VT: Ashgate.

Dunnett, Oliver. 2009. "Identity and Geopolitics in Hergé's *Adventures of Tintin*." *Social & Cultural Geography* no. 10 (5):583–599.

Emad, Mitra C. 2006. "Reading Wonder Woman's Body: Mythologies of Gender and Nation." *Journal of Popular Culture* no. 39 (6):954–984.

Es, Murat. 2011. "Frank Miller's *300*: Civilization Exclusivism and the Spatialized Politics of Spectatorship." *Aether: The Journal of Media Geography* no. 8 (2):6–39.

Foer, Franklin. 2009. *How Soccer Explains the World: An Unlikely Theory of Globalization*. New York: Harper Collins.

Foglesong, David S. 2007. *The American Mission and the "Evil Empire": The Crusade for a "Free Russia" Since 1881*. Cambridge, UK: Cambridge University Press.

Fox, Charity. 2014. "Paramilitary Patriots of the Cold War: Women, Weapons, and Private Warriors in *The A-Team* and *Airwolf*." In *How Television Shapes Our Worldview: Media Representations of Social Trends and Change*, edited by Deborah A. Macey, Kathleen M. Ryan, and Noah J. Springer, 171–191. Lanham, MD: Lexington Books.

Franklin, Daniel P. 2006. *Politics and Film in the United States*. Lanham, MD: Rowman & Littlefield.

George Daccache, Jenny, and Brandon Valeriano. 2012. *Hollywood's Representations of the Sino-Tibetan Conflict: Politics, Culture, and Globalization*. Basingstoke, UK: Palgrave Macmillan.

Grayson, Kyle, Matt Davies, and Simon Philpott. 2009. "Pop Goes IR? Researching the Popular Culture–World Politics Continuum." *Politics* no. 29 (3):155–163.

Gregory, Derek. 1994. *Geographical Imaginations*. Cambridge, MA: Blackwell.

——. 2004. *The Colonial Present: Afghanistan, Palestine, Iraq*. Hoboken, NJ: Wiley-Blackwell.

Güney, Aylin, and Fulya Gökcan. 2010. "The 'Greater Middle East' as a 'Modern' Geopolitical Imagination in American Foreign Policy." *Geopolitics* no. 15:22–38.

Hall, Lucy. 2015. "Making Feminist Sense of 'The Americans'." *E-International Relations*, available at www.e-ir.info/2015/05/20/making-feminist-sense-of-the-americans/ [last accessed 13 July 2015].

Hammett, Daniel. 2014. "Narrating the Contested Public Sphere: Zapiro, Zuma and Freedom of Expression in South Africa." In *Civic Agency in Africa: Arts of Resistance in the 21st Century*, edited by Ebenezer Obadare and Wendy Willems, 204–224. Rochester, NY: Boydell & Brewer.

Harvey, David. 2005. "The Sociological and Geographical Imaginations." *International Journal of Politics, Culture, and Society* no. 18:211–255.

Haugevik, Kristin M. 2008. "Politics Among Islanders: Anarchy, Power, Struggle, and Conflict in 'Lost'." Paper presented at the annual convention of the International Studies Association, San Francisco, CA (26–29 March).

Hendershot, Cynthia. 2001. *I Was a Cold War Monster: Horror Films, Eroticism, and the Cold War Imagination*. Madison, WI: Popular Press.

Higson, Andrew. 2002. "The Concept of National Cinema." In *Film and Nationalism*, edited by Alan Williams, 52–67. New Brunswick, NJ and London: Rutgers University Press.

Hogan, Jon. 2009. "The Comic Book as Symbolic Environment: The Case of Iron Man." *ETC.: A Review of General Semantics* no. 66 (2):199–214.

Höglund, Johan. 2014. *The American Imperial Gothic: Popular Culture, Empire, Violence*. Farnham, UK and Burlington, VT: Ashgate.

Holland, Edward C. 2012. "'To Think and Imagine and See Differently': Popular Geopolitics, Graphic Narrative, and Joe Sacco's 'Chechen War, Chechen Women'." *Geopolitics* no. 17 (1):105–129.

Hopkins, Jeff. 1994. "A Mapping of Cinematic Place: Icons, Ideology, and the Power of (Mis)representation." In *Place, Power, Situation, and Spectacle: A Geography of Film*, edited by Stuart Aitken and Leo E. Zonn, 47–66. Lanham, MD: Rowman & Littlefield.

Horten, Gerd. 2002. *Radio Goes to War: The Cultural Politics of Propaganda During World War II*. Berkley and Los Angeles: University of California Press.

How, James. 2000. "*2000AD* and Hollywood: The Special Relationship between a British Comic and American Film." In *Comics and Culture: Analytical and Theoretical Approaches to Comics*, edited by Anne Magnussen and Hans-Christian Christiansen, 225–241. Copenhagen: Museum Tusculanum Press.

Howie, Luke. 2011. *Terror on the Screen: Witnesses and the Reanimation of 9/11 as Image-Event, Popular Culture and Pornography*. New York: New Academia Publishing.

Hughes, Rachel. 2010. "Gameworld Geopolitics and the Genre of the Quest." In *Observant States: Geopolitics and Visual Culture*, edited by Fraser MacDonald, Rachel Hughes, and Klaus Dodds, 123–141. London and New York: I. B. Tauris.

Ingram, Alan, and Klaus Dodds. 2011. "Counterterror Culture." *Environment and Planning D: Society and Space* no. 29:89–97.

Inthorn, Sanna. 2010. "Europe Divided, or Europe United? German and British Press Coverage of the 2008 European Championship." *Soccer & Society* no. 11 (6):790–802.

Irr, Caren. 2013. *Toward the Geopolitical Novel: U.S. Fiction in the Twenty-First Century*. New York: Columbia University Press.

Ivakhiv, Adrian. 2011. "Cinema of the Not-Yet: The Utopian Promise of Film as Heterotopia." *Journal for the Study of Religion, Nature & Culture* no. 5 (2):186–209.

Jackiewicz, Edward, and James Craine. 2012. "Scales of Resistance: Billy Bragg and the Creation of Activist Spaces." In *Sound, Society and the Geography of Popular Music*, edited by Ola Johansson and Thomas L. Bell, 33–51. Farnham, UK and Burlington, VT: Ashgate.

Jarosz, Lucy. 1992. "Constructing the Dark Continent: Metaphor as Geographic Representation of Africa." *Geografiska Annaler* no. 74 (2):105–115.

Jay, Martin. 2011. *Essays from the Edge: Parerga and Paralipomena*. Richmond: University of Virginia Press.

Jazeel, Tariq. 2005. "The World Is Sound? Geography, Musicology and British-Asian Soundscapes." *Area* no. 37 (3):233–241.

Johnson, Derek. 2011. "Neoliberal Politics, Convergence, and the Do-It-Yourself Security of *24*." *Cinema Journal* no. 51 (1):149–154.

Kassabova, Kapka. 2010. "From Bulgaria with Love and Hate: The Anxiety of the Distorting Mirror." In *Facing the East in the West: Images of Eastern Europe in British Literature, Film and Culture*, edited by Barbara Korte, Eva Ulrike Pirker, and Sissy Helff, 67–78. Amsterdam: Rodopi.

Kaye, Sharon. 2010. *Ultimate Lost and Philosophy: Think Together, Die Alone*. Hoboken, NJ: John Wiley & Sons.

Kjellen, Rudolf. 1917. *Der Staat also Lebenform*. Leipzig: Hirzel.

Kong, Lily. 1995. "Popular Music in Geographical Analyses." *Progress in Human Geography* no. 19 (2):183–198.

Kruse, Robert J, II. 2012. "Geographies of John and Yoko's 1969 Campaign for Peace: An Intersection of Celebrity, Space, Art, and Activism." In *Sound, Society and the Geography of Popular Music*, edited by Ola Johansson and Thomas L. Bell, 11–32. Farnham, UK and Burlington, VT: Ashgate.

Kukkonen, Karin. 2011. "The Map, the Mirror and the Simulacrum: Visual Communication and the Question of Power." In *Images in Use: Towards the Critical Analysis of Visual Communication*, edited by Matteo Stocchetti and Karin Kukkonen, 55–68. Amsterdam and Philadelphia, PA: John Benjamin Publishing Co.

Kunczik, Michael. 1997. *Images of Nations in International Public Relations*. Mahwah, NJ: Lawrence Erlbaum Associates.

Kuper, Simon. 2006. *Soccer Against the Enemy: How the World's Most Popular Sport Starts and Fuels Revolutions and Keeps Dictators in Power*. New York: Nation Books.

Latham, Andrew A. 2001. "China in Contemporary American Geopolitical Imagination." *Asian Affairs: An American Review* no. 28 (3):138–145.

Lippmann, Walter. 1922. *Public Opinion*. New York: Harcourt, Brace and Company.

Lipschutz, Ronnie D. 2001. *Cold War Fantasies: Film, Fiction, and Foreign Policy*. Lanham, MD: Rowman & Littlefield.

MacDonald, Fraser. 2008. "Space and the Atom: On the Popular Geopolitics of Cold War Rocketry." *Geopolitics* no. 13:611–634.

MacDonald, Fraser, Rachel Hughes, and Klaus Dodds. 2010a. "Introduction." In *Observant States: Geopolitics and Visual Culture*, edited by Fraser MacDonald, Rachel Hughes, and Klaus Dodds, 1–19. London and New York: I. B. Tauris.

——. 2010b. *Observant States: Geopolitics and Visual Culture*. London and New York: I. B. Tauris.

Madison, Nathan Vernon. 2013. *Anti-Foreign Imagery in American Pulps and Comic Books, 1920–1960*. Jefferson, NC: McFarland.

Manzo, Kate. 2012. "Visualising Modernity: Development Hopes and the 2010 FIFA World Cup." *Soccer & Society* no. 13 (2):173–187.

Martin, Andrew, and Patrice Petro. 2006. *Rethinking Global Security: Media, Popular Culture, and the "War on Terror."* New Brunswick, NJ: Rutgers University Press.

McAllister, Matthew P., Jr., Edward H. Sewell, and Ian Gordon. 2001. "Introducing Comics and Ideology." In *Comics and Ideology*, edited by Matthew P. McAllister, Jr. Edward H. Sewell, and Ian Gordon, 1–14. New York: Peter Lang.

Merkel, Udo. 2013. "Bigger than Beijing 2008: Politics, Propaganda and Physical Culture in Pyongyang." In *Post-Beijing 2008: Geopolitics, Sport and the Pacific Rim*, edited by J. A. Mangan and Fan Hong, 135–160. London and New York: Routledge.

Minear, Richard H. 2012. "Dr. Seuss Went to War." *University of California San Diego*, available at http://library.ucsd.edu/speccoll/dswenttowar/index.html [last accessed 20 July 2015].

Mitchell, Donald. 2000. *Cultural Geography: A Critical Introduction*. New York: Wiley.

Morey, Peter, and Amina Yaqin. 2011. *Framing Muslims*. Cambridge, MA: Harvard University Press.

Morrissette, Jason. 2014. "Zombies, International Relations, and the Production of Danger: Critical Security Studies versus the Living Dead." *Studies in Popular Culture* no. 36 (2):1–28.

Muller, Benjamin J. 2008. "Securing the Political Imagination: Popular Culture, the Security *Dispotif* and the Biometric State." *Security Dialogue* no. 39 (2–3):199–220.

Murray, Chris. 2000. "*Pop*aganda: Superhero Comics and Propaganda in World War Two." In *Comics and Culture: Analytical and Theoretical Approaches to Comics*, edited by Anne Magnussen and Hans-Christian Christiansen, 141–156. Copenhagen: Museum Tuscalanum Press.

Myers, Garth, Thomas Klak, and Timothy Koehl. 1996. "The Inscription of Difference: News Coverage of the Conflicts in Rwanda and Bosnia." *Political Geography* no. 15 (1):21–46.

Nauright, John. 2004. "Global Games: Culture, Political Economy and Sport in the Globalised World of the 21st Century." *Third World Quarterly* no. 25 (7):1325–1336.

Neumann, Iver B. 2011. "'Religion in the Global Sense': The Relevance of Religious Practices for Political Community in *Battlestar Galactica* and Beyond." *Journal of Contemporary Religion* no. 26 (3):387–401.

Nexon, Daniel H., and Iver B. Neumann. 2006. *Harry Potter and International Relations*. Lanham, MD: Rowman & Littlefield.

Ó Tuathail, Gearóid. 1996a. *Critical Geopolitics: The Politics of Writing Global Space*. Minneapolis: University of Minnesota Press.

——. 1996b. "An Anti-Geopolitical Eye: Maggie O'Kane in Bosnia, 1992–93." *Gender, Place and Culture* no. 3 (2):171–185.

——. 2000. *Critical Geopolitics*. Abingdon, UK: Taylor & Francis.

——. 2006. "Thinking Critically about Geopolitics." In *The Geopolitics Reader*, edited by Gearóid Ó Tuathail, Simon Dalby, and Paul Routledge, 1–14. Abingdon, UK: Routledge.

Oremus, Will. 2012. "The World's Most Popular Online Newspaper." *Slate*, available at www.slate.com/articles/business/moneybox/2012/02/daily_mail_new_york_times_ how_the_british_tabloid_became_the_world_s_most_popular_online_newspaper_. html [last accessed 26 September 2014].

Orr, Christopher. 2006. "Kiefer Madness: The Politics of *24*." *New Republic* no. 234 (19):16–18.

Paik, Peter Y. 2010. *From Utopia to Apocalypse: Science Fiction and the Politics of Catastrophe*. Minneapolis and London: University of Minnesota Press.

Pain, Rachel 2009. "Globalized Fear? Towards an Emotional Geopolitics." *Progress in Human Geography* no. 33 (4):466–486.

Pain, Rachel, and Susan J. Smith. 2008. *Fear: Critical Geopolitics and Everyday Life*. Aldershot, UK: Ashgate.

Pinkerton, Alasdair, and Klaus Dodds. 2009. "Radio Geopolitics: Broadcasting, Listening and the Struggle for Acoustic Spaces." *Progress in Human Geography* no. 33 (1):10–27.

Porter, Lynnette R., David Lavery, and Hillary Robson. 2008. *Finding Battlestar Galactica: An Unauthorized Guide*. Naperville, IL: Sourcebooks, Inc.

Power, Marcus. 2007. "Digitized Virtuosity: Video War Games and Post-9/11 Cyber-Deterrence." *Security Dialogue* no. 38 (2):271–288.

Puar, Jasbir K. 2007. *Terrorist Assemblages: Homonationalism in Queer Times*. Durham, NC: Duke University Press.

Purcell, Darren, Melissa Scott Brown, and Mahmut Gokmen. 2010. "Achmed the Dead Terrorist and Humor in Popular Geopolitics." *Geoforum* no. 75:373–385.

Rawnsley, Gary D. 1996. *Radio Diplomacy and Propaganda: The BBC and VOA in International Politics, 1956–64*. London: Macmillan Press.

Ridanpää, Juha. 2009. "Geopolitics of Humour: The Muhammad Cartoon Crisis and the *Kaltio* Comic Strip Episode in Finland." *Geopolitics* no. 14 (4):729–749.

——. 2010. "Metafictive Geography." *Culture, Theory and Critique* no. 51 (1):47–63.

Rieder, John. 2008. *Colonialism and the Emergence of Science Fiction*. Middletown, CT: Wesleyan University Press.

Robinson, Nick. 2012. "Videogames, Persuasion and the War on Terror: Escaping or Embedding the Military–Entertainment Complex?" *Political Studies* no. 60:504–522.

Rose, Gillian. 2001. *Visual Methodologies: An Introduction to the Interpretation of Visual Materials*. Thousand Oaks, CA: SAGE.

Said, Edward W. 1979. *Orientalism*. New York: Vintage Books.

——. 1994. *Culture and Imperialism*. New York: Vintage Books.

Salter, Mark B. 2011. "The Geographical Imaginations of Video Games: Diplomacy, Civilization, *America's Army* and *Grand Theft Auto IV*." *Geopolitics* no. 16 (2):359–388.

Särmä, Saara. 2015. "Collage: An Art-Inspired Methodology for Studying Laughter in World Politics." In *Popular Culture and World Politics: Theories, Methods, Pedagogies*, edited by Federica Caso and Caitlin Hamilton, 110–119. Bristol, UK: E-International Relations.

Saunders, Robert A. 2008a. "Buying into Brand Borat: Kazakhstan's Cautious Embrace of Its Unwanted 'Son'." *Slavic Review* no. 67 (1):63–80.

——. 2008b. *The Many Faces of Sacha Baron Cohen: Politics, Parody, and the Battle over Borat*. Lanham, MD: Lexington Books.

——. 2012a. "Brand Interrupted: The Impact of Alternative Narrators on Nation Branding in the Former Second World." In *Branding Post-Communist Nations: Marketizing National Identities in the "New" Europe*, edited by Nadia Kaneva, 49–78. New York and London: Routledge.

——. 2012b. "Undead Spaces: Fear, Globalisation, and the Popular Geopolitics of Zombiism." *Geopolitics* no. 17 (1):80–104.

——. 2015. "Imperial Imaginaries: Employing Science Fiction to Talk about Geopolitics." In *Popular Culture and World Politics: Theories, Methods, Pedagogies*, edited by Federica Caso and Caitlin Hamilton, 149–159. Bristol, UK: E-International Relations.

Saunders, Robert A., and Vlad Strukov. Forthcoming. "The Popular Geopolitics Feedback Loop: Thinking beyond the 'Russia versus the West' Paradigm." *Europe-Asia Studies*.

Schutz, Charles. 1989. "The Sociability of Ethnic Jokes." *Humor* no. 2 (2):165–177.

Schwartz, Joan M., and James R. Ryan. 2003. "Introduction: Photography and the Geographical Imagination." In *Picturing Place: Photography and the Geographical Imagination*, edited by Joan M. Schwartz and James R. Ryan, 1–18. London and New York: I. B. Tauris.

Scott, Cord A. 2007. "Written in Red, White, and Blue: A Comparison of Comic Book Propaganda from World War II and September 11." *Journal of Popular Culture* no. 40 (2):325–343.

——. 2011. *Comics and Conflict: War and Patriotically Themed Comics in American Cultural History from World War II through the Iraq War*. Chicago: Loyala University Chicago.

Scott, Matthew. 2014. "Some Thoughts on Popular Geopolitics, Science Fiction, and Orson Scott Card's Novel *Ender's Game*." *Trurl and Klapaucius*, available at http://trurlandklapaucius.wordpress.com/2014/08/06/some-thoughts-on-popular-geopolitics-science-fiction-and-orson-scott-cards-novel-enders-game/ [last accessed 22 September 2014].

Shaheen, Jack. 1994. "Arab Images in American Comic Books." *The Journal of Popular Culture* no. 28 (1):123–133.

Shapiro, Michael J. 2008. *Cinematic Geopolitics*. Abingdon, UK: Taylor & Francis.

Sharp, Joanne P. 1993. "Publishing American Identity: Popular Geopolitics, Myth and *The Reader's Digest*." *Political Geography* no. 12 (6):491–503.

——. 1996. "Hegemony, Popular Culture and Geopolitics: *The Reader's Digest* and the Construction of Danger." *Political Geography* no. 15 (6–7):557–570.

——. 1998. "Reel Geographies of the New World Order: Patriotism, Masculinity, and Geopolitics in Post-Cold War American Movies." In *Rethinking Geopolitics*, edited by Gearóid Ó Tuathail and Simon Dalby, 152–169. London and New York: Routledge.

——. 2000a. *Condensing the Cold War: Reader's Digest and American Identity*. Minneapolis: University of Minnesota Press.

——. 2000b. "Refiguring Geopolitics: *The Reader's Digest* and Popular Geographies of Danger at the End of the Cold War." In *Geopolitical Traditions: Critical Histories of a Century of Geopolitical Thought*, edited by David Atkinson and Klaus Dodds, 332–352. London and New York: Routledge.

——. 2009. *Geographies of Postcolonialism*. Los Angeles: SAGE.

Shaw, Ian G. R., and Joanne P. Sharp. 2013. "Playing with the Future: Social Irrealism and the Politics of Aesthetics." *Social & Cultural Geography* no. 14 (3):341–359.

Shaw, Ian G. R., and Barney Warf. 2009. "Worlds of Affect: Virtual Geographies of Videogames." *Environment and Planning A* no. 41:1332–1343.

Shaw, Tony. 2007. *Hollywood's Cold War*. Amhearst: University of Massachusetts Press.

Shoemaker, David. 2013. *The Squared Circle: Life, Death, and Professional Wrestling*. New York: Penguin.

Singer, Marc. 2002. "'Black Skins' and White Masks: Comic Books and the Secret of Race." *African American Review* no. 36 (1):107–119.

Šisler, Vít. 2008. "Digital Arabs: Representation in Video Games." *European Journal of Cultural Studies* no. 11 (2):203–220.

Smith, Matthew J. 2001. "The Tyranny of the Melting Pot Metaphor: Wonder Woman as the Americanized Immigrant." In *Comics and Ideology*, edited by Matthew P. McAllister Jr., Edward H. Sewell, and Ian Gordon, 130–150. New York: Peter Lang.

Somerville, Keith. 2012. *Radio Propaganda and the Broadcasting of Hatred: Historical Development and Definitions*. Houndsmills, UK and New York: Palgrave Macmillan.

Stenger, Josh. 2004. "Consuming the Planet: Planet Hollywood, Stars, and the Global Consumer Culture." In *Hollywood: Cultural Dimensions: Ideology, Identity and Cultural Industry Studies*, edited by Thomas Schatz, 346–365. London: Taylor & Francis.

Street, John. 1997. *Politics and Popular Culture*. Cambridge, UK: Polity Press.

Sussman, Gerald. 2012. "Systemic Propaganda and State Branding in Post-Soviet Eastern Europe." In *Branding Post-Communist Nations: Marketizing National Identities in the "New" Europe*, edited by Nadia Kaneva, 23–48. New York and London: Routledge.

Thrift, Nigel. 2000. "It's the Little Things." In *Geopolitical Traditions: A Century of Geopolitical Thought*, edited by David Atkinson and Klaus Dodds, 380–387. London: Routledge.

van Munster, Rens, and Casper Sylvest. 2015. *Documenting World Politics: A Critical Companion to IR and Non-Fiction Film*. London: Routledge.

Vogel, Ezra F. 1979. *Japan as Number One: Lessons for America*. Cambridge, MA: Harvard University Press.

Volčič, Zala, and Mark Andrejevic. 2015. *Commercial Nationalism: Selling the Nation and Nationalizing the Sell*. London: Palgrave Macmillan.

Wallace, Dickie. 2008. "Hyperrealizing 'Borat' with the Map of the European Other." *Slavic Review* no. 67 (1):36–49.

Weinstein, Simcha. 2006. *Up, Up, and Oy Vey! How Jewish History, Culture, and Values Shaped the Comic Book Superhero*. Pikesville, MD: Leviathan Press.

Weldes, Jutta. 1999. "Going Cultural: Star Trek, State Action, and Popular Culture." *Millennium: Journal of International Studies* no. 28 (1):117–134.

Whitehall, Geoffrey. 2003. "The Problem of the 'World and Beyond': Encountering 'the Other' in Science Fiction." In *To Seek Out New Worlds: Exploring Links between Science Fiction and World Politics*, edited by Jutta Weldes. Houndsmills, UK: Palgrave Macmillan.

Williams, Alan. 2002. "Introduction: The Concept of National Cinema." In *Film and Nationalism*, edited by Alan Williams, 1–24. New Brunswick, NJ and London: Rutgers University Press.

Woodyer, Tara. 2013. "Action Figures, Cultures of Militarism and Geopolitical Logics." *Museum of Childhood*, available at www.museumofchildhood.org.uk/whats-on/exhibitions-and-displays/past-exhibitions-and-displays/past-exhibitions-and-displays-2013/war-games/war-games-perspectives/action-figures [last accessed 25 September 2014].

Wright, B. W. 2003. *Comic Book Nation: The Transformation of Youth Culture in America*. Baltimore, MD and London: The Johns Hopkins University Press.

4 A brand new Eurasia

Places, spaces, and peoples of the post-Soviet realm

Eurasia, a portmanteau of "Europe" and "Asia," has always been and remains a contentious geopolitical concept. From a purely geographical perspective, this term refers to the vast supercontinent that stretches from the Atlantic coasts of Portugal and Norway in the west to the Kamchatka and Indochina peninsulas along the Pacific Ocean in the east. The ancient Greeks established a historical division between the European and Asian continents based on a fallacious supposition that the two spaces were ultimately separated by large bodies of water. Unlike Africa, the third continent in this foundational triad, these two great expanses were named after pagan goddesses. Despite the absence of any genuine geographical barriers between Europe and Asia (Sullivan 2012) and the existence of millennia of sustained and expansive "Eurasian exchange" (Vinokurov and Libman 2012),[1] a variety of historical factors served to reify and then ossify this delimitation, a trend which has been reinforced for more than 2,000 years through political, economic, and other power structures (see van der Dussen and Wilson 2005). This imagined division between Europe and its Others—particularly its eastern Others—is one of the primary concerns of this study.

Russia's rise as a great power in the Early Modern period triggered a new understanding of what constituted the European continent. No longer just a peripheral set of duchies, the ascendant Romanov Empire commanded the attention of the wider European world. Sitting astride the continental divides of the Ural Mountains and the Caucasus, Russia became intrinsically linked to both Europe *and* Asia. After the conquest of Siberia and the annexation of territories with large populations of Turkic-speaking Muslims, the country never again fit neatly into either category, instead coming to serve as a permanent frontier zone between East and West, mirroring the ambiguity of Eurasia itself as a liminal space. From a geographical standpoint, situating Russia has remained a perennial obsession among theorists, both within and outside of the region. Russian nationalists and pan-Slavists propagated the idea of a Eurasian "middle realm" in the nineteenth century, rallying around the sacral—even mystical—nature of the lands between longitude 30 degrees East[2] and the Lena River in Siberia (Lewis and Wigen 1997, 222). Other scholars conceived of Russia-Eurasia as a "third continent" between (Western) Europe and (Eastern) Asia (Laruelle 2007, 24). At the turn of the last century, Halford Mackinder's famous essay, "The Pivot of

History" (1904) focused the attention of the Anglophone West (as well as a rising Germany) squarely on the region, permanently linking the fate of Russia *qua* the "heartland of Eurasia" with that of global power.

In terms of contemporary geopolitics, Eurasia is uniquely associated with the former Soviet Union, and Russia in particular. Yet, while the term encompasses the new constellation of states and polities that once formed the geopolitical universe known as the USSR, Eurasia is no longer "just another name for Russia" (Trenin 2002, 29); however, at the same time, it is the outgrowth of the geopolitical universe that once was the USSR. The word itself is pregnant with meaning and highly controversial in certain quarters, in both its application and usage. When Russia gained independence from the USSR in the early 1990s, there was some support for choosing the appellation "Eurasia" as the country's new name, before politicians finally settled on the less controversial "Russian Federation" (*Rossiyskaya Federatsiya*) favored by those citizens living in the ethnic republics (in lieu of the more ethnically exclusive "Russia" (*Rossiya*), a term which is used in everyday parlance). The prospect of adopting the official denomination of Eurasia had support from a group which has come to be known as the neo-Eurasianists, a cadre of thinkers who stress the multicultural and multiconfessional nature of the diverse peoples of contemporary Russia, with Lev Gumilëv, Aleksandr Panarin, and Aleksandr Dugin being the most influential voices (see Rossman 2007; Tsygankov 2007; Laruelle 2012).[3] With such a name, Russia would have effectively positioned itself as a homeland for all its peoples; this would have included not only the numerically dominant Slavs, but also the various Turkic, Finnic, Ugrian, Caucasian, Mongolic, Tungusic, and Paleoasiatic nations that are indigenous to the Russian realm (see Saunders and Strukov 2010; Blinnikov 2011). Likewise, the delinking of the *russki* ethnos[4] from the name of the state would have made more political space for the various faith groups that are native to the lands that constitute contemporary Russia, including practitioners of Islam, Judaism, and Lamaism (Tibetan Buddhism), as well as the rich array of animistic, shamanistic, and polytheistic beliefs systems of the region.

Outside of Russia, the Eurasian mantle is even more politicized. Those living in the Baltic States vehemently reject categorization of their countries as Eurasian due to the term's decidedly post-Soviet connotations. In Ukraine and Belarus, sentiments are mixed, whereas the Caucasian and Central Asian Republics generally embrace the term (or at least do not object to its application to their geopolitical profiles). Kazakhstan, in particular, promotes itself as a "Eurasian nation" as part of its national branding campaign (Saunders 2008); such framing functions both on a territorial level (like Turkey and the Russian Federation, Kazakhstan has both European and Asian territory) and from an ethnic standpoint (ethnic Kazakhs, the majority population, are Asiatic in their ethnogenesis, while most of the rest of the country's ethnic groups are European in ancestry, i.e., Russians, Ukrainians, Germans, etc.). Regardless of the controversies surrounding the nomenclature, the term "Eurasia" functions—at least for now—as a binding element for all the states that once composed the Union of Soviet Socialist Republics.

Russia: Eurasia's heartland

Over 17 million square kilometers in size, the Russian Federation is easily the largest country in the world. Though it dominates the northern half of the Eurasian supercontinent, the country contains less than 3 percent of its human population. From its most westerly point on the Baltic coast to its extreme eastern islands in the Bering Sea, Russia is more than 8,000 kilometers wide; from its Arctic fringes to its southern belt, the country is more than 4,000 kilometers long. According to George Feifer writing in *Reader's Digest* in 1984, "Although some inhabited places are further north on the map and actually as cold, Russia's lumbering heritage of isolation and backwardness makes it more frozen. Russia remains psychologically the most northern of nations" (qtd. in Sharp 2000, xv). Despite the dominant notion that the Russian lands are uniform and monotonous—largely the result of pervasive depictions of snowy wastes in Western films and other forms of popular culture (see Chapter 7)—the country lays claim to every major biome except for tropical rainforests. In the northern fringe, one finds tundra and taiga, but in the south there are large swaths of desert-like steppe lands; most of Russia's population of 142 million resides in areas dominated by temperate broadleaf forests, an undeniably "European" landscape.

Nearly a continent unto itself, Russia is geopolitically divided between its European core and four peripheries: the Northern Caucasus, the Far North, Siberia, and the Russian Far East. An asymmetrical federation, Russia is a complex agglomeration of dozens of politico-territorial sub-units, referred to as federal subjects. Possessing the highest level of autonomy are the country's twenty-two ethnic republics, which in most cases possess a president, constitution, parliament, flag, anthem, control of their own borders, and at least one official language other than Russian.[5] This is the result of a centuries-long history of territorial expansion in which the Romanov Empire incorporated hundreds of ethnic, religious, and linguistic groups into the state, although ethnic Russians always remained a plurality. As a "settler empire," Russia has an ethnic geography shaped by the historic expansion of Cossacks and other Eastern Slavs southward from the (Eastern Orthodox) Kyiv-Moscow ecumene towards the (Muslim) Caucasus and eastward into (polytheistic) Siberia and northeastern Eurasia, creating what has been called a "boreal empire" (Shaw 1999). Military conquest was the primary vehicle of territorial expansion, often through violent measures including "policies of scorched earth, deportation, and merciless reprisals against whole villages" (Zürcher 2007, 17); however, not all such interactions were as contested or bloody. During this period, the Eastern Slavic population was slowly distinguished into three separate groups, the Great Russians (ethnic Russians), White Russians (Belarusians), and the so-called Little Russians (Ukrainians), a rather artificial construct that only reached fruition with Soviet-era mass education and national delimitation projects. Imperial expansion also resulted in the inclusion of significant populations of Yiddish-speaking Jews from the defunct Polish-Lithuanian Commonwealth, as well as Buddhist populations in Kalmykia and southern Siberia. In the Volga-Ural region and the North Caucasus significant populations of "non-Russians" remain to this day, while in Siberia and the Far East, Russian/Soviet migration policies diluted the demographic dominance of indigenous peoples, resulting in an ethnic majority for the East Slavic population.[6]

These "countries within a country"—a situation that curiously mirrors the old Soviet structure of union republics and reflects tsarist-era gubernates—can be found in the North Caucasus-Caspian region (Adygea, Karachai-Cherkessia, Kabardino-Balkaria, North Ossetia, Ingushetia, Chechnya, Dagestan, and Kalmykia), the Volga-Ural basin (Mordovia, Chuvashia, Mari El, Udmurtia, Tatarstan, and Bashkortostan), southeastern Siberia (Altai, Khakasia, Tuva, and Buryatia), and the Far North (Karelia, Komi, and the Sakha Republic).[7] In the post-Soviet period, Russia's first democratically elected president, Boris Yeltsin, urged the leaders of the republics to take "all the sovereignty they could swallow" (Hahn 2003, 345). This hyper-federal regime allowed for the development of robust presidential systems in many republics, as well as complex, often inscrutable structures of political patronage. In the aftermath of Vladimir Putin's 2005 federal reforms, which suspended local elections of regional leaders, the power of these leaders sharply diminished (direct elections were reinstated after 2011–2012 protests, but with "municipal filters" put in place to ensure Kremlin-approved candidates). During his time in office, Putin has forced the republics to bring their constitutions into line with the federal constitution, signifying a re-imposition of Moscow's control over the republics. Besides the ethnic republics, Russia is divided into krais (frontier territories), oblasts, federal cities, and other political structures (eighty-five in total), all with differing levels of self-governance.[8] In these administrative units, ethnic Russians possess majority status and the Russian language tends to be the only recognized medium of "official" communication.

With over 60,000 kilometers of international borders, it is not surprising that Russia has more neighbors than any other state. Most of these countries are part of the so-called "near abroad" (i.e., former Soviet republics), including Estonia, Latvia, Lithuania, Belarus, Ukraine, Georgia, Azerbaijan, and Kazakhstan.[9] Russia also shares a land border with the Nordic countries of Finland and Norway, as well as with Poland via the Kaliningrad oblast, which is comprised of territory annexed from Germany after World War II. In East Asia, Russia neighbors China, Mongolia, and North Korea. Since its recognition of Georgia's breakaway republics of Abkhazia and South Ossetia in 2008, these states too can be considered to be on the Russian frontier, though they are de facto parts of the Russian Federation. Russia also possesses maritime boundaries with the United States, Sweden, and Japan. Russian territorial claims in the Arctic Ocean may result in future maritime borders with Canada and Greenland as well (see Laruelle 2013). The country shares jurisdiction of the Caspian Sea with Iran and Turkmenistan, as well as Kazakhstan and Azerbaijan (see Figure 4.1).

Since the breakup of the USSR, the Russian Federation has become embroiled in a number of geopolitical conflicts within post-Soviet space. During the 1990s, Boris Yeltsin implemented the so-called "Monroeski Doctrine" (Stewart-Ingersoll and Frazier 2012, 124), which retooled the United States' claim to be the only legitimate great power in the Americas in an effort to establish a Russian "sphere

Figure 4.1 The Russian Federation (Emily A. Fogarty/DIVA-GIS, Global Administrative
Areas and Natural Earth)

of exclusive influence" in the near abroad. Under this aegis, Moscow intervened in
the Transnistrian War (1990–1992) and the Tajikistan Civil War (1992–1997), and
played a partisan role in both the Nagorno-Karabakh War (1988–1994) between
Armenia and Azerbaijan and the Georgian Civil War (1991–1993). Within his own
country's borders, Yeltsin ordered military action to hold onto the breakaway North
Caucasus republic of Chechnya, resulting in the First Chechen War (1994–1996),[10]
while simultaneously supporting secessionists in Moldova (Transnistria) and
Georgia (Abkhazia, South Ossetia, and Ajaria). Beyond the use of military assets,
Yeltsin's government also pressured its neighbors on the issue of some 25 million
"ethnic countrymen" (*sootechestvenniki*), i.e., ethnic Russians, Russophones, and
those people who identified with the (defunct) Soviet state (see, for instance, Chinn
and Kaiser 1996; Kolstø 1999; Saunders 2005). Consequently, Russia was able to
exert influence in the peaceful states of Kazakhstan, Latvia, and Ukraine, where
sizeable populations of Russophones and self-identifying ethnic Russians resided.

Hoping to retain influence across Eurasia, Yeltsin established the
Commonwealth of Independent States (CIS) as a weak political union of the non-
Baltic post-Soviet republics.[11] In 1992, he also established the Collective Security
Treaty Organization (CSTO) to provide for common security among certain mem-
ber-states of the CIS. Certain countries, including Belarus, Armenia, Kazakhstan,

and Tajikistan saw great value in a close relationship with Moscow (in fact, the Border Service of Russia functioned as the de facto border guards of Tajikistan's frontier with Afghanistan until 2005), while other states including Turkmenistan, Ukraine, Uzbekistan, and Georgia were less enthusiastic about close ties to the Kremlin. On the international level, Yeltsin's first administration was dominated by an "Atlanticist" agenda, which promoted integration with Western Europe and North America as well as the normalization of relations with erstwhile enemies Turkey, Japan, and South Korea (see Tsygankov 2013). However, following a drop in popularity associated with economic decline (partly due to the US's failure to meet its promises to assist the country during economic transition), the conflict in Serbia/Kosovo, and NATO expansion in central Europe, Yeltsin embraced a new foreign policy doctrine which attempted to draw Russia closer to Iran, India, and China as a check on American global hegemony.[12]

Vladimir Putin's ascendency and the geopolitical tremors associated with the 9/11 attacks on the US reconfigured relations across Central Asia and the Caucasus. The Shanghai Cooperation Organization, originally established in 1996 to fight terrorism and separatism in Inner Asia, was strengthened in 2001 by the inclusion of Uzbekistan (which had tense relations with Yeltsin's government). US–Russian cooperation in the aftermath of 9/11 created an environment where American military assets were situated in several Central Asian countries as Moscow and Washington engaged in expanded intelligence-sharing and strategic cooperation. However, following the US-led invasion of Iraq in 2003 and the launch of George W. Bush's "freedom agenda" (see Stent 2014), Putin steadily distanced his country from the US, while advocating that his allies in the region do likewise (most significant was the break between Uzbekistan and Washington in 2005; however, the US removed its troops from Kyrgyzstan in 2014, continuing the trend). The US's support for the so-called "Color Revolutions" in Georgia (2003), Ukraine (2004), and Kyrgyzstan (2005) further complicated regional relations. According to Dmitri Trenin (2006), it was at this point that Russia decided to abandon integration into the West and to establish its own geopolitical "solar system" in Eurasia.

Emboldened by rising oil and natural gas revenues, Putin took an increasingly oppositional position on issues related to EU accession and NATO membership for the former Soviet republics. He also emerged as an advocate for the values of "managed democracy" (Colton and McFaul 2003), particularly in post-Soviet Eurasia. Those countries that have fallen afoul of Russia's geopolitical scheme (most notably Georgia and Ukraine) have often paid the price through interruption of gas supplies, bans on imports (especially foodstuffs), and other forms of geo-economic disciplining. Russia has also exerted greater influence over the trade affairs of its immediate neighbors under Putin, especially through the Eurasian Economic Community (EurAsEC) which was established in 2000 and the formation of the Eurasian Economic Union (EEC) in 2014,[13] as well as through government-controlled corporations like Gazprom and Transneft. Most recently, Russia banned imports of foodstuffs from the EU, the United States, Canada, Australia, and Norway, making the country even more reliant on trade with post-Soviet states and ensuring closer economic ties across much of Eurasia.

Whereas Yeltsin tended to be cautious in military interventions beyond the borders of the Russian Federation, Putin has been more aggressive, eschewing the need for international support or cloaking the activities under the veil of peace-keeping operations in the so-called "internal wars" of the former Soviet Union (Zürcher 2007). In 2008, simmering relations between Tbilisi and Moscow boiled over, resulting in a short war in which Russia invaded its southern neighbor and subsequently recognized two of Georgia's breakaway republics (Abkhazia and South Ossetia) as sovereign states. This was the first time since 1979 that the Russian military "crossed national borders to attack a sovereign state" (Cornell and Starr 2009, 3). Four years later, civil strife in Ukraine created a political crisis in the region wherein Crimea voted to secede from Ukraine; Russia immediately annexed what it claimed to be its "lost" territory of Crimea (the region was transferred from the RSFSR to the Ukrainian SSR in 1956).[14] Subsequently, Russia backed anti-government separatists in the east of the country (centered in Luhansk and Donetsk), resulting in significant internal displacement, destruction of urban infrastructure, and loss of life, including the downing of an international airliner traveling from Amsterdam to Kuala Lumpur. On a broader scale, Putin sent a submarine to stake a claim on the North Pole's seabed in 2007, effectively extending Russian sovereignty across half of the Arctic Ocean, while also resuming long-range bomber patrols (suspended since the end of the Cold War). The country's secret services have also been increasingly active under Putin, purportedly supporting cyber-attacks on Estonia (Katin-Borland 2012), engaging in assassinations of "enemy agents" in Qatar and the United Kingdom (Pringle 2010), and expanding espionage activities in Georgia, Ukraine, the US, and elsewhere (Sulick 2013). Most recently, Russian airpower has been deployed in Syria in an effort to destroy "terrorist groups" fighting the Assad regime, provoking criticism from Western powers already involved in the conflict.

Internationally, Russia has assumed much of the role once held by the USSR as the premier supplier of hydrocarbons to the thirsty markets of Europe, while establishing new commercial bridgeheads in East Asia. In the first decade of the twenty-first century, spiking oil and natural gas prices (buoyed by Anglo-American adventurism in the Greater Middle East) helped to rejuvenate the Russian economy and restore the country's position in the global geo-economic schema, with energy provision often functioning as a "critical motivating factor" in the country's foreign policy (Sussex 2012, 215). Russia's wealth stems from its natural resources, including oil, natural gas, uranium, and precious metals. Once a global force in heavy industry, the collapse of the command-and-control economy of the Soviet era has greatly reduced Russia's competitiveness in terms of manufactured goods (with the exceptions of its nuclear power and armaments industries); however, modest gains have occurred since the economy began to recover in the 2000s, particularly in the high technology sector.

Due to Soviet policies, the ecological situation is quite dire, with massive areas of the country plagued by air, water, and ground pollution. Despite such contamination and the fact that less than 10 percent of the land mass is used for farming, Russia remains one of the world's largest countries in terms of agriculture. In addition to the country's role as both a producer of petroleum and natural gas, Russia also functions as an indispensable transshipment zone for fossil fuels, owing to pipelines and transportation infrastructure built during the Soviet period. Russia's situation as a transit bottleneck between Central Asia and Western Europe has allowed the country to exert substantial political pressure on Kazakhstan, Turkmenistan, Uzbekistan, and Azerbaijan as these countries have sought to export their natural resources to the West and/or China.

Despite the construction of internationally funded alternative routes via the southern Caucasus, post-Soviet republics have found that circumventing Russia's geopolitical domination remains problematic. Under President Vladimir Putin, Russia became an adept at the "grammar of commerce" (Luttwak 1990) as applied to its geopolitical relations, particularly the countries of the near abroad, but also other states in Europe and Asia. Russia's relations with both Ukraine and Belarus have put the flow and subsidies of natural gas, ostensibly handled by Russian corporations (but in reality determined by the Kremlin), at the center of regional relations, with acute ramifications for countries as far afield as Bosnia and Italy. Through the close relationship between Russia's oligarchs and the government, the Kremlin has been able to influence markets and guarantee privileged access for its companies in nearly all of the former Soviet republics, thus reinforcing the nodality of Moscow in Eurasia. Through significant cross-border trade, arms exports, and nuclear power assistance, Moscow has remade its once contentious relationships with Iran and the People's Republic of China, while exports of oil and natural gas have allowed the country to develop burgeoning trade relations with its Cold War enemies, particularly Germany and Japan.

The Baltic States: From Russia's "window to the West" to Norden's "East"

The eastern shores of the Baltic rim have long been a place of both conflict and cultural exchange. Ruled by a succession of Teutonic Knights, Swedish kings, and Russian tsars, the region reflects this history in its ethnic (Germanic, Finnic, Baltic, and Slavic) and religious diversity (Paganism, Catholicism, Lutheranism, and Eastern Orthodoxy are all prevalent, as was Judaism prior to World War II), as well as strong trade links to other parts of the Baltic basin via the Hanseatic League (*Hansa*) that date to the late Middle Ages. In the context of post-Soviet Eurasia, the term "Baltic States" refers to the former Soviet republics of Estonia, Latvia, and Lithuania (see Figure 4.2). These three countries were all once part of tsarist Russia, but—along with Finland—gained their independence as a result of the Bolshevik Revolution and the changing political landscape triggered by the end of World War I. Following Nazi Germany's invasion of Poland, the Soviet Union occupied the Baltic States, ultimately incorporating them into the USSR as union republics; not surprisingly, these nations were the first to gain their independence with the dissolution of the USSR in the early 1990s.

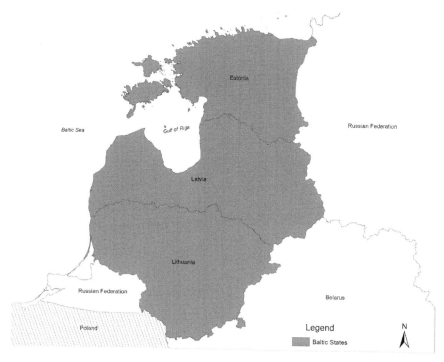

Figure 4.2 The Baltic States (Emily A. Fogarty/DIVA-GIS, Global Administrative Areas and Natural Earth)

Estonia is the smallest and northernmost of the Baltic States. Situated on the Gulf of Finland and the Baltic Sea, the country is less than 50 kilometers from its northern neighbor, Finland, with which it shares cultural, linguistic, and historical ties (the Estonian language is part of the Finnic family, and thus related to Finnish). Estonia is approximately 45,000 square kilometers in size and has a population of 1.3 million, of which two-thirds are ethnic Estonians (Russians account for approximately one-quarter of the population). The majority of the population is not religious, though Lutheranism and Eastern Orthodoxy are the legacy faiths of Estonians and Russians, respectively; pagan practices remain strong among rural Estonians. The country is renowned for its environmentally friendly policies and diversity of flora and fauna.

Latvia, sandwiched between Estonia and Lithuania, is second largest in terms of size and people, having a land mass of nearly 65,000 square kilometers and a population of 2.2 million. Nearly 60 percent of the population is ethnic Latvian, with Russians accounting for about one-third of the population. The Latvian language is one of only two surviving Baltic languages (the other being Lithuanian). Like the Estonians, most Latvians are not religious; however, Lutheranism is dominant among observant Latvians, while about 20 percent of Latvians are Roman

Catholics; Eastern Orthodoxy is practiced by ethnic Russian believers. Latvia is characterized by well-defined regional identities affiliated with the Kurzeme (western Latvia), Vidzeme (central Latvia), and Latgale (eastern Latvia) regions.

The largest and southernmost of the Baltic republics is Lithuania. The country has a population of 3.2 million and is 65,200 square kilometers in size. Lithuanians account for 84 percent of the population, with Poles and Russians being the largest minority groups. The national language is Lithuanian, held by many linguists to be the most archaic living language in the Indo-European family (Baldi 1983). Significantly more religious than the other Baltic nations, nearly 80 percent of Lithuanians profess Roman Catholicism. Lithuania's geography differs from its northern neighbors primarily as a result of its extremely short coastline, as well as its border with Russia's Kaliningrad Oblast (a problematic issue due to certain European Union restrictions on travel).

Never resigned to their incorporation in the USSR, the Baltic States maintained foreign consulates abroad and were politically supported by their diasporas in the West throughout the Cold War.[15] During perestroika, strong national revival movements emerged in each of the republics, often coordinating with one another on linguistic, cultural, and economic policies. Led by Lithuania, their combined struggle for independence (often centered on media-friendly displays of nationalism subsequently deemed the "Singing Revolutions") and Moscow's hard-fisted suppression of the nationalist movement hurt Mikhail Gorbachev's international standing in the late Soviet period. In 1991, the three republics voted to cut ties to Moscow, initiating the unraveling of the USSR. In the post-Soviet period, these states have had a particularly complicated relationship with the Russian Federation, evidenced by their resolute refusal to join the Commonwealth of Independent States or any other Russian-dominated bloc. Additional problems stemmed from the three countries' move to join the US-dominated NATO alliance in 2004, economic disputes related to oil and natural gas pipelines, and discrimination against ethnic Russians (particularly in Estonia and Latvia, where many locally born and longtime resident ethnic Russians and other "Soviet peoples" were not automatically granted citizenship upon independence, thus becoming legal residents but stateless persons).

Borders and frontiers, a consummate issue in regional geopolitics, also manifested as bones of contention. Trenin sums up the case quite succinctly:

> The very emergence of independent Estonia and Latvia . . . led to border claims by Tallinn and Riga. Whereas Moscow believed that it had generously *granted independence*, no strings attached, to the three former Soviet republics, the latter viewed this as a *restoration of their independence* unlawfully suspended by the Soviet Union—from 1940 through 1991—of which the Russian Federation claimed to be the successor. (2002, 157)

The Baltic States, along with Poland and Finland, have sought to use their membership in the European Union and NATO to constrain Russian actions, which they view as neo-imperial violations of their national sovereignty. Reaching back to a distant past, some commentators have branded this collective action a latter-day manifestation of the "Hansa spirit," a sort of mutual aid society for parvenu states in the twenty-first century (Smith 2001, 3). In many cases, the plucky Balts have been aided by their wealthier EU counterparts, particularly through the so-called Northern Dimension initiative, which seeks to "provide a common framework for the promotion of dialogue and concrete cooperation, strengthen stability and well-being, intensify economic cooperation, [and] promote economic integration, competitiveness and sustainable development in Northern Europe" (EC 2010). As Christopher Browning (2005) points out, the northern reaches of the European continent—and particularly the Baltic Sea's eastern fringe, once the site of fierce Cold War suspicions and intrigue—are now at the forefront of regional integration and the de-emphasizing of historical geopolitical divisions (though sometimes generating new ones in the process).

Despite such sanguine evaluations of the affairs of northeastern Europe, not all voices are singing in unison. While Estonia, Latvia, and Lithuania are often grouped together as the "post-Soviet Baltic Republics," certain political elites bristle at this characterization. Despite nearly half a century of Sovietization, many local leaders and opinion-makers decry the post-Soviet moniker *and* attempts to lump the three countries together (i.e., apart from other littoral states of the Baltic, such as Sweden and Finland). Reflecting historic-religious factors rather than geophysical traits, contemporary geographers have aided local attempts at splitting the Baltics. Robert C. Ostergren and Mathias Le Bossé (2011), for instance, include Estonia and Latvia in Nordic Europe, while Lithuania is confined to East-Central Europe alongside Poland and Hungary. While such a division may play well with political elites in Tallinn and Riga, it rends asunder the last remaining Baltic-speaking peoples (the Latvians and the Lithuanians) and ignores myriad factors that bind Lithuania to its northern neighbors. Other geographers including Harm J. de Blij (de Blij and Muller 2008) situate Estonia (though not Latvia) within Norden, thus linking (at least geo-conceptually) the small state more closely to Iceland than to either of its actual neighbors, Latvia or Russia. This ambiguity leaks far beyond the narrow confines of academia, often serving as grist in popular debates launched from the comfort of bar stools and the editorial boardrooms of news media outlets, thus reflecting the importance of discursive battles in the larger information wars that characterize the post-Cold War era in post-socialist Europe.

Regarding geo-economics, the Baltic States exhibit meaningful differences from their post-Soviet counterparts. As members of the EU, as well as other regional trade blocs, Estonia, Latvia, and Lithuania represent highly integrated states with regards to the global economy. Estonia emerged as a paragon of free trade and neoliberalism following its independence from the USSR, adopting a flat tax rate for businesses, thus attracting a large number of corporations from other parts of Europe, particularly the Nordic countries. Latvia and Lithuania have followed in the wake of Estonia, though often with less vigor. Certainly, Latvia's massive exposure to the real estate problems that afflicted the United States and Spain during the 2008–2009 global financial crisis give proof to the country's integration into the world economy, as does its miraculous recovery since the downturn.

The Baltic States, with their ready embrace of shared sovereignty, integration in the Baltic, European, and global marketplaces, and city-state-like political systems that value social mobility, economic freedoms, and technocracy more than territoriality, are paradigms of the post-Cold War geo-economic state.[16] The region's embrace of new media platforms perhaps best affirms this contention. Analyzing the various metrics for technology penetration, e.g., cellular phones, teledensity, Internet use, personal computers per person, etc., it quickly becomes obvious that the Baltics shared little in common with their post-Soviet peers, instead resembling Western European nations (see Saunders 2009b)—although this has certainly changed in recent years as Russia has emerged as an ICT dynamo. Estonia, in particular, has embraced the full suite of ICTs as a mechanism for the transformation of governance—from e-voting to live streaming of parliament in cyberspace—and for improving the commercial and social system of the entire country. The fact that the software for Skype, one of the world's leading instant messaging and voice/video messaging services, was written by Estonian engineers is a testament to the country's ongoing digital revolution (or at least we are led to believe so by Estonia's nation branders).

The Western Republics: The "new" Eastern Europe

Ukraine, Belarus, and Moldova are situated on the East European Plain between the steppes of Russia, the forested uplands of Central Europe, and the mountainous Balkan Peninsula (see Figure 4.3). Collectively, these three countries represent a transitional zone between Russia and Europe-proper, sometimes being labeled the "new Eastern Europe" (see Plokhy 2011). These states represent a geopolitical region that is undeniably European, yet one which remains inextricably linked to its Soviet past (much more so than the Baltic States discussed above). The region is a mixed agricultural-industrial zone which experienced massive development during the pre- and post-World War II periods of Soviet industrialization. According to Intourist's *Pocket Guide to the Soviet Union*:

> After centuries of stagnation, the sleepy towns are now being converted into huge industrial centers, new railways are being laid, ports and mines are being reconstructed, the greatest power plant in Europe is nearing completion, everywhere new factories, grain elevators, socialized cities are springing up. (Intourist 1932, 337)

However, there was a heavy price to pay for the modernization of the Soviet West: air, water and soil pollution are rampant across the region. Moreover, the aftereffects of the 1986 Chernobyl nuclear disaster continue to plague the regional ecology and cause birth defects as well as abnormally high rates of cancer and other debilitating diseases.

Ukraine, at over 600,000 square kilometers, is not only the biggest of these countries, it is also the largest wholly European country. Ukraine's population of 45 million includes the titular nationality, which make up two-thirds of the

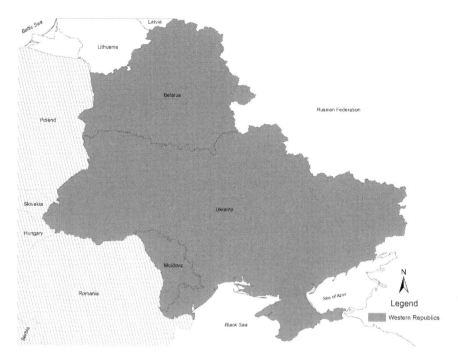

Figure 4.3 The Western Republics (Emily A. Fogarty/DIVA-GIS, Global Administrative
Areas and Natural Earth)

population, as well as a wide variety of ethnic minorities including Russians,
Ruthenes, Hungarians, Vlachs (Romanians), Poles, and Crimean Tatars.[17] The
Ukrainian language is an East Slavic language with substantial borrowings from
Polish. Russian is the preferred language of many in the eastern part of the country,
while Surzhyk, a Russian-Ukrainian creole, is also a popular idiom. Three-fourths
of the country is nominally Ukrainian Orthodox, splitting their allegiance between
the Kyiv and Moscow patriarchies and, alternately, the Ukrainian Autocephalous
Orthodox Church. Minority populations of Greek Catholics, Roman Catholics,
Protestants, Jews, and Muslims also live in the country. Religion and language
mark important dividing lines within Ukraine's domestic geopolitical identity,
often pulling Ukrainians westerly toward Poland and the European Union or east-
wards to Russia and post-Soviet Eurasia (see Knudson Gee 1995), creating a highly
dynamic and arguably cosmopolitan polity. True to its history as the "breadbasket
of Europe," Ukraine retains a heavy focus on agriculture, bee-keeping, and animal
husbandry, but is also a significant producer of manufactured goods.[18]

Ukraine's position between Russia and southeastern Europe, (former) control
of the former Russian territory of Crimea, and its long Black Sea and Azov coasts
made it a vital partner for Russia's trade relations and national security in the post-
Soviet period. However, following a series of governmental changes beginning with

the Orange Revolution in 2004, relations between Moscow and Kyiv fluctuated. Following President Victor Yanukovych's rejection of a proposed EU association agreement in late 2013, pro-European activists revolted in what came to be known as the Euromaiden Uprising, leading to his ouster. Subsequently, the majority of Crimea's citizens supported secession from Ukraine, and shortly thereafter acceded to the Russian Federation in a highly controversial annexation which has not been fully recognized by the international community. Throughout 2014, Russian troops amassed along Ukraine's eastern border, providing material support to separatist forces associated with the Donetsk People's Republic, as well as some other breakaway portions of the country. A bloody conflict ensued, claiming hundreds of lives, including nearly 300 people aboard Malaysia Airlines Flight 17, which was downed over disputed territory on 17 July 2014 in circumstances that remain mysterious, with both sides pointing the finger at the other. Throughout 2015, Russia became more deeply embroiled in the conflict, with revelations that its soldiers were dying in the Donbass becoming commonplace.[19] In late 2015, Vladimir Putin confirmed that Russian "military specialists" were working alongside the rebels, though he stated these individuals were not regular Russian troops (Khomami 2015).

Belarus, sometimes known as "White Russia," is a medium-sized, landlocked European country of just over 200,000 square kilometers. As a Soviet-era creation,[20] the Belarusian state lacks natural geographical borders, reflecting instead partially contrived ethnic distinctions which suited the geopolitical interests of Moscow after World War II. The initial site of the initial Axis invasion of the USSR, Belarus still exhibits the privations of World War II, combined with the damage wrought by the fallout from the 1986 Chernobyl nuclear reactor explosion in neighboring Ukraine. Belarus has a population of nearly 10 million; ethnic Belarusians enjoy clear majority status, with Russians, Poles, and Ukrainians collectively comprising less than 20 percent of the population. Belarusian is an East Slavic language, distinct from Ukrainian and Russian; however, many Belarusians speak Russian or Trasianka (a Russian-Belarusian creole) in their day-to-day activities; in fact, a higher percentage of Belarusians can speak Russian than is the case for the citizenry of the Russian Federation (see Saunders 2014). Eighty percent of Belarusians profess Eastern Orthodoxy, though regular practice of the faith is rather weak. Minorities of Catholics, Jews, and non-believers make up the remainder. Belarus retains much of the socialist system of the late Soviet period, having opted not operate according to the neoliberal economic reforms that have occurred elsewhere in the region, resulting in hyperinflation in recent years.

Moldova is a small, densely populated, landlocked country situated between Ukraine and Romania. Moldova has a land mass of 33,850 square kilometers and a population of 4.3 million. Nearly 80 percent of the population is Moldovan, an ethnic group with extremely close ties to the neighboring Romanians (Cash 2011); Ukrainians and Russians are the largest minorities, predominantly residing in the breakaway "Slavic" republic of Transnistria (officially called the Pridnestrovian Moldavian Republic) in the east of the country. A Romance language, Moldovan is nearly identical to Romanian, except for its historical use of the Cyrillic alphabet.[21] Assertions that it is in fact a separate language are disputed, with many linguists arguing that it is nothing more than a variant of the northeastern dialect of standard Romanian (King 1999). On 5 December 2013, Moldova's Constitutional Court ruled that the official language of the country was "Romanian," thus making the case that no distinction exists between the two languages. Many Moldovan citizens use Russian in their everyday lives, particularly in the areas of the country which were part of the USSR prior to 1940. Nearly all Moldovans come from Eastern Orthodox backgrounds (Romanian, Ukrainian, or Russian), though religious practice is not universal. Once a major center of European Jewry, the country now retains only a small Jewish community owing to the Holocaust and subsequent Jewish emigration. Europe's poorest country, Moldova is a major sender of economic migrants abroad and is quite dependent on remittances. Farming, particularly viniculture, drives the domestic economy, as much of the country's industrial sector has contracted since 1991.

As a regional collective, the Western Republics have the lowest level of international recognition among the four post-Soviet groupings. This is, in part, due to the relatively recent conceptualization of the three states as constituting a common geographical space. Beginning in the 1960s, Sovietologists in the United States and elsewhere began employing the rather amorphous descriptor of the "Soviet West" to refer to the SSRs of Ukraine, Belarus, and Moldova.[22] Today, this contrived description finds a corollary in the term "Western CIS States" (see, for instance, Isakova 2004). According to Serhy Yekelchyk (2006), the Soviet Union's occidental fringe played a particularly important role in the dissident movement during the 1960s, thus raising the possibility that the westernmost union republics might emerge as catalysts of nationalist unrest in the future. One-part description and one-part prescription, the term "Western Republics" still carries with it a tinge of Cold War politicization, reflecting the US military-academic complex's strategy to weaken the Soviet Bloc by driving in wedges wherever possible (see Robin 2001). Certainly, these regions, abutting staunchly Catholic Poland and the "maverick" Communist state of Romania, provided a channel for the spillover of democratic ideas from Eastern Europe and/or irredentist sentiment stemming from World War II-era border changes.

With the introduction of glasnost, these three republics steadily took on more substantive control of their destinies within the larger structure of the Soviet Union, with two of three (Ukraine and Belarus) participating in the Belavezha Accords in December 1991, which disbanded the USSR and established the Commonwealth of Independent States in its wake. Despite sharply differing popular alignments towards the idea of Europe, the three countries are treated rather uniformly in terms of their relationship with the rest of the continent and the European Union. As members of the CIS, the Western Republics are not generally considered viable candidates for admission to the EU in the foreseeable future, instead occupying a second-tier position alongside North African countries and other members of the Eastern Partnership within the European Neighborhood Policy (ENP). The 2007 expansion of the Schengen Agreement (the visa-free arrangement that exists between the continental members of the European Union) to include Latvia, Lithuania, Hungary, Poland, and Slovakia created what some have labeled a "new

Iron Curtain," erected and maintained by billions of dollars of "barbed wire, cameras and motion detectors" along the EU's new eastern frontiers, separating populations that once moved back and forth across Soviet and Eastern Bloc borders with ease (see Gomez and Dudikova 2010). Even with Ukraine's comparative openness to western investment and freewheeling-though-chaotic attempts at pluralism, the country is treated little differently in structural terms from "Europe's last dictatorship" Belarus (Rausing 2012) and the political and economic "black hole" of Moldova (Orlandi 2007), thus making the argument, at least for the purposes of this study, for grouping the three countries together.

Internally, these three republics generated their own engines of change, particularly Ukraine. As the site of the devastating *Holodomor* (Ukrainian: "death by hunger") famines of the 1930s, de-Stalinization in "Little Russia" (as the region was pejoratively known) led to increasing differentiation of Ukrainian and Russian identities (though both continued to exist under the Soviet umbrella). In terms of historical statehood, Belarus and Moldova enjoyed weaker legacies than their Ukrainian neighbor, which was site of the first Russian state, Kievan Rus, as well as a short-lived republic in the wake of the Bolshevik Revolution (Wilson 1997); yet, the vigor with which they embraced independence proved the ultimate viability of Leninist-Stalinist national delimitation conducted decades earlier. Following the catastrophic Chernobyl explosion in 1986, Ukraine and Belarus gravitated away from the Kremlin, realizing their precarious position in a surreptitiously imperial system that privileged the center over the periphery. Meanwhile, Moldova—carved out of Ukrainian and Romanian territory cribbed in 1940—began an inexorable slide into ethnic conflict between its Latinate- and Slavic-speaking residents. However, despite commonalities, grouping these states together sometimes proves problematic. As Angela E. Stent remarks:

> Although these three countries are referred to as the "Western Newly Independent States (NIS)," it is not clear that they form a region in terms of common goals or consensus on interaction with each other. Apart from their common Soviet legacy and geographical contiguity, it is premature to speak of them as a region. Moreover, they do not consider themselves as such. (2007, 2)

Consequently, there are as many reasons to consider these countries separately as there are to consider them as any sort of unit, despite current trends in geopolitical thought.

Reflecting their history as founding members of the USSR and their geographical position between Russia's oil and natural gas fields and energy-hungry countries in Western Europe, the Western Republics are intimately linked to Russia in the post-Soviet period, a fact which is crucial to understanding these countries' political geography. From the curious Union of Russia and Belarus to Russian peacekeepers' continued presence in Transnistria, Moscow's geopolitical dominance remains a central concern to both the foreign and domestic policies of these three states. The presence of large numbers of ethnic Russians in Ukraine, particularly Crimea (a region transferred from Russia to the Ukrainian SSR under Nikita

Khrushchev and recently annexed by the Russian Federation), allowed Moscow a great deal of influence over its neighbor after independence, as did the basing of the Russian Black Sea Fleet in Sevastopol under a lease that was to last until 2042; this deal was struck in 2010 under now-ousted President Viktor Yanukovych in return for guarantees on lower prices for Russian natural gas through 2019 (see Motyl 2010). Provision of and subsidies on natural gas have been and remain a lever for the exercise of Russian power, with Gazprom choosing to shut off supplies to Ukraine at crucial points in political negotiations between Moscow and Kyiv. Increasing numbers of Russian troops on Ukraine's eastern border (and, according to many intelligence reports, within Ukraine) are also important tools for Russia in its relationship with its western neighbor.

Belarus, as a primary route for the transshipment of natural gas to Germany and a strident critic of the EU and NATO, has long-served as a reliable Russian client state in the post-Soviet period. However, more recently, the country's authoritarian President Aleksandr Lukashenko has distanced himself from the Kremlin, seeking better deals from Russian gas companies and opening new doors to the West. Correspondingly, economic elites in Russia have begun to label Belarus as a parasitic entity, suggesting that Russia—as the host—needs to slough off its burdensome satellite (see Klinke 2008). Relations have deteriorated even further with Minsk's vocal support for Ukraine in the current regional crisis. Transnistria remains a headache for Europe and a tool of influence for Moscow over Moldova: the breakaway republic is essentially a mafia-dominated satrapy of Russia, where international law cannot reach. The European Parliament has described the breakaway republic as "a black hole in which illegal trade in arms, the trafficking in human beings and the laundering of criminal finance was carried on" (Wiersma 2002). In recent years, trade in illicit weapons (linked to the Russian military contingent of over 1,000 soldiers based in the region) has weighed heavily on Transnistria's image as an ungoverned and dangerous space. In fact, the region was once home base for the Russian arms dealer Viktor Bout, popularly known as the "Merchant of Death" and loosely depicted in the 2005 film *Lord of War* (see Chapters 5 and 7). Adding to this negative perception is the fact that a large number of pro-Russian cyber-attacks and other acts of Internet-based criminality have originated from the capital Tiraspol, most famously when a pro-Kremlin youth launched a series of devastating attacks on Estonian sites in 2007.

Transcaucasia: The Euro-Asian borderlands

Lying on the southern slopes of the Caucasus mountain range, Azerbaijan, Armenia, and Georgia are arguably the most continentally ambiguous countries in the world. From a purely geographical standpoint these three states are in Asia, given that the Caucasian watershed represents Europe's southernmost territorial frontier.[23] However, Transcaucasia's long association with Russia, particularly as part of the USSR, ties these countries to West through culture, trade, architecture, and numerous other markers of identity. As early (Christian) nations, Armenia and Georgia possess histories of statehood dating back to the classical period;

however, (Muslim) Azerbaijan is a creation of the modern period, resulting from tsarist territorial acquisitions made at the expense of Persia. Originally established as a tripartite Transcaucasian federation in the early days of the Soviet Union,[24] Armenia, Georgia, and Azerbaijan gained the status of union republics in 1936, thus paving the way for full independence more than half a century later. Situated between the Caspian and Black Seas and neighboring Russia, Turkey, and Iran (see Figure 4.4), the three post-Soviet republics of the South Caucasus transitioned from being a geopolitical "black holes" (Gachechiladze 2002) in the early 1990s to important players in contemporary geopolitics, particularly given the region's spiking importance in petroleum exports and international security.

Roughly shaped like an eagle in flight, Azerbaijan is a smallish country of 86,600 square kilometers. The country has a population of 8.3 million, 90 percent of whom are ethnic Azerbaijanis, a Turkic people closely related to the Turks. Approximately 15 million Azeris reside in neighboring Iran, a fact that has long complicated cross-border relations with the Islamic Republic (Souleimanov and Ditrych 2007). The Azeri language, a member of the Oghuz sub-family of Turkic, employs a Latin alphabet similar to that of Turkish, though the older Cyrillic script remains in use in certain quarters. Statistically, 99 percent of the country's population is Muslim, with Shi'ism prevailing (the country's Christian minority mostly fled at the end of the 1980s); however, Azerbaijan is considered to be the

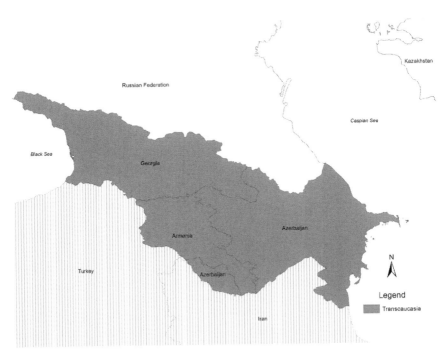

Figure 4.4 Transcaucasia or the Southern Caucasus (Emily A. Fogarty/DIVA-GIS, Global Administrative Areas and Natural Earth)

most secular Muslim state in the world (Ismayilov 2014). Azerbaijan's geopolitical identity is intertwined with that of Armenia because of the former's exclave of Nakhchivan, from which it is separated by Armenian territory, as well as the more problematic issue of Nagorno-Karabakh, an ethnically Armenian region that declared its independence from Azerbaijan more than two decades ago. The statelet is economically and militarily supported by Armenia, which in turn is increasingly propped up by Russia and Russian investors.

One of the earliest nations to embrace Christianity, Armenia prides itself on more than 2,000 years of statehood. Complicating such claims, the lands of the Armenians have long been under the rule of foreign powers, from the Romans to the Russians. Despite Allied plans for a large, independent Armenian state after the Great War, Kemalist Turkey and Bolshevik Russia prevented such an outcome, reducing Armenia to a rump state under Soviet suzerainty until independence in 1991. Armenia—or Hayastan as it called by the Armenians—is a mountainous land of approximately 30,000 square kilometers with a population of nearly 3 million, nearly all of whom are ethnic Armenians. With a unique alphabet and only distant links to other Indo-European tongues, the Armenian language stands as a key marker of national identity, both for the country's citizens and a diaspora of upwards of 9 million worldwide. Landlocked and bordered by the hostile states of Turkey and Azerbaijan, Armenia is highly dependent on sustaining good relations with Russia (Moscow possesses missile sites and a sizeable military contingent in the country) and Iran, as well as garnering economic support from the large expatriate Armenian communities in the US and France (Darieva 2011).

The Republic of Georgia, or Sakartvelo as it known to Georgians, is washed by the Black Sea and situated between southern Russia and northwestern Turkey. The country has a population of approximately 4.6 million and is nearly 70,000 square kilometers in size. Similarly to the Armenians, Georgians converted to Christianity in the early Middle Ages, and are historically affiliated with St. George (hence the country's name in the West and the use of the saint's cross on the current flag). Approximately 80 percent of the population are adherents of or are historically affiliated with the Georgian Orthodox Church; Muslims in the breakaway republic Abkhazia and the formerly autonomous region of Ajaria account for roughly 10 percent of the population. Ethnic Georgians dominate the country, representing 83 percent of the population. The Georgian language is indigenous to the Caucasus and unrelated to any of the Indo-European languages or the various Turkic languages of the region; the language's alphabet Mkhedruli is equally distinct. Georgia has long experienced secessionist struggles involving its three ethnically distinct regions of Ajaria, Abkhazia, and South Ossetia, all which were established in the Soviet period (see Kabachnik 2012). While Ajaria was successfully reintegrated in 2004, Abkhazia and South Ossetia have severed all ties with the rest of the country following recognition as sovereign states by Moscow.[25]

Russia's diplomatic relations with the South Caucasus are complex and almost always contentious. Moscow's support of Armenian secessionists in the Nagorno-Karabakh War (1988–1994) led to poor relations with Azerbaijan for much of the 1990s. Similarly, the Kremlin's backing of Georgia's breakaway republics of Abkhazia and South Ossetia, both of which share a border with the Russian Federation, has complicated relations with Tbilisi following its independence, as have Russia's real and perceived involvement in the country's domestic politics. Moscow, conversely, has strongly criticized Georgia's actions during its two conflicts with Chechen rebels, who often found sanctuary within poorly policed areas of Georgia's mountainous frontier with the Russian Federation. The Kremlin's relations with its southern neighbor steadily deteriorated after the election of Columbia-educated, US-backed Mikheil Saakashvili following the so-called Rose Revolution in 2003. In the following years, Moscow cut natural gas exports to Georgia, banned key imports from the country, and suspended transportation and postal links. Following attempts by Georgian military units to reassert authority in South Ossetia, Russia invaded Georgia, occupying large parts of the country (including Stalin's birthplace, Gori) before international arbitration resulted in a ceasefire and withdrawal. Russia's warmest relations in the region have been with Armenia, which is highly dependent on Russian subsidies. The United States has challenged Russian hegemony is the region, particularly through military support to Georgia and investment in Azerbaijan's oil sector. Western support for a pipeline connecting Caspian oil exporters to the Mediterranean, as well as tentative plans for the inclusion of Georgia in the NATO, proved to be the most visible displays of US attempts at influence in the region.

While Azerbaijan, Armenia, and Georgia are divided by language, faith, culture, and strategic interests, the three countries are united in the desire to be included within the European realm rather than be linked to the Middle East. Regional geo-economics affirm this normative position, with a strong commercial orientation towards the West and/or Russia, though each country is also careful to maintain positive relations with Iran, an increasingly important player in the larger region. Of the three countries, Azerbaijan is easily the wealthiest as a result its large oil reserves. The capital Baku emerged as one of the world's first "oil towns" in the late nineteenth century, and has continued to serve as a major site of petroleum extraction ever since. The country has invested in massive development projects in Baku, as well as becoming a premier hub for international sporting and cultural events, from Eurovision (2012) to the European Games (2015). Georgia is developing its tourism infrastructure, as well as continuing to focus on mineral water and wine exports, though the 2008 war has proved a major setback; however, reforms put in place before the global financial crisis have provided the country with a remarkable level of resiliency. Armenia, the poorest country in the region, continues to suffer from a lack of self-sustaining industries in the post-Soviet period, and is partially dependent on remittances, international aid, and Russian investment.

The Central Asian Republics: Europe in Asia

As a geopolitical region, Central Asia has historically been demarcated as the space between Russia, Europe, China, and the Indian subcontinent. However, since the dissolution of the USSR, the term has become synonymous with the five post-Soviet republics of Kazakhstan, Uzbekistan, Turkmenistan, Kyrgyzstan, and Tajikistan (see Figure 4.5); Afghanistan is sometimes included as well, but is more often affiliated with the "Greater Middle East" (see Güney and Gökcan 2010). In the seventeenth century, Russia began exerting influence in the region, steadily extending its frontier southwards throughout the 1800s. Following the assertion of Bolshevik power in the region, Turkestan—as it was then known—was divided into five ethnic republics, which today are the sovereign states listed above. During the period of Soviet rule, Moscow subsidized the development of industry, agriculture, and education throughout the region, resulting in standards roughly equivalent to those in parts of Eastern Europe, though the price to be paid was the destruction of traditional culture and economic systems through intense Sovietization. In the wake of independence from the USSR, living standards plummeted across most of the region, with the exception of Kazakhstan, where oil revenues and a small population allowed the country to buck regional trends (Brill Olcott 2002). Lingering regional instability associated with Afghanistan, the international drugs trade, the rising influence of Islamist groups, and the Tajik Civil War (1992–1997) also plagued the southernmost republics of Central Asia. In the wake of 9/11, Central Asia moved to center stage in global politics as the US—with Russian approval—established a sizeable military presence in the region. China and Russia, however, have since worked together to dislodge Washington from this privileged position, with almost complete success (U.S troops were forced out of Uzbekistan in 2005 following condemnation of a violent crackdown on dissidents in the Fergana Valley and were asked to leave Kyrgyzstan in 2014).

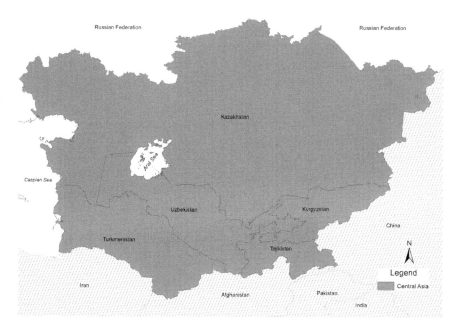

Figure 4.5 The Central Asian Republics (Emily A. Fogarty/DIVA-GIS, Global Administrative Areas and Natural Earth)

The ninth largest country in the world, Kazakhstan is also the biggest of the Central Asian Republics (CARs) at 2.7 million square kilometers, yet it only has a population of 16.5 million. Kazakhs are the titular majority, accounting for 65 percent of the total population (up sharply from 40 percent in the late 1980s). Ethnic Russians make up nearly one-quarter of the population; numerous other minorities (Ukrainians, Germans, Tatars, Koreans, etc.) also call the republic home, due to World War II-era deportations which consigned many ethnic groups to partial or total relocation to the steppes and cities of Kazakhstan. Kazakhs are mostly Sunni Muslims, though the level of religiosity in the republic remains quite low, and the practice of other religions is vigorously protected, an outcome of Sovietization and active management of inter-faith relations by the Nazarbayev government. While many Kazakhs continue to speak Russian, their language—a member of the Kipchak sub-family of Turkic languages—is on the rise in use and prestige. Kazakhstan possesses the world's longest contiguous border with its northern neighbor Russia. The two countries maintain a friendly and economically robust relationship, including long-term leasing rights to Kazakhstan's Baikonur cosmodrome, making Kazakhstan one Moscow's most loyal allies in the post-Soviet realm (Rumer 2000). Once the site of the Soviet Union's nuclear testing facilities, Kazakhstan rapidly de-nuclearized upon independence, and campaigns for full nuclear disarmament worldwide (EoK 2006). The country possesses prodigious oil and natural gas reserves, as well as significant mineral resources, all of which are currently being developed by a host of international petroleum companies including Agip, ENI, ExxonMobil, Total, and Shell.

In terms of population, Uzbekistan is the largest country in the region, with nearly 29 million inhabitants; however, at 447,000 square kilometers, it is dwarfed by its northern neighbor Kazakhstan. Ethnic Uzbeks, a Turkic-speaking, Sunni Muslim people, make up about 75 percent of the population, with Russians constituting the largest ethnic minority. Unlike the Kazakhs, many Uzbeks are religious, particularly in the Fergana Valley, a region where political Islam has gained ground in recent decades (Brill Olcott and Babajanov 2003). Uzbeks form sizeable minorities in all of the country's neighbors including Afghanistan, thus creating a constant source of tension in the region. Due to Stalinist gerrymandering of its borders, Uzbekistan is home to many of the great cities of the Silk Road (e.g., Bukhara, Samarkand, and Kokand), which were major centers of learning and commerce during the Islamic Golden Age (750–1250). However, Uzbekistan's doubly landlocked status complicates its geopolitical position in the era of neoliberal globalization. Since the mid-nineteenth century, cotton has been a major cash crop in the country, but ill-conceived irrigation programs and heavy use of pesticides has wreaked havoc on the national ecology, including the depletion of two-thirds of the Aral Sea (shared with Kazakhstan). Gold and natural gas are important export products for the national economy. Uzbekistan includes one autonomous region, Karakalpakistan, which—unlike its peers in Moldova and Georgia—has not opted for secession from its parent state in the post-Soviet era.

Situated on the Caspian Sea, but with almost 90 percent of its land mass composed of deserts, Turkmenistan is a country of 491,000 square kilometers.

Its population of 5.1 million is dominated by ethnic Turkmen, who are closely related to the Turks and Azeris; small minorities of Uzbeks and Russians also reside in the country. Possessing long borders with Iran and Afghanistan and enjoying the world's fourth-largest proven natural gas reserves, the country's first post-Soviet president, Saparmurat Niyazov (Türkmenbaşy), opted for a policy of neutrality, thus distinguishing Turkmenistan from all other former Soviet republics, which either embraced Russia via CIS membership or gravitated towards Western-backed alliances. Turkmenistan also distinguished itself by the cult of personality of the "President for Life," being a "human rights black hole" (Stoudmann 2003), and imposing a near total ban on Internet use through the late 1990s and the first part of the new millennium. Reflecting an unwillingness to move beyond Stalinism, Niyazov's cult of personality was only rivalled by that of Kim Jong-il: the autocrat renamed the months of the year (his mother's name replaced "April"), erected monumental statues to his leadership of the country, and—most bizarrely—penned a "poorly written, chaotically organized, and factually dubious" revisionist history entitled *Ruhnama* (Saunders 2009a). Niyazov claimed the book was more important than the Quran, and required all citizens to have a copy and students to read from it on a daily basis. Under his rule, the country began to exploit its natural gas reserves; however, economic reforms impoverished the county's rural populations, denying them basic services including healthcare. The result has been massive immigration to the capital, with many Turkmen living at the edge of society in the gleaming metropolis of Ashgabat, forced to abandoned traditional social structures associated with tribalism (though there is a cultural renaissance occurring among those who have chosen to remain in the villages). Additionally, there has been a major brain drain as many Russophones and other skilled workers have left for other countries. Since Niyazov's death in 2006, the country has made moderate political reforms and significantly opened up to the outside world, though Turkmenistan still remains one of the world's most reclusive states.

Kyrgyzstan, formerly known as Kirghizia, is a small, mostly mountainous republic of less than 200,000 square kilometers in area. Its population of 5.6 million consists of two-thirds ethnic Kyrgyz—a Muslim, Turkic people closely related to the Kazakhs—and Uzbeks, who account for 14 percent of the population, as well as ethnic Russians and a number of other minorities. Unlike its lowland neighbors Uzbekistan and Kazakhstan, Kyrgyzstan lacks significant hydrocarbon resources, and has instead opted to use its geography to improve its economic situation in the post-Soviet period, permitting basing rights for both the Russian Federation and the United States; however, Bishkek abruptly ended its 22-year-old bilateral treaty with the US in mid-2015 over the awarding of a human rights prize to jailed activist Azimjon Askarov. Additionally, the country has sought to take advantage of its stunning alpine landscapes, especially around Issyk Kul, by marketing itself as an eco-tourism destination. Once framed as an "island of democracy" (Juraev 2008) in the region, Kyrgyzstan descended into authoritarianism, followed by a series of popular revolutions and thinly-veiled political coups which have debilitated the country over the past decade.

With few natural resources and an infelicitous border that deprived the Persian-speaking Tajiks of their historic urban centers of Bukhara, Samarkand, and Kokand (Soucek 2000), Tajikistan is the poorest of the former Soviet republics. The country's poverty is due in part to its vertiginous, mostly rural geography: at 143,000 square kilometers, Tajikistan is home to the 7,495-meter Qullai Ismoili Somoni, formerly known as Communism Peak, the highest point in the former USSR. The economic situation was made much worse by the five-year civil war that pitted Islamist insurgents and democratic reformers against the neo-Soviet, pro-Russian old guard (the ultimate victors in the conflict). Drug trafficking has also taken its toll on the nation, as has emigration, with nearly 1 million Tajiks (mostly men) working abroad. The country has a population of 8 million, four-fifths of whom are ethnic Tajiks, a people of Indo-Iranian descent and thus distinct from their Turkic neighbors in Kyrgyzstan and Uzbekistan. Sunni Islam is dominant in the country, although there is a small contingent of Ismaili Shi'a Muslims among the country's residents. Tajikistan has long been ruled by Emomali Rahmon, who commands strong Russian support (Moscow enjoys basing rights in the country and long oversaw border security). China is an important investor in the republic, financing mining and transportation projects.

In terms of its geopolitical and geo-economic position, the region of Central Asia is ascendant. Situated on the northern fringe of the volatile Middle East and South Asia and linking Europe to East Asia, the five Central Asian republics are both vital to stability on the super-continent as well as a zone of potential instability. Zbigniew Brzezinski (1997) once described the region as the "Eurasian Balkans," intimating that the CARs might follow Yugoslavia's descent into ethnic cleansing and endemic conflict. However, despite a few minor flare-ups of Islamist insurrection, the region has remained relatively stable since the late 1990s. Through cooperation with China and Russia, separatist and terrorist threats have been quashed. With rapid development of oil and natural gas reserves, particularly in Kazakhstan and Turkmenistan, as well as the construction of new pipelines linking hydrocarbon exports directly to the Mediterranean and East Asia, the region has also emerged as an important player in global energy (Tazhin 2008). Kazakhstan has differentiated itself through a major investment in sport (cycling teams, international games, etc.), cultural promotion, and economic-technology conferences, as well as building a world-class, twenty-first-century showcase capital in the form of Astana. In political terms, however, the region has become synonymous with managed democracy and suffers from the "President for Life" syndrome (Collins 2009). According to Martha Brill Olcott (2005), the region enjoyed a "second chance" at political pluralism following 9/11; however, this opportunity was squandered by dominant political elites, who used the US's "war on terror" as cover for the elimination of rivals and diminution of a burgeoning civil society.

As will be discussed in the ensuing chapters, the complex geographical and protean geopolitical space of post-Soviet Eurasia allows it continue to be framed as a mystery—or, to reprise Churchill's term, an *enigma*—to many in the West, a sort of European Orient. Russia, Ukraine, Eurasia, the Caucasus, and Central Asia are terms which, generally speaking, evoke almost the exact same associations despite robust nation branding efforts. The following analysis aims to explore why this is the case, focusing on popular cultural representations of post-Soviet space. In order to accomplish this, we will turn our attention to the vagaries of national-image production, both by states and by those acting outside the state, in an effort to understand why geographical imaginaries of the post-Soviet realm remain stagnant and why "Soviet ghosts" (and even older revenants) continue to haunt the Western mind when it turns to thoughts of the Eurasian realm.[26]

Notes

1 The Bosporus and the Dardanelles do not present any practical barrier to movement between the two continents. Moreover, the territorial delimitation of the Caucasus and the Ural mountain ranges are later additions to the delimitation structure. Even in the case of Africa, we see few geographical reasons for separating the continents, a contention that is strengthened by the historical shift of the border between Asia and Africa being shifted from the Nile River to the western edge of the Sinai Peninsula.

2 St. Petersburg lies on this line, as does the occidental border of the pre-WWII Soviet Union; following territorial annexations in the 1940s, the western fringe of the USSR extended farther into Europe-proper.

3 Despite inclusionary claims towards Russian/Eurasian religious traditions, many neo-Eurasianists espouse a form of "veiled anti-Semitism" (Laruelle 2012), condemning the influence of the Khazars, medieval "nomadism," Zionism, and other forms of "Jewish influence" as anathema to the Eurasian *esprit*.

4 Ethnic Russians are known as *russki*, while the term *rossiski* is used to connote all citizens of the Russian Federation regardless of ethnic background.

5 This number includes the recent addition of Crimea, though the annexation of the former Ukrainian province is not recognized by much of the international community.

6 An important exception is the southern Siberian Republic of Tuva (formerly known as Tannu Tuva and Urianghai before that), an independent state until it was quietly annexed to the USSR in 1944. Today, the indigenous Tuvans make up three-quarters of the republic's population.

7 Importantly, Russia's newest ethnic republic, Crimea, does not fit neatly into any of these geographical categories as it lies on the Black Sea.

8 Once again, this figure takes into account Ukrainian territories annexed in 2014, including the Republic of Crimea and the federal city of Sevastopol.

9 The term "near abroad" carries strong post-imperial connotations, discursively structuring these states as part of Russia's regional sphere of influence based on past inclusion in the USSR. Other countries included in the near abroad, but which do not share a territorial border with the Russian Federation, are as follows: Moldova, Armenia, Uzbekistan, Turkmenistan, Kyrgyzstan, and Tajikistan.

10 Yeltsin would initiate the Second Chechen War (1999–2000) with the help of his heir-apparent, Prime Minister Vladimir Putin. During its insurgency phase (2000–2009), this conflict spread well beyond the borders of the Chechen Republic, engulfing much of the North Caucasus, sowing chaos in Georgia, and prompting terrorist attacks in Moscow and other parts of Russia (most spectacularly the attack on School No. 1 in Beslan, North Ossetia, in September 2004).

11 Estonia, Latvia, and Lithuania eschewed any ties to Russia upon independence, opting instead for incorporation into the European Union and NATO.

12 These dramatically different foreign policy vectors were largely influenced by Yeltsin's foreign ministers, the Western-oriented Andrei Kozyrev (1991–1996) and hardliner

Yevgeny Primakov (1996–1998), who argued for the creation of a multipolar world to counter US hegemony.

13 EurAsEC includes Russia, Belarus, Kazakhstan, Kyrgyzstan, and Tajikistan; Uzbekistan withdrew from the organization in 2008.

14 Interestingly, both conflicts broke out during the course of the Olympic Games; the first during the Beijing Summer Olympics and the second as Russia hosted the winter games at Sochi.

15 The United States, Great Britain, and a significant number of other Western powers refused to recognize the wartime annexation of the Baltic States during the Cold War.

16 In their contrasting of geopolitics with geo-economics, John Agnew, Katharyne Mitchell, and Gearóid Ó Tuathail (2003) characterize the latter as associated with deter-ritorialized post-Fordist political economies where deregulation and decentralization of governance are the norm, the state and market are "networked" through public–private partnerships, borders and foreign/domestic distinctions are lessened, and nodal-ity within regional and global economic systems are treated as superlative.

17 The vast majority of Crimean Tatars (*Qirimlar*) now live under Russian control follow-ing the annexation of the province in 2014. The Tatars, who were deported en masse to Central Asia during World War II, on the whole disapproved of Crimea's secession from Ukraine, fearing a loss of autonomy within the republic as well as a return to the tender mercies of the Kremlin.

18 Though, importantly, the majority of the country's industrial base is in the chaotic Donbass region.

19 On 27 February 2015, Russian opposition politician Boris Nemstov was gunned down near Red Square, purportedly for his role in preparing a report on the role of Russian troops in the fighting in eastern Ukraine, despite the Kremlin's official denials to the contrary.

20 According to the country's official web site (http://www.belarus.by/en/), Belarus was part of a series of grand duchies, commonwealths, and empires, not being its own state until 1918 when a People's Republic was declared in response to the German occu-pation, followed a year later by the establishment of the Belarusian Soviet Socialist Republic under Soviet authority.

21 The Romanian language employed a variant of the Cyrillic alphabet until orthographic reforms in the nineteenth century. During Soviet times Moldovan was written in Cyrillic, but since the late 1980s the Latin alphabet has been used in official documen-tation (outside of Transnistria).

22 Occasionally, the Baltic States were included under this label as well, though Lithuania, Latvia, and Estonia were generally kept separate due to their interwar independence.

23 Maritime borders separate the Balkan Peninsula from Asia Minor, while the Ural Mountains, Ural River, and Caspian Sea demarcate the easterly edge of European Russia.

24 The historical merging of the South Caucasus under Russian/Soviet rule presents a variety of difficulties for the three republics as they seek to differentiate themselves on the international stage (see Chapter 8).

25 The list of other countries recognizing independence is relatively short: Nicaragua, Venezuela, and Nauru.

26 I borrow the term "Soviet ghosts" from the title of Rebecca Litchfield's (2014) photo-graphic essay on abandoned Soviet sites.

References

Agnew, John, Katharyne Mitchell, and Gearóid Ó Tuathail. 2003. *A Companion to Political Geography*. Hoboken, NJ: Blackwell.

Baldi, Philip. 1983. *An Introduction to the Indo-European Languages*. Carbondale, IL: Southern Illinois University Press.

Blinnikov, Mikhail S. 2011. *A Geography of Russia and Its Neighbors*. New York: The Guilford Press.

Brill Olcott, Martha. 2002. *Kazakhstan: Unfulfilled Promise*. Washington, DC: Carnegie Endowment for International Peace.

Brill Olcott, Martha. 2005. *Central Asia's Second Chance*. Washington, DC: Carnegie Endowment for International Peace.

Brill Olcott, Martha, and Bakhtiyar Babajanov. 2003. "The Terrorist Notebooks." *Foreign Policy* no. 135 (135–140).

Browning, Christopher S. 2005. " Westphalian, Imperial, Neomedieval: The Geopolitics of Europe and the Role of the North." In *Remaking Europe in the Margins: Northern Europe after the Enlargements*, edited by Christopher S. Browning. Aldershot, England and Burlington, VT: Ashgate.

Brzezinski, Zbigniew. 1997. *The Grand Chessboard: American Primacy and Its Geostrategic Imperatives*. New York: Basic Books.

Cash, Jennifer R. 2011. *Villages on Stage: Folklore and Nationalism in the Republic of Moldova*. Münster, Germany: Verlag.

Chinn, Jeff, and Robert J Kaiser. 1996. *Russians as the New Minority: Ethnicity and Nationalism in Soviet Successor States*. Boulder, CO: Westview Press.

Collins, Kathleen. 2009. "Economic and Security Regionalism among Patrimonial Authoritarian Regimes: The Case of Central Asia." *Europe-Asia Studies* no. 61 (2):249–281.

Colton, Timothy J., and Michael McFaul. 2003. *Popular Choice and Managed Democracy: The Russian Elections of 1999 and 2000*. Washington, DC: Brookings Institution Press.

Cornell, Svante E., and S. Frederick Starr. 2009. *The Guns of August 2008: Russia's War in Georgia*. Armonk, NY: M. E. Sharpe.

Darieva, Tsypylma. 2011. "Rethinking Homecoming: Diasporic Cosmopolitanism in Post-Soviet Armenia." *Ethnic & Racial Studies* no. 34 (3):490–508.

de Blij, Harm J., and Peter O. Muller. 2008. *Geography: Realms, Regions, and Concepts*. Hoboken, NJ: John Wiley & Sons.

EC. 2010. "The Northern Dimension Policy." *European Commission Department of External Relations*, available at: http://ec.europa.eu/external_relations/north_dim/index_en.htm [last accessed 13 September 2010].

EoK. 2006. *Kazakhstan's Nuclear Disarmament: A Global Model for a Safer World*. Washington DC: Embassy of the Republic of Kazakhstan to the United States of America and the Nuclear Threat Initiative.

Gachechiladze, Revaz. 2002. "Geopolitics in the South Caucasus: Local and External Players." *Geopolitics* no. 7 (1):113–138.

Gomez, James M., and Andrea Dudikova. 2010. "'New Iron Curtain' Descends in EU Free-Travel Split." *Bloomberg*, available at: www.bloomberg.com/apps/news?pid=ne wsarchive&sid=aKQMJPPrOifU [last accessed 20 December 2010].

Güney, Aylin, and Fulya Gökcan. 2010. "The 'Greater Middle East' as a 'Modern' Geopolitical Imagination in American Foreign Policy." *Geopolitics* no. 15:22–38.

Hahn, Gordon M. 2003. "The Past, Present, and Future of the Russian Federal State." *Demokratizatsiya* no. 11 (3):343–362.

Intourist. 1932. *A Pocket Guide to the Soviet Union*. Edited by L. A. Block. Moscow and Leningrad: Vneshtorgisdat.

Isakova, Irina. 2004. *Russian Governance in the 21st Century: Geo-Strategy, Geopolitics and New Governance*. London and New York: Routledge.

Ismayilov, Elnur. 2014. "Islam in Azerbaijan: Revival and Political Involvement." In *Religion, Nation and Democracy in the South Caucasus*, edited by Alexander Agadjanian, Ansgar Jödicke and Evert van der Zweerde, 96–111. London and New York: Routledge.

Juraev, Shairbek. 2008. "Kyrgyz Democracy? The Tulip Revolution and Beyond." *Central Asian Survey* no. 27 (3–4):253–264.

Kabachnik, Peter. 2012. "Wounds That Won't Heal: Cartographic Anxieties and the Quest for Territorial Integrity in Georgia." *Central Asian Survey* no. 31 (1):45–60.

Katin-Borland, Nat. 2012. "Cyberwar: A Real and Growing Threat." In *Cyberspaces and Global Affairs*, edited by Sean S. Costigan, 3–21. Farnham, UK and Burlington, VT: Ashgate.

Khomami, Nadia. 2015. "Putin Holds Annual Press Conference in Moscow." *The Guardian*, available at www.theguardian.com/world/live/2015/dec/17/vladimir-putins-annual-press-conference-live [last accessed 30 December 2015].

King, Charles. 1999. "The Ambivalence of Authenticity, or How the Moldovan Language Was Made." *Slavic Review* no. 58 (1):117–142.

Klinke, Ian. 2008. "Geopolitical Narratives on Belarus in Contemporary Russia." *Perspectives: Central European Review of International Affairs* no. 16 (1):109–131.

Knudson Gee, Gretchen. 1995. "Geography, Nationality, and Religion in Ukraine: A Research Note." *Journal for the Scientific Study of Religion* no. 34 (3):383–390.

Kolstø, Pål. 1999. "Territorialising Diasporas: The Case of the Russians in the Former Soviet Republics." *Millennium: Journal of International Studies* no. 28 (3):607–631.

Laruelle, Marléne. 2007. "The Orient in Russian Thought at the Turn of the Century." In *Russia between East and West: Scholarly Debates on Eurasianism*, edited by Dmitry Shlapentokh, 9–37. Leiden, Netherlands and Boston, MA: Brill.

——. 2012. *Russian Eurasianism: An Ideology of Empire*. Baltimore, MD: Johns Hopkins University Press.

——. 2013. *Russia's Arctic Strategies and the Future of the Far North*. Armonk, NY: M. E. Sharpe.

Lewis, Martin W., and Kären E. Wigen. 1997. *The Myth of Continents: A Critique of Metageography*. Berkeley: University of California Press.

Litchfield, Rebecca. 2014. *Soviet Ghosts: The Soviet Union Abandoned: A Communist Empire in Decay*. London: Carpet Bombing Culture.

Luttwak, Edward. 1990. "From Geopolitics to Geoeconomics: Logic of Conflict, Grammar of Commerce." *The National Interest* no. 20:17–23.

Mackinder, Halford J. 1904. "The Geographical Pivot of History." *The Geographical Journal* no. 23 (4):421–437.

Motyl, Alexander J. 2010. "Ukrainian Blues." *Foreign Affairs* no. 89 (4):125–136.

Orlandi, Fernando. 2007. "Moldova: A Black Hole in Europe." *East* no. 16:137–141.

Ostergren, Robert C., and Mathias Le Bossé. 2011. *The Europeans: A Geography of People, Culture, and Environment*, 2nd ed. New York: The Guilford Press.

Plokhy, Serhii. 2011. "The 'New Eastern Europe': What to Do with the Histories of Ukraine, Belarus, and Moldova." *East European Politics & Societies* no. 25 (4):763–769.

Pringle, Robert W. 2010. "The Intelligence Services of Russia." In *The Oxford Handbook of National Security Intelligence*, edited by Loch K. Johnson, 774–789. Oxford: Oxford University Press.

Rausing, Sigrid. 2012. "Belarus: Inside Europe's Last Dictatorship." *The Guardian*, available at www.theguardian.com/world/2012/oct/07/belarus-inside-europes-last-dictatorship [last accessed 7 October 2012].

Robin, Ron. 2001. *The Making of the Cold War Enemy: Culture and Politics in the Military-Intellectual Complex*. Princeton, NJ and Oxford: Princeton University Press.

Rossman, Vadim. 2007. "The Orient in Russian Thought at the Turn of the Century." In *Russia between East and West: Scholarly Debates on Eurasianism*, edited by Dmitry Shlapentokh, 121–191. Leiden, Netherlands and Boston, MA: Brill.

Rumer, Boris Z. 2000. *Central Asia and the New Global Economy*. Armonk, NY: M. E. Sharpe.

Saunders, Robert A. 2005. "A Marooned Diaspora: Ethnic Russians in the Near Abroad and Their Impact on Russia's Foreign Policy and Domestic Politics." In *International Migration and Globalization of Domestic Politics*, edited by Rey Koslowski, 173–193. Abingdon, UK and New York: Routledge.

——. 2008. "Buying into Brand Borat: Kazakhstan's Cautious Embrace of Its Unwanted 'Son'." *Slavic Review* no. 67 (1):63–80.

——. 2009a. "Turkmenistan: Rage Against the Ruhnama." *Transitions*, available online at www.tol.org/client/article/20408-rage-against-the-ruhnama.html?print [last accessed 3 March 2010].

——. 2009b. "Wiring the Second World: The Geopolitics of Information and Communications Technology in Post-Totalitarian Eurasia." *Digital Icons: Studies in Russian, Eurasian and Central European New Media* no. 1:1–24.

——. 2014. "The Geopolitics of Russophonia: The Problems and Prospects of Post-Soviet 'Global Russian'." *Globality Studies Journal* (40):1–22.

Saunders, Robert A., and Vlad Strukov. 2010. *Historical Dictionary of the Russian Federation*. Lanham, MD: Scarecrow Press.

Sharp, Joanne P. 2000. *Condensing the Cold War: Reader's Digest and American Identity*. Minneapolis: University of Minnesota Press.

Shaw, Denis J. B. 1999. *Russia in the Modern World: A New Geography*. Oxford: Blackwell Publishers.

Smith, David J. 2001. *Estonia: Independence and European Integration*. London and New York: Routledge.

Soucek, Svat. 2000. *A History of Inner Asia*. Cambridge: Cambridge University Press.

Souleimanov, Emil, and Ondrej Ditrych. 2007. "Iran and Azerbaijan: A Contested Neighborhood." *Middle East Policy* no. 14 (2):101–116.

Stent, Angela E. 2007. "The Lands In Between: The New Eastern Europe in the Twenty-First Century." In *The New Eastern Europe: Ukraine, Belarus, Moldova*, edited by Daniel Hamilton and Gerhard Mangott, 1–24. Washington, DC: Center for Transatlantic Relations.

——. 2014. *The Limits of Partnership: U.S.–Russian Relations in the Twenty-First Century*. Princeton, NJ: Princeton University Press.

Stewart-Ingersoll, Robert, and Derrick Frazier. 2012. *Regional Powers and Security Orders: A Theoretical Framework*. London and New York: Routledge.

Stoudmann, Gérard. 2003. "Turkmenistan, a Human Rights 'Black Hole'." *Helsinki Monitor* no. 14 (2):117–124.

Sulick, Michael J. 2013. *American Spies: Espionage against the United States from the Cold War to the Present*. Washington, D.C.: Georgetown University Press.

Sullivan, Rob. 2012. *Geography Speaks: Performative Aspects of Geography: Performative Aspects of Geography*. Farnham, UK and Burlington, VT: Ashgate.

Sussex, Matthew. 2012. "Strategy, Security and Russian Resource Diplomacy." In *Russia and Its Near Neighbours*, edited by Maria Raquel Freire and Roger E. Kanet, 223–245. Basingstoke, UK: Palgrave Macmillan.

Tazhin, Marat. 2008. "The Geopolitical Role of the Main Global Players in Central Asia." *American Foreign Policy Interests* no. 30 (2):63–39.

Trenin, Dmitri. 2002. *The End of Eurasia: Russia on the Border between Geopolitics and Globalization*. Washington, DC: Carnegie Endowment for Global Peace.

——. 2006. "Russia Leaves the West." *Foreign Affairs* no. 85 (4):87–96.

Tsygankov, Andrei P. 2007. "Finding a Civilisational Idea: 'West,' 'Eurasia,' and 'Euro-East' in Russia's Foreign Policy." *Geopolitics* no. 12 (3):375–399.

——. 2013. *Russia's Foreign Policy: Change and Contniuity in National Identity*. Lanham, MD: Rowman & Littlefield.

van der Dussen, Jan, and Kevin Wilson. 2005. *The History of the Idea of Europe*. London and New York: Routledge.

Vinokurov, Evgeny, and Alexander Libman. 2012. *Eurasian Integration: Challenges of Transcontinental Regionalism*. Houndsmills, UK: Palgrave Macmillan.

Wiersma, Jan Marinus. 2002. "Report from the Chairman of the European Parliament (Ad Hoc Delegation To Moldova)." *European Parliament*, 5–6 June, pp. 1–22. Available at www.europarl.europa.eu/meetdocs/committees/afet/20021007/473437EN.pdf [last accessed 10 September 2010].

Wilson, Andrew. 1997. "Myths of National History in Belarus and Ukraine." In *Myths and Nationhood*, edited by Geoffrey Hosking and George Schöpflin, 182–197. London: Hurst & Company.

Yekelchyk, Serhy. 2006. "The Western Republics: Ukraine, Belarus, Moldova and the Baltics." In *The Cambridge History of Russia, Volume 3: The Twentieth Century*, edited by Ronald G. Suny, 522–548. Cambridge: Cambridge University Press.

Zürcher, Christoph. 2007. *The Post-Soviet Wars: Rebellion, Ethnic Conflict, and Nationhood in the Caucasus*. New York and London: New York University Press.

5 The post-Soviet bogeyman

A guide to the dangerous personae of the former USSR

In a March 2012 interview with CNN, Republican Party frontrunner for the presidential nomination Mitt Romney stated that Russia was, "without question," America's "number one geopolitical foe" (Willis 2012). Several months later, the future secretary of state John Kerry, speaking at the 2012 Democratic National Convention, used this dubious assertion in his endorsement of Barack Obama for a second term, sardonically noting that "Mitt Romney talks like he's only seen Russia by watching *Rocky IV*" (DNC 2012). In what some political pundits described as a Kerry's "best speech ever" (Schultz qtd. in Shahid 2012), the senior Senator from Massachusetts and foreign policy maven invoked popular culture in a highly effective riposte, deftly challenging Romney's credentials for managing the world's sole superpower. *Rocky IV* (1985) represents the zenith of late-Cold War filmic propaganda through the medium of professional boxing, pitting the rugged, self-made, underdog—Rocky Balboa (Sylvester Stallone)—against a pharmacologically and technologically enhanced Soviet foe—Ivan Drago (Dolf Lungren). The film taps a host of geopolitical codes familiar to anyone who lived through the Reagan administration (1981–1989), when the USSR was popularly referred to as the "Evil Empire." Rooted in the pop-culture mythology of George Lucas' *Star Wars*, this powerful rhetorical and geopolitical framing agent was extremely effective in structuring the Cold War as an existential battle between "good (US/NATO) and "evil" (USSR/Warsaw Pact). According to film historian William J. Palmer, *Rocky IV*, which was marketed with the tagline "When East Meets West, the Champion remains standing" (IMDB 2013), employs all of the ideologized stereotypes and hackneyed structures of "right-wing military fantasies" and "the need for cold war confrontation" (Palmer 1995, 219), epiphenomena which continue to manifest in certain conservative political circles to this day.

Foregrounding his quip about Romney's Russophobia, Kerry referenced former Republican vice presidential candidate Sarah Palin's 2008 comment that "one could see Russia from Alaska," thus quixotically burnishing her own foreign policy qualifications as the former governor of America's northernmost state. *Saturday Night Live* veteran Tina Fey's subsequent parody of Palin initiated a multifaceted grassroots popular-cultural response, including faux McCain-Palin advertisements showing a comically celestial Vladimir Putin peering at Alaska over the Earth's Arctic horizon, among other popularly generated spoofs of U.S.–Russian relations in the North Pacific Basin. Needless to say, "strategic fictions" (Davis 2006) in popular culture—from films and novels to Facebook memes and bumper stickers—play a palpable role in how Americans conceive of geopolitics in the twenty-first century. While Americans (and Britons) are reading fewer newspapers and watching fewer traditional news programs, media consumption continues to grow, with online media and television consumption currently at parity (Anderson 2010). Thus, quotidian understandings of geopolitical events, international relations, and global affairs are increasingly influenced by the images (re)presented in popular media forms. This is particularly important in the case of what has been deemed "mediated evil" (Altheide 2009), which results from the interplay between political communication (discourses produced by the political elite) and visual representation (imagery created by cultural producers). The resulting nexus produces a powerful metaphoric structuration that informs how security is actually understood and acted upon in the current world system (Brown 2009). While pop culture has long played a role in everyday conceptualizations of place and space, there is increasing academic analysis of the relationship between geopolitics, international relations, and popular-culture production and consumption, a reflection of the new speed of transmission, cultural hybridity, novel forms of consumption, the rise of global fandoms, and a DIY culture of entertainment production. In an effort to explore the popular-cultural turn in contemporary East–West geopolitics, this chapter engages one aspect of this phenomenon: the representation of the post-Soviet/Russian bogeyman.

The fall of the "Red Giant" and Russia's "resurrection"

In late 1991 the USSR dissolved, bringing an end to the political, social, and economic experiment that was the Soviet Union. With this act, Russian President Boris Yeltsin and his counterparts in the other republics wrought a geopolitical transformation the likes of which history has rarely—if ever—witnessed. The Red Giant, once a source of international fear, fascination, and inspiration, was not only humbled: it hacked itself to pieces, producing an outcome which Western Sovietologists had failed to predict even a few years prior. This messy breakup remade the world map in the process, a fact that Russia's long-serving President Vladimir Putin has bemoaned as a geopolitical tragedy. In some ways, this macro-scale geopolitical realignment solved certain problems of (popular) conceptual ambiguity, e.g., the rather inappropriate use of the term "Russia" for the whole of the Soviet Union—a tendency that never abated during the Cold War—became moot. Interestingly, speaking of Russia's image, Vladimir Lebedenko (2008), deputy director at the Ministry of Foreign Affairs, points out that French leader Charles de Gaulle never once referred to the Soviet Union, only "Russia." Certain tendencies to this effect remain, as many citizens in the Anglophone West continue to treat post-Soviet space as an undifferentiated mass, roughly equivalent to "Russia." This is especially true of Ukraine, which until the events of 2014–2015, was often subsumed within the conceptual hegemony of "Russia." Only through civil war and the annexation of Crimea did the idea of an independent Ukraine find purchase in everyday American geopolitical imagination(s). The same might be said for Kazakhstan, given that the country only found "fame" through the "shame" of Sacha Baron Cohen's Boratistan parody (see Saunders qtd. in Motyl 2015). However, in theory at least, the last decade of the second millennium was a further stage in the worldwide process of decolonization and the ultimate realization of national self-determination, something promised by both Woodrow Wilson and Vladimir Lenin a century ago.

The so-called "Balkanization" of northern Eurasia has, however, proved challenging for the "average American" to master, thus replicating the previously mentioned problem of ambiguity, i.e., for many casual observers of international politics, all of the former USSR was converted into a new, non-Soviet Russia. In addition to ambiguity, there was ambivalence towards Russia and the other post-Soviet republics. Long perceived as a seething maelstrom of anti-Americanism, anti-Westernism, and anti-capitalism, the Newly Independent States now represented potential security partners—perhaps even possible allies in the North American Treaty Organization (even Russia was considered as a candidate at one point in the early 1990s). At the very least, these "new" countries were ripe, often eager, converts to democracy and the free market (though the initial estimations of this enthusiasm proved to be overly optimistic, as post-Soviet presidents like Aleksandr Lukashenko of Belarus and the megalomaniacal Saparmurat Niyazov would prove). Despite the tectonic shift in geopolitical alignments across post-Soviet Eurasia, the Cold War died hard.

In their own ways, the military-industrial complex, academia, and the media and entertainment industries each reflected an palpable unwillingness to recognize what Francis Fukuyama (1992) deemed the "end of history," harboring acute concerns that Russia, while laid low, had not changed—and that it remained a threat to order and stability in the post-Cold War era. This was particularly true of Hollywood, which possesses an inelastic "ideological matrix" (Irr 2013), and one which despite conventional wisdom is almost always supportive of US foreign policy goals (see Alford 2010). As certain scholars have argued, the end of the Cold War and the digital networking of information has actually led to a situation where the boundaries between these sectors have diminished, creating what some have called the Military-Industrial-Media-Entertainment network (Der Derian 2009) or simply "The Complex" (Turse 2009). Clearly, the normative trajectories of the military-industrial complex and academia lie outside the scope of this study, as the representation of post-Soviet space and its citizenry by culture producers, particularly filmmakers, is my focus. However, recent events in Eastern Europe (namely Russia's military posture vis-à-vis Ukraine and by extension, the West), economic sanctions on Russian and Crimean businessmen, and the downing of Malaysia Airlines Flight 17 have been greeted with a certain level of self-destructive enthusiasm in these quarters, fulfilling decades of cinematic prophesizing that Russia would ultimately pick up its preordained mantle and become the West's "Geopolitical Enemy Number 1."

The villainous "post-Soviet" archetype in Western popular culture

My analysis examines more than a dozen popular films and one film-like television series, *24* (2001–2010), that use Russian (or generically post-Soviet, i.e., Russianesque) bogeymen as villains key to their plot structures. All these media products can be categorized as action-thrillers, a Western genre exported to other countries during the twentieth century. Some are based on potentially plausible scenarios where realism is sought and occasionally achieved, including *Crimson Tide* (1995), *The Peacemaker* (1997), *The Bourne Supremacy* (2004), *Eastern Promises* (2007), and *Salt* (2010). Others, including *Air Force One* (1997), *The Saint* (1997), *Rollerball* (2002), *The Sum of All Fears* (2002), *xXx* (2002), *A Good Day to Die Hard* (2014), and two James Bond films, *GoldenEye* (1995) and *The World Is Not Enough* (1999), require a significant suspension of disbelief on the part of the viewer—as works of science fiction, the last few even more so: *Star Trek VI: The Undiscovered Country* (1991), *Babylon A.D.* (2008), *Iron Man 2* (2010), *Chernobyl Diaries* (2012), and *Captain America: The Winter Soldier* (2014). While the focus of this chapter is on film as an "ideology machine" (Kellner 2013, 81), I also make references to other media formats, including videogames and comic books, in order to provide a more encompassing analysis of the post-Soviet villain in contemporary popular culture.

The ensuing analysis is, in part, a response to James Donald's timely and biting question "[W]hy are there so many grotesque foreign villains in popular fiction?" (1988, 36).[1] Through a critical reading of these representations of post-Soviet space and citizens, I have constructed a tentative topology of what I call the *post-Soviet bogeyman*. This schematic is divided into five separate, though often linked, archetypes: gangsters, mercenaries, revanchists, terrorists, and "mad" scientists (see Figure 5.1). Contra the idea that Hollywood has undergone a "crisis of representation" when it comes to depictions of post-Soviet Russians (Goering and Krause 2005–2006; Robinson 2008) and that the post-Soviet East is "no longer a thrilling foe" without Communism (Takors 2010, 230), these archetypes scaffold an interrelated "threat paradigm" that reinforces "common sense-based views" (see Anderson 2009) about the whole of the former USSR in the post-Cold War American geographical imagination. Moreover, these cinematically embedded binaries of the friend/enemy meta-structure that continue to pervade international relations thinking, with Russia still be seen as the quintessential Other in the West (see Semeneko, Lapkin, and Pantin 2007), provide proof that "power struggles and rivalries are embedded in identity debates . . . [Mental maps of the post-Cold War era] are now fluid, multidimensional, almost 'holographic' projections of this geopolitical discourse" (Debrix 2004, 162).

While the "Muslim/Arab/Middle Easterner" threat matrix certainly occupies a greater share of Anglophone pop-culture production in the post-1991 period, there is a palpable sense of loss associated with the end of the Cold War in geopolitical thrillers. Without a respectable (though ideologically antithetical) and geopolitically formidable (though ultimately vincible) foe like the USSR, the Western alliance

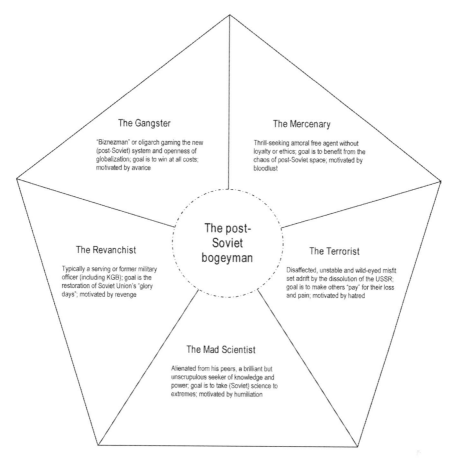

Figure 5.1 A topology of the "post-Soviet bogeyman" in contemporary film

(i.e., the US/UK/NATO/ANZUS monolith) rings hollow as a geopolitical standard-bearer of light and goodness. Consequently, there has been a tangible reticence to let go of the Cold War past. As a result, the frequency with which Russians are negatively portrayed in film, television series, and videogames has even allowed certain (non-Russian) actors to make lucrative careers leveraging their expertise at affecting "Russianness" (see Table 5.1). Such bogeyman-style images represent the norm seen by North Americans, Britons, Irish, and Antipodean Anglophones in celluloid representations of the post-Soviet "East," with a few notable exceptions, specifically films that play on the post-Soviet buffoon (see Chapter 6), the big-budget Russian exports *Night Watch* (2004) and *Day Watch* (2006), and what I deem the "existential exception" (discussed at the end of this chapter). By constantly engaging in a social construction of a reality in which the whole of post-Soviet space is a danger zone (see Chapter 7), Hollywood reinforces and in some

Table 5.1 Prominent post-Soviet "baddie" actors

Actor	Nationality	Film/TV series/ videogame	Character traits
Rade Šerbedžija	Croatian	*The Saint* (1997)	"Ivan Tretiak," corrupt Russian oligarch bent on restoring the USSR
		Snatch (2000)	"Boris the Blade," Russian-Uzbekistani gangster in London
		24: Season 6 (2008)	"Dmitri Gredenko," Russian general plotting to use a suitcase bomb against the US
		Red Widow: Season 1 (2013)	"Andrei Petrov," Russian mob boss in San Francisco
Karel Roden	Czech	*The Bourne Supremacy* (2004)	"Yuri Gretkov," murderous Russian oligarch
		Hellboy (2004)	"Grigori Rasputin," revivified tsarist-era wizard fomenting an apocalypse
		RocknRolla (2008)	"Uri Omovich," crooked Russian tycoon in England
		Grand Theft Auto IV (2008)	"Mikhail Faustin," cocaine-addicted mafia boss in the U.S.
		Orphan (2009)	"Dr. Värava," head of a sinister mental hospital in Estonia
Marton Csokas	New Zealander	*xXx* (2002)	"Yorgi," disaffected Chechen-war veteran and terrorist operating in the Czech Republic
		Covert Affairs: Season 5 (2014)	"Ivan Kravec," former FSB agent and terrorist
		The Equalizer (2014)	"Teddy," cleaner for the Russian mob
Jürgen Prochnow	German	*Air Force One* (1996)	"General Ivan Radek," genocidal dictator of Kazakhstan
		Dark Sector (2008)	"Yargo Mensik," former GRU officer turned criminal agent
		24: Season 8 (2010)	"Sergei Bazhaev," Russian mafia kingpin backing the assassination of a world leader

cases reifies everyday geopolitical attitudes about Russia and the post-Soviet republics which in turn have shaped elite geopolitical codes, homeland security initiatives, and foreign policy, including the Cooperative Threat Reduction (CTR) Program, the expansion of NATO, massive investment in cybersecurity, and intergovernmental cooperation to combat organized crime. Moreover, there is an increasing tendency among American politicians to play on popular culture in their own electioneering, from Ronald Reagan's use of *Star Wars* terminology to besmirch the USSR (Smith 1987) to John McCain's lauding of the patriotism of *24*'s Jack Bauer (Verheul 2010) to Barack Obama's hearty defense of satirists to impugn the image of Kim Jong-un (Saunders 2014).

The revanchist: Restoring the lost empire by any means necessary

I will begin, somewhat paradoxically, not in post-Soviet Eurasia but in the distant future, by focusing on the Klingon warmonger General Chang (Christopher Plummer), the main antagonist in *Star Trek VI: The Undiscovered Country* (1991). The character of Chang established a resonant paradigm for understanding the new post-Soviet villain just as the USSR was about to enter the "dustbin of history" (Fukuyama 1992). Released on 6 December 1991, the sixth installment in the *Star Trek* film series presented audiences with a fantastical yet "thoughtful critique" of the coming post-Cold War order (Alford 2010, 116), on the very weekend that Boris Yeltsin, Stanislau Shushkevich, and Leonid Kravchuk—the three heads of the core Soviet republics—convened at a state dacha in Belovezhkaya Pushcha in Belarus to dissolve the Union (an act which came to full fruition on [the West's] Christmas Day later that month). Hobbled by an unexpected and catastrophic explosion on the power-generating moon Praxis, the Klingon homeworld is placed in such environmental and economic jeopardy that the Empire submits to permanent normalization of relations with the Federation. Reflecting the real-world events of the Chernobyl disaster (1986) and the end of the Cold War (1989), *The Undiscovered Country* holds up a mirror to changes going on between Moscow and Washington, but also delivers a cautionary tale of latent dangers to come. In keeping with Star Trek's hallowed role of helping the West imagine its place in the international constellation of power and powers, this geopolitical "pivot" is important, as Jason Dittmer underscores when he writes: "While no ever told me the Klingons were the Soviets and the Federation was a loosely defined 'us,' nobody needed to" (Dittmer 2010, xi).

General Chang, who alludes to contemporary politics with his deeply meaningful assertion that "in space, all warriors are cold warriors," is part of cabal of Klingon (read Soviet) and Federation (read NATO) officers who seek to undermine the embryonic peace in an effort to keep the grand conflict alive. Chang, who had long trained the Klingons' best and brightest for the coming conflagration with the Federation, cannot bring himself to accept the new world order. In a nod to the late Cold War strategic fiction *The Hunt for Red October* (novel, 1984; film, 1990), Chang's warship can remain cloaked from radar, firing on its targets while remaining invisible: the ultimate "first-strike" weapon. However, unlike

Captain Marko Ramius (Sean Connery), Chang is intent on using this game-changing technology to prompt a war, thus inverting the somewhat hopeful message of *Red October*. This shift towards the revanchist Russian as the unredeemed villain, just on the cusp of independence, is characteristic of a larger reordering wherein Hollywood's binary framing of "good" and "bad" Soviets during the last years of the Cold War is generally replaced with singular representation of "dangerous" Russians after 1991.

In a melodramatic soliloquy, which quotes his enemy's greatest poet, William Shakespeare, Chang asks, "And if you wrong us, shall we not revenge?" thus rhetorically manifesting the unreconstructed Cold Warrior whose thirst for vengeance is unquenchable. Ultimately, good prevails over evil. Through ingenuity and gumption (rather than technological superiority), the crew of the Starship *USS Enterprise*, under the tenacious leadership of Captain James T. Kirk (William Shatner), outwits their foe and restores the peace, thus heralding a new era of cross-border cooperation with the Klingons. This outcome effectively transforms the Klingons (who, from Star Trek's inception in 1966 until *Star Trek 6*, served as a popular geopolitical stand-in for the Soviets) from a foe into a potential ally (i.e., post-Soviet Russians), though one that must be monitored carefully.

Echoing the theme of the recalcitrant Cold Warrior, *Crimson Tide* (1995) pits conservative (white) US submariner Captain Frank Ramsay (Gene Hackman) against his progressive (black) second-in-command, Lieutenant Commander Ron Hunter (Denzel Washington). The catalyst for confrontation occurs when Vladimir Rochenko (off-screen), a "rabid nationalist at the head of an army of defecting Russian soldiers," starts a rebellion against Moscow in protest against Russia's "acquiescence to NATO intervention in Chechnya." The viewer learns that Rochenko has taken control of a missile base near the North Korean border, threatening that if he or any of his men are harmed by the Russian army, he will launch a nuclear strike against Japan and the United States. Following a certain amount of confusion regarding orders, a contest of wills begins aboard the boat, in which the cool-headed Hunter ultimately prevails, averting World War III.

Film historian Ronnie D. Lipschutz (2001) begins his analysis of Cold War strategic cinema by unpacking this film, remarking that the ethos of the genre signals a nagging desire to return to the "good old days" where enemies were easy to identify. However, much to the chagrin of Cold Warriors like Ramsay, the "end of the Cold War has brought with it the end of the old game" (Lipschutz 2001, 4). His section on the film is appropriately titled by Ramsay's trenchant quote from the film: "Gentlemen, we are back in business!" (Lipschutz 2001, 1). While most of the motion picture delves into the US's generational, social, and racial issues that are overly familiar to those Americans living in the "Obama Era" (see Taylor 2010), the (phantasmal) post-Soviet revanchist Rochenko provides an evocative backdrop for exposing important fault lines in American society. However, we must also consider the role of this film in sustaining the notion that Russia (or at least its nukes) continued to present a "clear and present danger" to the US in the 1990s. Popular adherence to such fears sustained the policies of a Republican-dominated Congress (buttressed by hawkish Democrats) throughout the first decade of post-Soviet independence, ultimately scuttling President Clinton's efforts to take full advantage of the so-called "Peace Dividend" (MacLean 2006). Rather than substantively reducing strategic military assets as was planned, the US instead held onto much of its hardware, "just in case" the Russians backtracked.

A number of other films during the 1990s played on lingering fears that disgruntled elements within post-Soviet society would seek to turn back the clock on history, triggering a resumption of the Cold War. However, since 9/11, the Russian revanchist has taken on what might be labeled a postmodern patina. In 2006, Rade Šerbedžija joined the cast of the highly successful "faux geopolitical" (Der Derian 2010) thriller *24*. In Season 6, Šerbedžija plays General Dmitri Gredenko, an evil mastermind who employs Arab terrorists to wreak havoc on the US and avenge a ruined Russia in a grand scheme to provoke Washington into a "civilizational war" (Huntington 1993) with the Muslim world that will leave only (Eurasian) Russia standing. This narrative conveniently ignores the fact that Russia is deeply entrenched in its own religio-civilizational conflict along its Caucasian borderlands (occasionally spilling onto the streets of Moscow). However, as François Debrix points out, real-world politics have little bearing on such geopolitical drama: "Instead, tabloid geopolitics intends to spread a sense of panic by providing spectacular scenarios and doomsday prophecies about the realities that are generally not tangible or even directly meaningful to the reader/ viewer's experience" (2004, 158). In a pivotal scene that makes American fear a hybrid Soviet–Islamic blowback of existential proportions, Gredenko states: "Our country lost the Cold War because we were afraid to use [nuclear] weapons against the Americans. Today we will correct that mistake, and the Arabs will take all the blame." Both playing both on contemporary fears of "dirty bombs" (stemming from the arrest of would-be American terrorist José Padilla in 2002) and dredging up historical threats to the homeland (namely, the erroneously-named suitcase nukes deployed in Cuba during the 1962 missile crisis), *24* deftly linked the Cold War and the War on Terror in a bubbling witches' brew of fear, loathing, and xenophobia, one which subtly informs popular geopolitical attitudes towards Russian–Arab relations vis-à-vis US interests in the Middle East and deepens the sense of urgency associated with securitization.

The wild-eyed terrorist: Filling the ideological hole with nihilism

As the United States emerged from the decades-long struggle against Soviet totalitarianism, a fresh enemy was a requisite for the new world order (or at least Hollywood's scripting of it). Certainly, Saddam Hussein and Slobodan Milošević made handy foils for Washington, but they were individuals who posed no immediate or existential threat to the American homeland (nor did they make compelling filmic villains, despite various attempts at *reductio ad Hitlerum*). International terrorists, on the other hand, represented an increasingly potent threat, which—when catalyzed and multiplied by the flows of globalization unleashed at the end of the Cold War—could function as a stand-in for the "Red Menace." Consequently, terror-themed strategic cinema steadily grew in popularity in the post-Cold War era, and in several important instances—*The Peacemaker* (1997), *The Sum of All Fears* (2002), *The Dark Knight Rises* (2012)[2]—the Eurasian realm figured prominently in its deadly calculus through the prism of "loose nukes" and nuclear knowhow (i.e., rogue scientists who are ineluctably rendered as world-destroyers through the "natural" outcome of the former Soviet Union's economic decline after 1991) emanating from the dark corners of the derelict, ideologically bereft behemoth that is the former USSR (see Kotkin 2002), a politico-geographical trope that will be explored in greater depth in Chapter 7. Hollywood trained on this frame with surprising alacrity, churning out more than a dozen pieces of (post-)Sovietophobic strategic cinema in the first decade after independence.[3]

The action-thriller *Air Force One* (1997) revolves around a terrorist attack on the US president's airplane following a joint US–Russian action against the leader of Kazakhstan.[4] In the film, the German actor Jürgen Prochnow plays General Ivan Radek, a genocidal dictator responsible for the deaths of hundreds of thousands in the Central Asian republic. This is a curious plot device given the comparatively warm relations between ethnic Kazakhs and Russians in the republic since the dissolution of the USSR; however, it does "rhyme" with contemporary prognostications of post-Sovietologists such as Zbigniew Brzezinski (1997), who predicted that Central Asia would inevitably become a bloody cauldron of ethnic cleansing. As Helena Vanhala (2011) points out, President James Marshall (Harrison Ford) provides a simulacrum of Clinton in his hard stance against international terrorism, while at the same time reminding the viewer of the absence of action on the part of the administration towards the Rwandan genocide in 1994. Given Radek's possession of Kazakhstan's nuclear arsenal (all nuclear-tipped missiles were actually decommissioned and shipped to Russia by 1995), he represents a threat to both Russia and the world through a "new Cold War." Following Radek's apprehension, a group of Russian terrorists loyal to the general take over Air Force One; led by Ivan Korshunov (played by English actor Gary Oldman), the terrorists hold the president and his family captive in an effort to gain the release of their leader (see Figure 5.2). Korshunov defends his actions in geo-economic terms, castigating the West for turning his country into a cesspool of "prostitutes and gangsters," neatly encapsulating the overall mediatic representation (Weber 2009) of post-1991 Russians, while also reinforcing gendered stereotypes about the nation (see Williams 2012).

While Radek sits comfortably within the category of "revanchists" discussed above, Korshunov is something altogether different: a maniacal terrorist, though one with a motivation rather familiar (though nonetheless repulsive) to those who lived through the Cold War. Unlike the revanchist, he kills for no reason, does not have control of his emotions, delights in causing others pain, and shows no respect for his enemies: his zealotry places him beyond the pale of politics. However, unlike others of his ilk, Korshunov readily and repeatedly declares his love for "Mother Russia," which—in the context of the mid-1990s—means the Soviet Union. In a piquant manifesto delivered to the US vice president, he declares:

Figure 5.2 The post-Soviet terrorist as depicted by Gary Oldman (Columbia Pictures/
Photofest)

What arrogance to think you could ever understand my intentions? . . . What do I want? When Mother Russia becomes one great nation again, when the capitalists are dragged from the Kremlin and shot in the street, when our enemies run and hide in fear at the mention of our name and America begs our forgiveness on that great day of deliverance, you will know what I want.

However, not all post-Soviet terrorists will show such allegiance to the country of their birth, instead railing against systems and structures in equal measure regardless of whether they are Anglo-American or (post-)Soviet. A case in point is *The Peacemaker* (1997), starring A-list actors George Clooney and Nicole Kidman, which played on fears of ideologically-adrift terrorists in the 1990s, while deftly combining this concern with elements of revanchism and rampant criminality. The often impossible-to-follow plot takes the protagonists, both US government agents, on a tour of the disintegrating socialist federations of the USSR and Yugoslavia, pitting them against the Russian mafia, rogue Red Army generals, Chechen insurgents, and a disgruntled Yugoslav bent on bombing the United Nations.

Scottish actor Robert Carlyle's character Renard in the James Bond film *The World Is Not Enough* (1999) is an even purer form of the post-Soviet madman. In typical Bond fashion, Renard, a.k.a. Victor Zokas—once a KGB assassin, now a freelance killer-for-hire—is scarred both inside and out. He boasts:

"You can't kill me. I'm already dead," referring to the bullet lodged in his head that will ultimately bring about his demise, though perhaps also subtly alluding to the Soviet Union itself, a sort of geopolitical revenant. Renard is an ersatz zombie, as he is both immune to pain and incapable of pleasure. Despite (or perhaps because of) this, he ruthlessly continues his now seemingly meaningless "post-life" as a Soviet hitman (though the character was initially conceived as deterritorialized Bosnian) (see Dodds 2005). The often preposterous plot of the film revolves around Renard's nihilistic desire to attack the Bosporus as a macabre gift to his former lover Elektra Vavra King, an Anglo-Azeri oil tycoon played by Sophie Marceau, with a nuclear weapon compiled from plutonium stolen from Kazakhstan, nicely playing on the familiar "loose nukes" trope of post-Soviet Central Asia.[5]

Equally psychotic and hell-bent is Yorgi, played by New Zealand actor Marton Csokas, in *xXx* (2002). A former military officer in the Russian Army, Yorgi now resides in Prague, Czech Republic, where he leads Anarchy Ninety-Nine, a group of debauched terrorists who divide their time between discotheques, ski slopes, and scientific labs where they are developing a new chemical weapon. Disaffected by the "senseless" loss of so many comrades in the Chechen War, the group mutinied and formed a criminal enterprise/militia. The ultimate aim of Anarchy Ninety-Nine is to bring about the collapse of the world's governments by striking at major cities with a chemical agent (codenamed "Silent Night"), delivered via an ultra-fast submarine drone. In one of the greatest geographical plot blunders of all time, the script writers posited that the submersible—laughably christened *Ahab*—could get from the Czech capital to the open ocean in a matter of minutes, conveniently ignoring the series of locks between the Vltava River and the North Sea. This preposterous conceit, however, is surprisingly trenchant; in their collapse of time and space, the scriptwriters are able make it seem that the (deadly) post-Soviet frontier is but minutes away from the "civilized world." Once again, the notion of uncontrolled WMD appears, with the basic ingredients of Silent Night coming from materiel that has gone missing from Russia and other parts of the former USSR in the wake of the Soviet collapse, thus underlining both the merits and failures of the Cooperative Threat Reduction Program (see Davis 2006). Interestingly, Yorgi and his fellow anarchists represent an elision between the terrorist and the mercenary, in that they are really a criminal gang of thrill-seekers, taking advantage of the chaos of the economic transition and the libertine culture of post-totalitarian Eastern Europe, darkly mirroring the persona of the "American hero" Xander Cage (Vin Diesel), a scofflaw and extreme sport athlete turned reluctant spy for the National Security Agency (NSA).

"Real-world" politics have also shaped the depiction of the post-Soviet terrorist in contemporary media, as Season 5 of the hit television thriller *24* (2005–2006) saw the introduction of Chechen-like terrorists operating on American soil. However, rather than contextualizing the so-called "Dawn Brigade" as Islamists from the Russian-ruled Caucasus, the series played it a bit safer, depicting the group led by veteran British actor Julian Sands simply as "anti-Russian separatists." These terrorists act in the name of a free Kaukistan, a breakaway republic "somewhere" in Central Asia, positing an improbable geopolitical pretzel that the viewer need not disentangle to "understand." The (ethnically European) Kaukistanis are simply stand-ins for Chechens, seeking revenge on Russia (and the world) for "violence inflicted on our sons and daughters," demanding "national sovereignty" and promising to meet "terror with terror" (Season 5, Episode 3). Inexplicably, the terrorists hope to turn Moscow into a no-man's-land via a chemical weapons attack that will somehow implicate the United States (a curious turnabout, given that the CTR Program was established, in part, to rid Central Asia of its WMD but in this case the Americans are the suppliers of said weapons). Ultimately, the terrorist network, led by Vladimir Bierko (Sands), gets distracted from its goal of attacking Moscow and killing the Russian president, opting instead to inexplicably pursue American targets, before being foiled by counter-terrorism agent Jack Bauer (Kiefer Sutherland). Thus, *24* presents a securitization dilemma that would have never presented had the USSR stayed together, an undeniable manifestation of the desire to return to the halcyon days of the Cold War when things were "simpler." The narrative of Season 5, while convoluted and often confounding, does make a fairly cogent case for non-involvement in Russian domestic politics, suggesting that entanglements will ultimately lead to attacks on the US homeland—although Russia's "poor" governance of its still-vast empire (Longworth 2006), its "restless frontier" (Trenin and Malashenko 2010), and the volatile "Eurasian Balkans" (Brzezinski 1997) are also framed as a cornucopia of reasons for continuing global instability and an iron-clad geopolitical justification for continuing to fear Russia decades after the cessation of the Cold War.

On that point, the last artefact to include would be the big-budget action film *Salt* (2010), in which Angelina Jolie plays a CIA operative (wrongly) suspected of being a Russian double-agent. While her character (an Anglophone deep-cover implant in the US who has thwarted her programming as a KGB agent) counteracts the standard structure for employing post-Soviet baddies on one level, the film still found a way to "resurrect the Russians as movie villains" (Ebert 2010). Controlled by Orlov (Daniel Olbrychski), a revanchist KGB agent who seems to have ignored the end of the Cold War, the film presents a frightening scenario wherein the US and its NATO allies' security apparatuses are filled with Soviet-trained sleeper agents who are ready to engage in suicide terrorism, nuclear blackmail, and mass killings at a moment's notice, although it is never explained why they remain loyal to a system and ideology that has long since disappeared. The viewer's only clue is that they are "orphans" who cling to the charismatic Orlov father-figure decades after he sent them out into the world on their various missions.

The ruthless mercenary: Killing for cash and the rush in the new Eurasia

While usually functioning as the henchmen for fuming revanchists or avaricious oligarchs, the blood-thirsty mercenary has emerged as a standard Russia "type" in contemporary action cinema. Whereas the mad terrorist was orphaned by the failure of the old system, the modern Russian mercenary is more like a latchkey-kid, gleefully at home in the anomie and lawlessness of the post-Soviet milieu. Such characters tend to revel in the (projected) amorality of the post-Soviet realm, carefully navigating the vagaries of a world where money and power can open any door and everything is for sale.[6] The corrupt Russian Federal Security Service (FSB) agent Kirill (Karl Urban) in *The Bourne Supremacy* (2004) is a perfect example. While ostensibly an agent of the Russian state, he is—to all intents and purposes—a hireling of shady oil mogul Yuri Gretkov (Karel Roden).[7] Kirill, like the members of Anarchist Ninety-Nine, seemingly spends much of his leisure time in thumping dance clubs, yet is always on call for the necessary "wet work" that goes with running a successful enterprise in Putin's (post-)Russia. He fabricates fingerprints at Berlin crime scenes, travels to Goa to kill spies, and robotically plows through pedestrians in the streets of Moscow in search of his prey. Kirill never shows remorse, nor does he espouse any percep-tible brand of politics. Unlike his master, he does not even seem to be motivated by power. He is the darkest reflection of the "deceptive" and "dissatisfied" *Homo post-Sovieticus* (Levada 2001), totally lacking in ideology or honor, a true agent of the now.[8]

The most recent installation of the long-running *Die Hard* series, *A Good Day to Die Hard* (2013), also employs a quasi-state operator as its principal vil-lain. Played by Serbian actor Radivoje Bukvić, Alik acts as an enforcer for the prominent Russian politician and president-in-waiting Viktor Chagarin (Sergei Kolesnikov). With its plot set in contemporary Moscow (as well as Chernobyl, Ukraine), the American hero John McClane (Bruce Willis) and his estranged son Jack (Jai Courtney) are targeted for death following the latter's botched CIA operation to retrieve an incriminating file on Chagarin. With its tagline, "Yippee Ki-Yay Mother Russia," it is not surprising that the American cowboy motif is quite prominent. In classic film villain camp, Alik delays executing the his foes as he waxes eloquent about international geopolitics, soliloquizing: "Do you know what I hate about Americans? Everything—especially the cowboys . . . You guys are so arrogant. It's not 1986, you know. Reagan is dead." Predictably, Alik's grandstanding results in the two escaping, thus precipitating a second (failed) attempt on their lives in which Alik uses a helicopter gunship to lay waste to a Moscow skyscraper (discussed in greater depth in Chapter 6). The film's rather tongue-in-cheek reference to the politics of Reagan's "Second Cold War" (Halliday 1986) bring into focus a number of issues related to globalization and the United States as a geopolitical somnambulist, continuing to act as if the old rules still apply. This is reinforced by the fact of Alik's ease at navigating multiple realms, both in terms of profession (he appears to be at once a respected entrepre-neur and a cold-blooded killer) and language (he only speaks English in the film, though he receives his orders in Russian).

Thus far I have spoken mostly of men, but there is one character that represents the post-Soviet bogeyman in feminine form: Xenia Onatopp. Portrayed with high camp by the Dutch actress, Famke Janssen, the Russo-Georgian *femme fatale* Onatopp is a pivotal character in 1995's *GoldenEye* (see Figure 5.3). In the film,

James Bond (Pierce Brosnan) attempts to prevent the hijacking of a high-energy, electromagnetic pulse (EMP) space weapon by a rogue agent of the British government; using the (old) Communist realm as its backdrop, the film begins in Cold-War Archangelsk and ends in Castro's Cuba. While there is not ample space to dwell on the symbolism of the character's nomenclature, it should be noted that her given name implies foreignness (from the Greek *kseno* or "of a stranger") and sexual emasculation ("on the top"). Like a praying mantis, we witness Xenia—a former Soviet fighter pilot from the Republic of Georgia—lure a Canadian admiral into a sexual liaison which culminates with the seductress reaching climax as she crushes her lover/victim. Onatopp, though clearly marked as a post-Soviet baddie, is actually a member of Janus, an international crime syndicate. As such, she personifies the international mercenary trope while reminding the viewer that the Eurasian realm remains a prime area for recruitment of such agents of destruction.

The second installment of Marvel Comics' Captain America film franchise, *The Winter Soldier* (2014), plays with the Russian/post-Soviet mercenary trope as well. The blockbuster film adapted elements of a decades-old narrative in which Captain America's WWII-era sidekick Bucky is believed killed in action, but in actuality was captured by Soviet special forces and trained as the USSR's "most lethal assassin" (Costello 2009, 232), with no memory of his earlier life. The motion picture, which was extremely popular in Russia, pits a Russian-speaking cyborg, Bucky Barnes (the eponymous "Winter Soldier," played by Romanian-American actor Sebastian Stan), against his old ally, Captain America/Steve Rogers (Chris Evans). His name reflects the frigid climate of the USSR

Figure 5.3 A deadly embrace from which only James Bond can escape (MGM/Photofest)

(now Russia), while he sports insignia reminiscent of the Soviet Union (a red star on his left arm, a bionic replacement). For much of the film, his face remains covered by a black mask, coding his dark intentions. Such geopolitical iconography is part and parcel of the character, both during and after the Cold War; in fact, the title logo for the *Winter Soldier* spin-off comic book series (2012–) features a stylized hammer and sickle.[9] Michael Busby (2013) argues that Marvel's use of "cognitive maps" and Soviet symbols in the series demonstrates an acute obsession with the lingering effects of the Cold War on the American psyche, particularly in relation to the still-resonant fear of Soviet/Russian sleeper agents in the United States.[10] Given the tendentious political situation between the US and the Russian Federation following the annexation of Crimea a month earlier, Disney-owned Marvel opted to drop "Captain America" from the film's title, marketing it instead as *The First Avenger: Another War* in an effort to "minimize antipathy towards the superhero's patriotic leanings" (Child 2014). The film premiered at number one at the Russian box office, thus complexifying the relationship between popular geopolitical spectants and the spected.

Between oligarch and mafioso: The new "biznezman" and monstrous capitalism unleashed

Perhaps most familiar to movie-goers is the effigy of the Russian "new biznezman": one part oligarch, one part mafia boss, one part sociopath. Some cases, such as the aforementioned Russian oil oligarch Yuri Gretkov in *The Bourne Supremacy* and Ivan Tretiak in *The Saint*, are captains of industry, ostensibly legitimate actors in the "Wild West" capitalism of post-Soviet Russia—though, as we soon learn, they are deeply embroiled in various intrigues, willing to kill at a moment's notice to ensure the continued success of their economic fiefdoms. Despite Putin's taming (or co-opting) of the *mafiya* since taking office (Steen 2004), Hollywood continues to project depictions of those 1990s-era Russian *mafiosi* Stephen Handelman has labeled "Comrade Criminals." This self-made brand of neoliberal monster has, in Handelman's words, been pivotal in the downfall of Russia since 1991.

> Newspaper headlines chart Russia's descent from superpower pride to third-world embarrassment. Ministry offices sell contracts. Stolen-weapons finds are discovered in army barracks, and uranium thieves are in the military-industrial complex. Millions of dollars' worth of timber, oil, gold, and drugs are smuggled out of Russia and the Commonwealth of Independent States every month. Since Russia is now a comparatively free economy, government does not have a monopoly on crime. In a murderous parody of free-market competition, mobsters fight open battles over territory in the streets of Russian cities, leaving their victims riddled with bullets in Chicago-style gangland assassinations. Criminal cartels, believed by polices to control as much of 40 percent of Russia's wealth, infiltrate stock exchanges and the real estate market. Gangsters not only open bank accounts; they open banks. (Handelman 1995, 3)

Not surprisingly, the victims of this predictable pathology are usually Brits or Americans (at least on the silver screen), though the "average Russian" is almost always collateral damage as these villains go after bigger fish in the West. While they are somewhat beyond the scope of this analysis, such representations of "chaos capitalism" reflect a psychological purging of Western guilt associated with the forced liberalization of the country's markets in the Yeltsin years. By force-feeding Eurasia Anglo-Saxon-style free marketism, the West designed a dangerous architecture that would eventually produce some level of "blowback," arguably presented in the form of Vladimir Putin, who some Western (and Russian) critics paint as the ultimate gangster-in-charge (see Siddiqui 2015).

Such is the case with *The Saint* (1997), wherein fears of a resurgent Russia are tapped as the international (though Anglophone)[11] spy Simon Templar goes head-to-head with Ivan Tretiak, played by frequent Eastern European baddie Rade Šerbedžija, an oil tycoon in the "new Russia."[12] Frighteningly prescient of Russia's coming post-2000 patriot-traditionalism, Tretiak even affects something of an intellectual air, having penned the "best-selling book" *The Third Rome*, thus evoking shades of Eurasianist thinkers like Aleksandr Dugin and Aleksandr Prokhonov. Combining the corrupt oligarch and revanchist tropes, Tretiak's ploy is to assume power in a coup d'état, which will supposedly restore the country's "lost glory," end Russia's destructive experiment with democracy, and allow the country to "rearm" (sic) and begin to "dictate terms to the West." Consequently, *The Saint* darkly foreshadowing the real life events which characterized the transition from Yeltsin to Vladimir Putin in the new millennium. Tretiak is secretly hording oil during Russia's coldest winter on record, thus setting up the elected president for a downfall. He constantly stokes anti-American sentiment, while publicly "fantasizing" about restoring the Russian Empire to "former might and size" (never once making mention of the USSR). The CNN-type reporter on the street succinctly frames the situation: "As the bitter chill descends on Moscow, Russians are warming to the angry rhetoric of former Communist boss Ivan Tretiak." Like many Western action-thrillers set in post-independence Russia, *The Saint* makes use of the impoverished Soviet-era nuclear physicist who will serve any master as long as he gets a paycheck. Tretiak's enforcer and son Ilya (Valery Nikolaev) also evokes a number of familiar bogeyman tropes, inexplicably carrying a walking stick, snorting cocaine, and cavorting with prostitutes. In one scene, he and his cronies are entertained by a literal rat race, betting tens of thousands of US dollars on the rodents as a balalaika band plays on.

Similarly fitting the mold is director John McTiernan's 2002 remake of the 1975 dystopian classic *Rollerball*. Set in the near future and situated in various locations across post-Soviet Eurasia, the film explores the emergence of a new blood sport called rollerball, which is headquartered in Kazakhstan. Russian entrepreneur Alexis Petrovich (played by French actor Jean Reno) has established a commercial empire with his sport-cum-media empire. We see the world over which Petrovich rules as the opening credits roll: a perplexing mix of crass U.S-style professional-sport commercialism juxtaposed with a bazaar-like atmosphere filled with turbaned, kalpaked, kaftaned, and hijabed Asians. The film is a paragon of "biased" or "empty" geographical knowledge (see Harvey 2005): Kazakhstan is an impoverished "shit hole" where the only professions are mining (male) and prostitution (female); Azerbaijan is depicted as a fundamentalist Muslim country where the Arabic alphabet is used and female newscasters are in full Islamic covering; Mongolia is a land where Chinese pictograms are dominant and Japanese street fashion prevails. While these sporting events take place in the post-Second World, the audiences are global. Rollerball's international audience includes Russophones, Mandarin speakers, Arabs, and the Anglophone world. The action centers around two rival teams: "Zolotop Tolpa" (likely a garbled rendering of "Yellow Horde," given the yellow logo which depicts a skull in a Mongol helm) and Petrovich's own team "Kazakhstan Vsadnik" ("Kazakhstan Horseman," though they are referred to as the "red horsemen"). As he gathers investors from around the world, the magnate avers: "I will not allow this game to become corrupted like so many other things in this part of the world," eerily echoing Putin's throaty defense of FIFA during the 2015 scandals.

However, we soon learn that all aspects of the game are corrupt. In one scene, Petrovich publicly threatens the Azeri minister of communications after he learns rollerball has been relegated to channel 109. He pulls a gun in front of dozens of journalists and promises to "erase" the minister's entire family, screaming, "We are channels one through five!" However, the rot goes deeper as the American protagonist (and the world's most famous rollerballer) uncovers a conspiracy to raise ratings by orchestrating ever-more gruesome events in the ring, including the murder of a mentally challenged Mongolian. We learn that "In the Old Regime, Petrovich was above the official rank in Spetsnaz before the age of 30. KGB colonels used to cross the street when they saw him coming. Now they all line up to grovel in the dust." Later the *soi-disant* strongman Petrovich brags, "I don't need a political position because I own the men who do." In the film's final scene, we are reminded of Petrovich's Russian origins once again via the symbolism of an Orthodox patriarch as his guest of honor at the international championship match. Petrovich, like all Russian mobsters, get his comeuppance in the end, with the rollerballers triggering a popular revolution in Kazakhstan.

Remarkably similar in its representation of Russian hyper-gangsters is the 2008 British-French-American film *Babylon A.D.*, starring Vin Diesel as Hugo Toorop, a begrudgingly moral mercenary hired to transport a miraculous young woman across Eurasia to North America to fulfil her destiny. Purportedly set in Russia, Kazakhstan, and Mongolia (Uzbekistan is also referenced, but only as the site of the attack of viral bio-weapons that killed almost everyone in the country), the film presents an utterly dilapidated, partially irradiated, and chaos-ridden realm where violence rules the day (see Chapter 7). Toroop's paymaster is the loathsome mobster Gorsky (Gérard Depardieu), who dons face makeup, including a "hilariously large fake nose" (Snider 2010), to project a particularly grotesque visage of post-communist/über-capitalist villainy. There is a peculiar irony at work here, connecting the real world to its mediatic Other, given that Depardieu would be granted Russian citizenship in 2013 as part of his very public taxation-based spat with the incoming socialist government of France.[13] Ultimately, Gorsky gets his due, suffering a smart-bomb nuclear attack by the mysterious Noelite sect (the plot details are so convoluted that a proper explanation would consume an entire chapter).

Darkly complementing the ostensibly "respectable" (but thoroughly corrupt and ultimately murderous) mogul is the familiar trope of the Russian mobster. The Russian (or often Chechen) mafia plays a background element in the plots of countless films released since 1991, including *Terminal Velocity* (1994), *Little Odessa* (1994), *The Jackal* (1997), *Boondock Saints* (1999), *Training Day* (2001), *Be Cool* (2005), *We Own the Night* (2007), *The Dark Knight* (2008), *Safe* (2012), *The Drop* (2014), and *The Equalizer* (2014). In nearly all of these films, Russian/ Chechen gangsters are represented as brutal criminals, lacking the "organized" habitus typically associated with the classic Italian/Sicilian mobster of old. While Semyon (Armin Mueller-Stahl), a UK-based Russian "thief in law" (*vor v zakone*) in *Eastern Promises* (2007), represents one of the most iconic depictions of this stereotypical "Russian" type, David Cronenberg's deft direction and Viggo Mortensen's sympathetic portrayal of the enigmatic and ethically-complex fixer Nikolai Luzhin won the film critical praise for its bold choice to represent Russians as something more than mindless killing machines. However, despite Cronenberg's more nuanced representation, even *Eastern Promises* promotes a similar *Weltbild* wherein the Russian gangster is destroying the West from within, profiting from the "freedom" of US/UK society. While rarely explicit, such films subtly reinforce an anti-immigrant sentiment while at the same time characterizing the post-Soviet man as naturally inclined towards violence, misogyny, and rapaciousness (much like Italian mafia films of an earlier day).

The mad scientist: Abandoned geniuses and orphaned technologies

Perhaps the most interesting post-Soviet bogeyman comes in the form of super-villain Ivan Vanko (Whiplash), played by Mickey Rourke, in *Iron Man 2* (2010). Vanko represents a pitiable piece of Cold War detritus, hollowed out by the chaos of Russia's shift to a market economy. His father Anton—a scientific genius of the first order—had once worked with Howard Stark, father of the film's hero Tony Stark (Robert Downey, Jr.), a.k.a. Iron Man, on the "arc reactor," a source of perpetual clean power. However, Anton succumbed to greed and was deported to his native Soviet Union following unethical attempts to profit from the technology. When the father dies, the son's desire for revenge is sparked. Enraged by his own country's decline (evinced by the ramshackle Moscow environs where Vanko keeps his lab and visual cues to the presence of organized crime in Russia) and Stark's fabulous wealth as an international weapons manufacturer (and, by implication, a thief of Soviet technology), Vanko begins to work on his own post-Soviet "super suit" to challenge Iron Man. After failing to destroy Stark at the Monaco Grand Prix (see Figure 5.4), Vanko takes advantage of a corrupt rival weapons-monger to build his own army of drone warriors to destroy Stark, therein reprising the familiar Cold War cinematic propaganda of soulless techno-warriors versus American individualism and personal gumption (spoiler alert: Iron Man emerges victorious as a defeated Vanko activates a self-destruct function on his body armor). Vanko, attempting to validate his criminal actions against civilian populations, seeks to contextualize Iron Man as the true villain: "You come from a family of thieves and butchers, and like all guilty men, you try to rewrite your history, to forget all the lives the Stark family has destroyed." *Iron Man 2* effectively scripts Russia as a belligerent and self-defeating entity, destined to fail because of its inability to engage in *Vergangenheitsbewältigung* or "dealing with the past," while simultaneously making a distinction between "good" American weapons manufacturers like the Stark family versus "bad" ones, personified by the avaricious Justin Hammer (Sam Rockwell).

Though a minor character, the unbalanced Boris Grishenko (Alan Cumming) in *GoldenEye* (1995) also fits well within the gestalt of the Russian mad scientist, though in this case, the gloss of "insane" is just as applicable as is the one of "angered." According the 007 Wiki (http://jamesbond.wikia.com/) fan web site:

> [Grishenko] is a brilliantly talented computer programmer and hacker; he is also a backstabbing, arrogant misogynist, who sexually harasses his female coworkers, has a nervous habit of clicking spring loaded ballpoint pens, and twirling before resuming the clicking motion while concentrating on a task, and shouts 'I am invincible!" whenever he succeeds.

In the Anglo-American geopolitical imagination, Grishenko symbolizes the shifting of the default enemy frame from KGB assassin to a new techno-scientific villain. In the words of columnist Mark Galeotti (2011), "The stereotype of the

Figure 5.4 A post-Soviet cyborg ready to wreak havoc (Paramount Pictures/Photofest)

Russian hacker has become such a common media trope that it gets recycled again and again. It also offers a handy update for those looking for new ways to perpetuate the 'Russian threat.'" In 1998, while announcing the creation of a new high-tech federal crime-fighting agency, Janet Reno, US President Bill Clinton's Attorney General, stated: 'The attack can come from anywhere in the world. You can sit in a kitchen in St. Petersburg, Russia, and steal money from a bank in New York' (qtd. in Alderson et al. 1998), thus designating Russia as the source of a new digital menace in the wake of the disappearance of the older thermonuclear threat (a fact underscored by the hacker attacks on Estonian servers in 2007, which triggered the establishment of a NATO-supported cyberwar training center in the Baltic republic). Since the mid-1990s, popular culture has regularly depicted disgruntled and tech-savvy Russians turning their ire on the West, wreaking havoc on computer systems, databanks and sensitive network architectures of all types. Grishenko, however, remains the most emblematic of this cybernetic menace.

The most chilling manifestation of the "mad scientist" frame is the depiction of (post-)Soviet scientists in the final minutes of the American horror film *Chernobyl Diaries* (2012). The backdrop of the story takes place in Pripyat, Ukraine, where a group of six foreign tourists (four Americans, a Norwegian, and an Australian) enter the abandoned city where the workers of the Chernobyl nuclear power plant resided before the 1986 meltdown. After their tour van mysteriously breaks down, the "extreme tourists" and their local guide come under attack by flesh-eating mutants, who—like the walking dead—hunt normal humans (though unlike zombies, they are evidently capable of reproduction). The radiation-mutated creatures pick off the Western interlopers one-by-one until the last two make their escape, only to be met by military personnel who shoot and kill one of them. The last survivor, Amanda (Devin Kelley), then passes out. When she awakens, she is confronted by Russian-speaking doctors and scientists in hazmat suits who are capturing the mutants and returning them to medical detention. In order to keep their nefarious experiments on Pripyat's carnivorous population clandestine, the scientists place Amanda in a cell with the monsters who immediately consume her in the darkness. In its last minutes, *Chernobyl Diaries* slyly transitions from a geographical to a geopolitical tale of horror by situating the post-Soviet military-industrial-science complex at the root of this cannibalistic nightmare. By "covering up" their perfidy, the (Western) viewer is reminded of the veil of secrecy that has long characterized life behind the "Iron Curtain."

Similarly horrifying is the representation of a "modern-day Doctor Frankenstein" in the second X-Files movie, *I Want to Believe* (2008). The cult television series *X-Files*, upon which the movie is based, regularly reference Russia during its decade-long run on FOX (1993–2002), including alien cover-ups by the USSR, biohazardous experiments gone wrong, and other techno-scientific intrigues. With Dana Scully (Gillian Anderson) and Fox Mulder (David Duchovny) back together again (this time in a sexual relationship), the plot revolves around a Russian researcher who employs chimerical Soviet transplantation science to affix the head of his gay lover onto a female victim he has abducted. Such dark fantasies of (post-)Soviet science continue to resonate across the decades, often finding their mythical fountainhead in Ilya Ivanovich Ivanov's human–chimpanzee hybrid experiments in the early days of the USSR, spuriously linked to a (non-existent) "super-warrior" proposed by Joseph Stalin (see Johnson 2011). Ensconced in a world of pederasty, organ harvesting, and amoral science, the use of the Russian Other (contrasted against the [now] reliably heterosexual American duo) seems only par for the course; yet, at the same time, both *Chernobyl Diaries* and *X-Files* serve to remind the viewer of Soviet science and technology gone wrong, and how the collapse of the USSR only pushed the practitioners of the scientific arts further towards the proverbial "dark side."

The existential exception: A deviation from the norm?

Within the genre of action films, the Russian bogey is alive and well. While some scholars claim that the end of the Cold War brought on a crisis in Hollywood's representation of the Russian, this is far from the case. Western cultural producers have generally maintained hoary stereotypes of KGB killers, while simply adding new villainous gestalts to the mix. Whereas Arabs once had a monopoly on terrorists in film, they must now share the celluloid with unhinged Russian bombers. At one time, the Italian mobster ruled the cinematic roost, but now—at the very least—he has to vie with the *vor* whose time in a Siberian prison has given him a leg-up on his Sicilian counterparts. While disaffected American vets used to have a corner on the market when it came to the mediated mercenary business, now they compete with former Spetsnaz. Russians have even started to displace the Germans as the prototypical "mad" (read amoral) scientists on the big screen. However, there is one arena where Russians tend to provide a helping hand to well-meaning Americans (and Brits): the existential crisis. In films where the Earth (or humanity as a whole) is threatened, Russians step in to help save the day (though they tend to be very "Russian" about it).

During the Cold War, science fiction depictions of the Soviets followed a fairly predictable pattern. In cases of explicit threats to the humanity/Earth/galaxy, Soviets would function as allies (of a sort). The most obvious examples are drawn from the first two *Star Trek* series (1966–1969/1987–1994), each of which featured a key supporting character from Russia, namely Pavel Chekov (Walter Koenig) and Worf (Michael Dorn).[14] In these and similar treatments, the Cold War is a historical footnote and Earth's governments have merged (typically under Anglophone hegemony centered in the United States) to meet the myriad challenges presented by interstellar travel and exploration. However, in other instances, the Soviet menace lurked in alien forms, subtly worming its way into America in such films as *It Came from Outer Space* (1953), *Them!* (1954), and *Invasion of the Body Snatchers* (1956) (see Hendershot 2001). Allegorical representations of the dangers of the Communist system continued throughout the Cold War, though they reached their apex during the "Red Scare" of the 1950s. Regardless, Russians tended to receive unfavorable treatment in nearly every filmic representation during the period, in both American and British film (see Shaw 2006, 2007).

Little has changed since the end of the US–Soviet conflict for global hegemony. In *Armageddon* (1998), cosmonaut Lev Andropov (played by Swedish actor Peter Stormare) assists an American crew of NASA scientists and deep-sea miners in their efforts to destroy a Texas-sized comet hurtling towards Earth. Andropov is always drunk and disheveled (quite ironic given Soviet Premier Yuri Andropov's anti-alcoholism campaign in the early 1980s), and frequently violent. He personifies the lapse of Russia from a once-great power into a ghostly shadow of its former self (Goering and Krause 2005–2006). A host of other post-1991 films similarly showcase Russians as bumbling, self-serving, and/or recalcitrant allies in the defense of Earth/humanity (almost invariably to be saved by an American/Anglophone "hero") from *Independence Day* (1996) to *Terminator: Salvation* (2009) to *Predators* (2010) to *The Darkest Hour* (2011) to *Resident Evil: Retribution* (2012).

In a very recent example of the trend, the summer blockbuster *Pacific Rim* (2013) depicts a cataclysmic scenario where the militaries of Earth must unite to fight giant Godzilla-like monsters (Kaiju) that are destroying the world's coastal cities. Four teams man the last of the Jaegers, giant human-controlled war machines that can match the gargantuan beasts punch-for-punch. One of these is the husband-and-wife duo of Sasha and Aleksis Kaidonovsky. The pair resembles a steam-punk version of Ivan Drago and with his wife Ludmilla (Brigitte Nielsen) from *Rocky IV*, with each sporting bleached blond crew cuts. Their Mark-1 Jaeger, *Cherno Alpha*, is the "oldest Jaeger still active in combat," relying on nuclear power and running on analog signaling, slyly reinforcing the anachronistic nature of (post-)Soviet technology. Ultimately, the Russians perish (alongside a team of Chinese triplets manning their own Sinic-themed Jaeger), paving the way for a combined American-Australian team to bring salvation to the planet. This framing of Russians as reluctant friends in time of existential crisis, though, is nothing new. When faced with a world-ending scenario or even the threat of aliens, Russians and Americans have been able to get along in past big-screen representations. This exception only proves the rule: the Russian remains the bogeyman despite the end the Cold War.

With Putin continuing to orient his country in an ever more oppositional position towards the West just as anti-Russian discourse increases in conservative political circles in the US and UK, we are likely to see a flurry of increasingly darker representations of Soviet/Russian/Eurasian "baddies" on the big screen. Wrapping up this book in the summer months of 2015, I had the opportunity to take in two of the season's blockbusters: *Ant-Man*, based on a minor Marvel Comics superhero, and *Mission: Impossible Rogue Nation*. Not coincidently, both films employ prologues associated with Russia and its WMD capacities. In *Ant-Man*, military scientist Hank Pym (Michael Douglas) resigns from the (US-government stand-in) S.H.I.E.L.D. in 1989 because of fears about his technology being used for evil and the recent loss of his wife (we later learn that Pym's wife, the superhero Wasp, sacrificed herself in the line of duty responding to "separatists [who] had hijacked a Soviet missile silo in Kursk and launched an ICBM at the United States"). The Cold War bubbles up through-out the film, a curious narrative thread given that most of the film's viewing audience was born after the Wall came down. In the most recent installment of the *Mission: Impossible* franchise, we find Ethan Hunt (Tom Cruise) and his crack team of counter-espionage agents in a field outside of Minsk attempting to intercept a cargo plane loaded down with VX nerve agent. Prohibited under United Nations resolutions, the production and stockpiling of VX (exceeding 100 grams per year) was outlawed by the Chemical Weapons Convention of 1993; however, the viewer quickly learns that Belarus not only has scores of kilograms of the deadly toxin, it is willing to sell it the highest bidder (subtle clues suggest Syria). These are just two examples of what portends to be a tidal wave of geopolitical thrillers in which Russia (or the old USSR)[15] is imagina-tively pitted as America's (and the West's) geopolitical enemy number one. Reprising a quote from earlier in the chapter, it seems that we are now truly "back in business."

Notes

1 He goes on to state: "'Racism' is no doubt the short answer, but it does not in itself reveal much about the dynamics of identification and differentiation in fiction, about our subjective investment in cultural hierarchies and symbolic boundaries."
2 Regarding the last of these films, it is ironic that despite Hollywood's fearmongering about ideologically laden post-Soviet WMD threats, in actuality, this film produced—in a roundabout way—its own terroristic outcomes when James Eagan Holmes, an unstable former neuroscience student, walked into a crowded Aurora, Colorado theater on the night of the film's premiere and shot dead twelve audience members, nightmarishly mim-icking the exploits of the "postmodern terrorist" known as the Joker (see Canavan 2010).
3 The terrorist frame of the 1990s (and beyond) actually retools the "Crazy Ivan" trope that emerged in Western popular culture during the 1970s, whereby Russians were presented as unpredictable and prone to "irascible violence, even brutality" (Brandt 2003, 52).
4 Condescendingly, the narrative presents a post-Cold War phantasy of a (Yeltsinite) Russia too weak to act on its own in the near abroad, which must thus couch its military intervention in Central Asia alongside that of the global police force, the US.
5 Interestingly, the notion of destroying—rather than capturing—the Bosporus turns an age-old Russian geopolitical objective on its head, though effectively achieving similar strategic goals.
6 Curiously absent in Hollywood depictions of such Russian mercenaries is a filmic repre-sentation of actual Russian mercenaries in real-life warzones like Bosnia and Kosovo (and most recently southeastern Ukraine), though Russian soldiers-of-fortune do figure promi-nently in *Mercenaries: Playground of Destruction* (2005) and similar videogame titles.
7 Like Šerbedžija, Roden has made a career of playing Eastern European villains, includ-ing a thrill-killer in *15 Minutes* (2001), a resurrected Rasputin in *Hellboy* (2004), and the corrupt oligarch Uri Omovich in *RocknRolla* (2008), as well as voicing a Russian gangster in *Grand Theft Auto IV* (2008), in which he uttered the timeless maxim: "This American greed takes over everyone . . . It's like a disease! Only I am still sane!"
8 An interesting contraposition to Kirill is presented in the film *Lord of War* (2005), a thinly veiled parody of infamous weapons monger Viktor Bout, currently serving a 25-year sentence in a Maryland correctional facility. In the film, Yuri Orlov (played by Nicolas Cage) is a Ukrainian-American arms dealer who maximizes the chaos of the collapsing Soviet Union to emerge as one of the world's most successful purveyors of

illicit heavy artillery. While the audience can empathize with the complex character of Orlov, Ukraine is rendered as a space where any and all things are for sale; however, the only important commodities emanating from the country are relics of the Cold War, i.e., heavy weapons that now are to be pumped into the neoliberal world system (ultimately resulting in dead, raped, maimed, and civilizationally negated Africans, at least according to the filmic arc).

9 A curious textual manifestation of the (utterly absent) Islamo-Russian alliance referenced earlier in the section on *24* also manifests in the comic book series, with a renegade Russian scientist prepping the Winter Soldier to wipe out the "Great Satan," i.e., America, a term that subtly evinces the ayatollahs of Iran without making specific geopolitical reference to the US enemy (see Busby 2013).

10 Though set in the early 1980s, the FX television series *The Americans* (2013–) has gone a long way to (re-)stoking these fears through its portrayal of two deep-cover KGB agents living in the Washington, DC area, raising a family while surreptitiously engaging in espionage and political assassinations (see Hall 2015).

11 In the prologue, we witness the young antagonist being raised in an orphanage in the "Far East," where the language of instruction is English, as is the head priest who torments young Simon.

12 In addition to his performances discussed in this chapter, the Croatian actor's roles have included a philandering, renegade Soviet general in *X-Men: First Class* (2011) and a monomaniacal Albanian gangster in *Taken 2* (2012).

13 Russian president Vladimir Putin, purportedly a good friend, bestowed a variety of honoraria upon the French actor, including making him the cultural ambassador to Montenegro. He subsequently moved to Saransk, Mordovia, an ethnic republic in the Ural region known mostly for its prison camps, including one that housed a member of the punk band Pussy Riot, where he settled at No. 1, Democracy Street (*Demokraticheskaya Ulitsa*) and declared he wished to become the "tsar" of the city.

14 In the latter case, the series played with the ambiguity of Russian/Soviet foreignness (as well as race) in multiple ways, scripting the character as a (black) Klingon adopted into a (white) Terran Russian family, and played by an African American actor. Worf often functioned as a liminal presence—or, as Yonassan Gershom puts it, a "marginal man" (2009, 101)—linking the Federation with the Klingon Empire, thus promising the coming rapprochement between the West and the (soon-to-be) post-Soviet East.

15 Recognizing the geopolitical-thriller content of *M:I*, trailers shown prior to the film included Guy Ritchie's forthcoming adaptation of the 1960s television series *The Man from U.N.C.L.E.* (2015) and Tom Hanks' historical biopic *Bridge of Spies* (2015), detailing events related to the downing of a US spy plane in 1960. Undeniably, Hollywood's enthusiasm for resurrecting the nuclear-charged brinksmanship of the early 1960s taps a growing sense of unease in the West about its relationship with the Russian Federation.

References

Alderson, David, David Elliott, Gregory D. Grove, Timothy Holliday, Stephen J. Lukasik, and Seymour E. Goodman. 1998. *Center for International Security and Cooperation Workshop on Protecting and Assuring Critical National Infrastructure: Next Steps*. Stanford, CA: Stanford University.

Alford, Matthew. 2010. *Reel Power: Hollywood Cinema and American Supremacy*. New York: Pluto Press.

Altheide, David L. 2009. *Terror Post-9/11 and the Media*. New York: Peter Lang.

Anderson, Jacqueline. 2010. *Understanding the Changing Needs of the US Online Consumer*. Cambridge, MA: Forrester Research, Inc.

Anderson, Jon. 2009. *Understanding Cultural Geography: Places and Traces*. Abingdon, UK: Taylor & Francis.

Brandt, Peter. 2003. "German Perceptions of Russia and the Russians in Modern History." *Debatte* no. 11 (1):39–59.

Brown, Neville. 2009. *The Geography of Human Conflict: Approaches to Survival*. Eastbourne, UK: Sussex Academic Press.

Brzezinski, Zbigniew. 1997. *The Grand Chessboard: American Primacy and Its Geostrategic Imperatives*. New York: Basic Books.

Busby, Michael. 2013. "Winter Soldier: A Geographical Perspective " *Constant Geography*, available at: http://constant-geography.blogspot.com/2013/06/winter-soldier-geographical-perspective.html [last accessed 19 August 2014].

Canavan, Gerry. 2010. "Person of the Year: Obama, Joker, Capitalism, Schizophrenia." In *Politics and Popular Culture*, edited by Leah A. Murray, 2–13. Newcastle upon Tyne, UK: Cambridge Scholars Publishing.

Child, Ben. 2014. "*Captain America: The Winter Soldier* Deals Stunning Blow at China Box Office." *The Guardian*, available at www.theguardian.com/film/2014/apr/09/captain-america-winter-soldier-movie-china-box-office [last accessed 26 March 2016].

Costello, Matthew J. 2009. *Secret Identity Crisis: Comic Books and the Unmasking of Cold War America*. London: Bloomsbury.

Davis, Doug. 2006. "Future-War Storytelling: National Security and Popular Film." In *Rethinking Global Secuirty: Media, Popular Culture, and the War on Terror*, edited by Andrew Martin and Patrice Petro, 13–44. New Brunswick, NJ: Rutgers University Press.

Debrix, François. 2004. "Tabloid Realism and the Revival of American Security Culture." In *11 September and Its Aftermath: The Geopolitics of Terror*, edited by Stanley D. Brunn, 151–190. London and Portland, OR: Frank Cass.

Der Derian, James. 2009. *Virtuous War: Mapping the Military-Industrial-Media-Entertainment Network*. London: Routledge.

——. 2010. "Imagining Terror: Logos, Pathos, and Ethos." In *Observant States: Geopolitics and Visual Culture*, edited by Fraser MacDonald, Rachel Hughes, and Klaus Dodds, 23–40. London and New York: I. B. Tauris.

Dittmer, Jason. 2010. *Popular Culture, Geopolitics, and Identity*. Lanham, MD: Rowman & Littlefield.

DNC. 2012. "Senator Kerry Comes Alive, Endorses Obama Foreign Policy." *2012 Democratic National Convention*, 6 September.

Dodds, Klaus. 2005. "Screening Geopolitics: James Bond and the Early Cold War Films (1962–1967)." *Geopolitics* no. 10: 266–289.

Donald, James. 1988. "How English Is It? Popular Literature and National Culture." *New Formations* no. 6:31–47.

Ebert, Roger. 2010. "Salt." *RogerEbert.com*, available at www.rogerebert.com/reviews/salt-2010 [last accessed 13 July 2015].

Fukuyama, Francis. 1992. *The End of History and the Last Man*. New York: Free Press.

Galeotti, Mark. 2011. "Why are Russians Excellent Cybercriminals?" *The Moscow News*, available at http://themoscownews.com/siloviks_scoundrels/20111121/189221309.html [last accessed 23 October 2014].

Gershom, Yonassan. 2009. *Jewish Themes in Star Trek: Where No Rabbi Has Gone Before!* Raleigh, NC: Lulu Press.

Goering, Elizabeth, and Andrea Krause. 2005–2006. "Still Enemies at the Gate? The Changing Iconography of Russia and Russians in Hollywood Films." *International Journal of the Humanities* no. 3 (7):13–19.

Hall, Lucy. 2015. "Making Feminist Sense of 'The Americans'." *E-International Relations*, available at www.e-ir.info/2015/05/20/making-feminist-sense-of-the-americans/ [last accessed 13 July 2015].

Halliday, Fred. 1986. *The Making of the Second Cold War*. New York: Verso.

Handelman, Stephen. 1995. *Comrade Criminal: Russia's New Mafiya*. New Haven, CT: Yale University Press.

Harvey, David. 2005. "The Sociological and Geographical Imaginations." *International Journal of Politics, Culture, and Society* no. 18:211–255.

Hendershot, Cynthia. 2001. *I Was a Cold War Monster: Horror Films, Eroticism, and the Cold War Imagination*. Madison, WI: Popular Press.

Huntington, Samuel P. 1993. "The Clash of Civilizations." *Foreign Affairs* no. 72 (3):22–49.

IMDB. 2013. "Rocky IV." *Internet Movie Database*, available at www.imdb.com/title/tt0089927/ [last accessed 2 February 2013].

Irr, Caren. 2013. *Toward the Geopolitical Novel: U.S. Fiction in the Twenty-First Century*. New York: Columbia University Press.

Johnson, Eric Michael. 2011. "Scientific Ethics and Stalin's Ape-Man Superwarriors." *Scientific American*, 10 November, avaliable at http://blogs.scientificamerican.com/primate-diaries/2011/11/10/stalins-ape-man-superwarriors/ [last accessed 14 July 2013].

Kellner, Douglas. 2013. "Film, Politics, and Ideology: Toward a Multiperspectival Film Theory." In *Movies and Politics: The Dynamic Relationship*, edited by James E. Combs, 55–91. Abingdon, UK and New York: Routledge.

Kotkin, Stephen. 2002. "Trashcanistan." *New Republic* no. 226 (14):26–38.

Lebedenko, Vladimir. 2008. "Country Case Insight—Russia: On National Identity and the Building of Russia's Image." In *Nation Branding: Concepts, Issues, Practice*, edited by Keith Dinnie, 107–129. Oxford: Butterworth-Heinemann.

Levada, Iurii A. 2001. "Homo Post-Sovieticus." *Sociological Research* no. 40 (6):6–41.

Lipschutz, Ronnie D. 2001. *Cold War Fantasies: Film, Fiction, and Foreign Policy*. Lanham, MD: Rowman & Littlefield.

Longworth, Philip. 2006. *Russia: The Once and Future Empire From Pre-History to Putin*. New York: Macmillan.

MacLean, George Andrew. 2006. *Clinton's Foreign Policy in Russia: From Deterrence and Isolation to Democratization and Engagement*. Farnham, UK: Ashgate Publishing, Ltd.

Motyl, Alexander J. 2015. "The Winners and Losers of 'Nation-Branding'." *World Affairs*, available at www.worldaffairsjournal.org/blog/alexander-j-motyl/winners-and-losers-%E2%80%98nation-branding%E2%80%99 [last accessed 30 December 2015].

Palmer, William J. 1995. *The Films of the Eighties: A Social History*. Carbondale, IL: Southern Illinois University Press.

Robinson, Harlow. 2008. "Russians in American Movies: Imagining the Enemy." *Russian Life* no. 51 (6):52–57.

Saunders, Robert A. 2014. "'The Interview' and the Popular Culture-World Politics Continuum." *E-International Relations*, available at www.e-ir.info/2014/12/23/situating-the-interview-within-the-popular-culture-world-politics-continuum/ [last accessed 3 December 2015].

Semeneko, Irina, Vladimir Lapkin, and Vladimir Pantin. 2007. "Russia's Image in the West (Formulation of the Problem)." *Social Sciences* no. 38 (3):79–92.

Shahid, Aliyah. 2012. "Lean Forward: John Kerry Gleefully Skewers Romney on Foreign Policy." *MSNBC*, available at www.msnbc.com/the-ed-show/john-kerry-gleefully-skewers-romney-f [last accessed 29 December 2013].

Shaw, Tony. 2006. *British Cinema and the Cold War: The State, Propaganda and Consensus*. London and New York: I. B. Tauris.

——. 2007. *Hollywood's Cold War*. Amhearst: University of Massachusetts Press.

Siddiqui, Sabrina. 2015. "Vladimir Putin is 'a Gangster and Thug', Says US Presidential Candidate Marco Rubio " *The Guardian*, available at www.theguardian.com/us-news/2015/oct/03/vladimir-putin-is-a-gangster-and-thug-says-us-presidential-candidate-marco-rubio [last accessed 30 December 2015].

Smith, Jeff. 1987. "Reagan, Star Wars, and American Culture." *Bulletin of the Atomic Scientists* no. 43 (1):19–25.

Snider, Eric D. 2010. "Eric's Bad Movies: *Babylon A.D.* (2008)." *Film.com*, available at www.film.com/movies/erics-bad-movies-babylon-a-d-2008 [last accessed 18 August 2014].

Steen, Anton. 2004. *Political Elites and the New Russia: The Power Basis of Yeltsin's and Putin's Regimes*. London and New York: Routledge.

Takors, Jonas. 2010. "'The Russians Could No Longer Be the Heavies': *From Russia with Love* and the Cold War Bond Series." In *Facing the East in the West: Images of Eastern Europe in British Literature, Film and Culture*, edited by Barbara Korte, Eva Ulrike Pirker, and Sissy Helff, 219–232. Amsterdam: Rodopi.

Taylor, Keeanga-Yamahtta. 2010. "Race and the Obama Era." *New Politics* no. XII (1), available at http://newpol.org/content/race-and-obama-era [last accessed 29 November 2013].

Trenin, Dmitri V., and Aleksei V. Malashenko. 2010. *Russia's Restless Frontier: The Chechnya Factor in Post-Soviet Russia*. Washington, DC: Carnegie Endowment for International Peace.

Turse, Nick. 2009. *The Complex: How the Military Invades Our Everyday Lives*. New York: Henry Holt and Company.

Vanhala, Helena. 2011. *The Depiction of Terrorists in Blockbuster Hollywood Films, 1980–2001*. Jefferson, NC: McFarland & Co.

Verheul, Jacobus. 2010. "Paranoia and Preemptive Violence in *24*." In *Politics and Popular Culture*, edited by Leah A. Murray, 137–147. Newcastle-upon-Tyne, UK: Cambridge Scholars Publishing.

Weber, Cynthia. 2009. *International Relations Theory: A Critical Introduction*. Abingdon, UK: Taylor & Francis.

Williams, Kimberly A. 2012. *Imagining Russia: Making Feminist Sense of American Nationalism in U.S.–Russian Relations*. Albany, NY: SUNY Press.

Willis, Amy. 2012. "Mitt Romney: Russia is America's 'Number One Geopolitical Foe'." *The Telegraph*, 27 March.

6   Laughable nations

Parodying the post-Soviet republics

In early 2012, just days after the twentieth anniversary of the dissolution of the USSR, NBC aired an advertisement during Superbowl XLVI promoting its Thursday-night comedy lineup, including the highly successful programs *30 Rock* and *The Office*. The 30-second ad depicts a squat, grayish building filled with stereotypical Eastern Europeans: men in fur hats, babushka-clad women, and plenty of goats and silver-capped teeth to go around. We learn that in this "remote village," almost everyone gets to enjoy the network's sitcoms, the lone exception being the unfortunate soul who must turn the hand crank that powers the hamlet's only television set. When this downtrodden fellow stops his work for a moment to catch a glimpse of *30 Rock*'s corporate tycoon Jack Donaghy (Alec Baldwin) whining "I wanna be a baby again!" his fellow villagers curse him in an unidentifiable, but nonetheless recognizably "Slavic" tongue for interrupting their mirth as the power cuts out. The villager quickly returns to his unsung labors, allowing the rest of the townsfolk to get back to their televisual pursuits, thus escaping their implied misery though Western-abetted "humor therapy." Herein, we witness the role that the post-socialist masses generally have come to occupy in the minds of the American public: impoverished, laughable urchins, content to occupy themselves with trickle-down American media. The fact that the ad aired during the most-watched television event in American television history underscores how pervasive the image of the broken-down post-Soviet East has become, as well as the West's lack of empathy for peoples enduring the hardships of transition to a market-based economy.

Through a close reading of three parodic geographies of post-Soviet space, this chapter interrogates the lingering prejudices of the Anglophone West towards the nations of the former Second World, and the impact of such farce on "imagined geopolitics" (Ridanpää 2009). Additionally, this chapter considers derisive humor as a geopolitical act and seeks to tease out the ramifications of such comedic geographing on real-world politics by contextualizing them in their relevant historical, cultural, and social milieux (Purcell, Scott Brown, and Gokmen 2010). These case studies include the section on the former Soviet Union from *The Onion*'s book of fake world geography, *Our Dumb World: Atlas of the Planet Earth* (2007), JetLag's fictive travel guide *Molvanîa: A Land Untouched by Modern Dentistry* (2004), and Sacha Baron Cohen's farcical primer on Kazakhstan *Borat: Touristic Guidings to Glorious Nation of Kazakhstan* (2007), authored under the pseudonym Borat

Sagdiyev. The concluding section of this chapter touches on a number of other popular send-ups of post-Soviet space and the (negatively) idealized citizens that occupy or hail from such places. This analysis focuses on the dynamics of power and the exploitation of the "raw resources" of current history (Goldsworthy 1998) as they manifest in international relations, and how spoof satirical, and parodic geographies perpetuate long-standing prejudices and interrupt or negate contemporary state-based efforts to burnish the images of the former Soviet republics by portraying these countries and their denizens as "goofy, weak, and feminized" (Bardan and Imre 2012, 173). The primary aim of this chapter is to interrogate the predilection among humor-based cultural producers to treat the post-Soviet realm as an unknown but instantly recognizable hermeneutic—lacking distinction or difference—and how this trend shapes popular geographical imagination of the region as a whole and certain individual countries in particular.

In doing so, I contend that there is a comparably stable post-Second World stereotype that exhibits neither the "dangerous" attributes of an enemy state (e.g., Iran, North Korea, or even China) nor the "unknowable" qualities of the non-West (e.g., Morocco, Nigeria, or Malaysia), but instead presents a mutated simulacrum of the West as a manageable (and managed) Other, a sort of "comic relief" to Euro-American civilization. This rather recent trend demonstrates a temporal shift from archetypes presented in the previous chapter, though such representations of the post-Soviet bogeyman continue to appear with regular frequency (especially since Russia's intervention in Ukraine). Through an examination of these geopolitical chimeras, I attempt to demonstrate how post-Soviet peoples are consistently represented as naïve, backward rubes and the core of the former Second World is portrayed as a ridiculous neverland littered with the detritus of the failed socialist experiment (a gestalt which is layered upon older views of the region and its people formed through a raft of earlier "fake" travel narratives dating back to the Enlightenment Era). The effect of this imaginative manipulation—or, to use the language of Debord (1983 [1967]), "spectacle"—is not neutral; in fact, it is part of a larger disciplining pathology. In their treatment of the former Eastern Bloc, popular cultural producers function as a *dispositif* (apparatus) of Western governments, intergovernmental organizations, and transnational corporations seeking influence in the decomposition and recomposition of the region and its systems (see Muller 2008). In contrast to the neo-Wallersteinian notion that the former Soviet Bloc is breaking up, with some portions moving to the core, others to the periphery, and the rest remaining in the semi-periphery (see, for instance, Berglund 2001), this chapter argues that the entire region—at least from a popular geopolitical vantage—is being "fixed" in a single, semi-peripheral location vis-à-vis the developed West, a sort of quasi-Orientalized farrago.

(En)visioning the post-Second World: Distorted representations and perspectival prejudices

As detailed in the Introduction, the last two decades have seen the emergence of a plethora of new, relatively unknown nations between the Oder River and the Sea of Okhotsk. The dissolution of the federal states of the Soviet Union, Yugoslavia, and Czechoslovakia produced 24 new countries (26 if one counts the Russian-recognized republics of Abkhazia and South Ossetia, and more still if one considers the likes of the breakaway statelets Transnistria and Nagorno-Karabakh). This figure is especially profound when one considers that across the post-Second World, only a handful of states are not "new," namely Poland, Hungary, Romania, Bulgaria, and Mongolia.[1] Following in the wake of the previous waves of decolonization in the twentieth century, this third wave of newly independent states face a host of challenges which were undreamed of by their counterparts in earlier periods of imperial dissolution and national liberation. As previously discussed, these new countries must grapple with the difficulties presented by the triple challenge of global economic interdependence, cultural hybridization, and the ebbing of state sovereignty wrought by the increasing power of non-state actors, including multinational corporations, inter-governmental organizations, transnational financial institutions, private military contractors, supranational judicial systems, and global civic networks (see Sperling 2009).

Turning to the region specific to this book, i.e., the former Soviet republics, we find that the battle over national image is in full swing. While Russia possesses one of the world's most deeply ingrained national images, most other post-Soviet countries possess weak or easily confused referents as independent states. As such, they have suffocated under the post-socialist frame, which carries with it a number of unattractive tropes including "inscrutability," "backwardness," "irrationality," and "xenophobia" (Semeneko, Lapkin, and Pantin 2007; Naarden and Leerssen 2007; Lipovetsky and Leiderman 2008). Adding insult to injury, the region is seen/"scened" (see Groth and Bressi 1997) as a "literal and figurative wasteland" (Gille 2000, 242), where the transitological paradigm has resulted in life which is inhospitable, insecure, and interstitial to the rest of the world. While we should be careful not to overemphasize the impact of parody on the international system, the quasi-political content of films such as *Borat: Cultural Learnings of America for Make Benefit Glorious Nation of Kazakhstan*, Steven Spielberg's *The Terminal* (2004), and the James Bond films of the late 1990s has an important role to play in shaping the geographical imagination and imagined geopolitics of the fifteen nations that once constituted the Soviet Union.

While the Cold War officially ended more than two decades ago, cultural biases associated with the ideological conflict continue to linger. This tendency is reinforced by preexisting structures embedded in the framework of a "power imbalance between East and West, where the East is clearly placed in a dominated subject position" (Kaneva 2007, 22). If this subversion is funny it proves even more effective (and affective), since, as Klaus Dodds states, "The spatial symbolic of humor can often challenge and even subvert dominant boundaries of national sovereignty and the nationalist scripting of place" (2005, 96), an assertion which is particularly relevant in the case studies addressed below. These prejudicial predilections appear as frames, i.e., political guidance embedded in media content via contextual clues, which in turn "evoke distinct patterns of judgment and opinions surrounding the issue" (Lenart and Targ 1992, 341). As another scholar argues, "Comedy is (serious) politics. Comedy reflects the social and historical contingencies of our individual and collective impulses to think (differently), both with(in) and against a particular set of hierarchies (and pathologies)" (Brassett 2015, 2).

Cold War tropes continue to dominate Anglophone cultural production; through these comforting, if inaccurate, lenses, Americans are able "to make sense of the former Soviet Union, the ex-socialist expanse from Brest to Vladivostok" (Condee 2006). This phenomenon is particularly visible in popular culture, which Grayson, Davies, and Philpott describe as an "important site where power, ideology and identity are constituted, produced and/or materialized" (2009, 155–156). Using Stephen Daniels' (2011) concept of geopolitical "imagineers"—i.e., those cultural producers whose work focuses on the formation of images that target the geographical imagination—I argue that such agents of imaginative power have transitioned from their roles as ersatz "Cold Warriors" to what might be termed "Cold Victors." In the wake of 1989, the triumphant West began a sort of neoliberal conquest of the post-socialist East, with the media functioning as an important adjunct to the process (see Imre 2009). An early example of this phenomenon was the spoof punk band-cum-art project known as the Leningrad Cowboys. The creation of Finnish film director Aki Kaurismäki, the Leningrad Cowboys was conceived as a joke that played both on the waning power of the USSR and Finnish stereotypes about Russians. From the beginning, the group, with their iconic unicornesque bouffant hairstyles and elven shoes, intended being the "worst rock 'n' roll band in the world" (Landry 2012, 411). The farcical project reached its apex in 1989 with the film *Leningrad Cowboys Go America*, which saw the band, the manager, and the "village idiot" Igor traveling across North America and Europe, before returning to the "promised land" of their old collective farm with the pilfered nose of the Statue of Liberty in hand, a sardonic representation that true freedom can only exist in the "fantastical Soviet countryside" (Nestingen 2013, 38). In Kaurismäki's depiction of the culture clash stemming from Siberian peasant-rockers' experiences with Western consumerism, we are presented with a harbinger of the coming triumphalism of Western, particularly Anglophone, popular cultural treatments of *Homo post-Sovieticus*.

Humor thus became and continues to function as an important part of this process as a form of political communication, wherein it "articulates group perspectives, expresses concerns and provides an outlet for aggression" (Purcell, Scott Brown, and Gokmen 2010, 377). As Sergey Armeyskov (2013) has argued, Russia and Russians were ripe for such targeting after 1991, given the deep well of cultural associations in the Western geographical imagination linking the people and land with images of bears, borscht, and banyas. He refers to this as "klyukvification," or turning Russia(ns) into a cranberry; *klyukva* is often used to describe "a combination of foreign cultural, historical, linguistic and lifestyle stereotypes about Russia(ns) and *lubok*-like depiction of Russia(ns)," what might be labeled as anti-branding. Armeyskov thus defines klyukvification as "a process of creating a peculiar stereotypical narrative using Russian cultural objects and concepts in a certain manner (exaggeration, putting them in a different context, etc.)"

(2013). By playing on these exaggerations, humor is more easily produced. Keeping in mind the admonition that laughter is rarely innocent, particularly when it involves questions of politics, geography, national identity (Ridanpää 2009), I now turn to my three case studies.

Dumbing down the world: *The Onion*'s post-Soviet landscape

We shall begin with *The Onion*'s *Our Dumb World: Atlas of the Planet Earth*. Facetiously marketed as the seventy-third edition, the cover of *Our Dumb World* promises "better-veiled xenophobia," "30% more Asia," and a "free globe inside" (Dikkers 2007). Published in 2007, this satirical book of world geography provides a fruitful field of visual and textual data for analysis through the lens of popular geopolitics, and, on a more granular level, a précis of the challenges facing the varied states of the post-Second World as they attempt to differentiate themselves in the highly pluralistic geopolitical landscape of the twenty-first century. *The Onion*, which sardonically bills itself as "America's Finest News Source," began as a student newspaper at the University of Wisconsin-Madison in 1988, but has since been transformed into a diverse media organization that includes a print newspaper with a national circulation of 400,000, a web site with 7.5 million unique visitors per month, an IFC television series, and a series of popular hard-cover books on topics ranging from sports to world history (see Johnson 2011). With the exception of *The A/V Club*, *The Onion*'s entertainment newsletter, the media organization's content is "fake news," or, more accurately, satirical news-like content. Like *The Daily Show* and *The Colbert Report*, *The Onion*'s news products are directed at a postmodern audience that is "conscious of the constructed nature of meaning and of its own participation in the appearance of things," thus resulting in the "self-referential irony that characterizes most of our cultural output today" (Colletta 2009, 856). Consequently, the tortured skein of lies, half-truths, and misanthropy that constitute *Our Dumb World* is intended to be read as ironic, though not necessarily without political meaning or bereft of geopolitical agendas.

Spread across the Europe and Asia chapters, the entries on post-Soviet republics evoke the multivalent stereotypes embedded in the West's contemporary geopolitical imagination. Most of these are contemporary, playing on hegemonic fantasies which manifested during the period of transition from totalitarianism and socialism to the current state of political and economic affairs in the region; however, others are purely historical, atavistic markers of hoary pasts. The Baltic States are the first post-Soviet countries to be profiled in the text. Lithuania, a proud country that was once the largest state in medieval Europe and which spearheaded the dissolution of the USSR in 1991, is marked by its recent United Nations designation as the nation with the highest suicide rate. Latvia, renowned for its beautiful beaches and picturesque uplands, is "nestled in a ditch between Lithuania and Estonia . . . home to Europe's most breathtaking swamps, picturesque marshes, and an endless horizon that glimmers with several different shades of runoff" (Dikkers 2007, 166). Estonia—a technology powerhouse and economic dynamo—is characterized as a peasant backwater, where the main university specializes in "folk dancing and ox rearing" (Dikkers 2007, 165). The topographic maps of the Baltic region further include such farcical sites as the Hitler Memorial Campground, soggiest town, and world's largest meatloaf.

Turning southeast, Belarus is in the section on Central Europe, effectively eschewing a more traditional geopolitical approach to the region that would locate the former within post-Soviet Eurasia. In a rare *beau geste*, Belarusians are lauded for refusing to be wiped off the face of the Earth, with the bloody history of Nazism and Stalinism recounted in ludicrous ways. As the site of the worst effects of the Chernobyl disaster, Belarus is declared "inhabitable by 2307," while many of its citizens have developed superhuman powers such as the ability to "liquefy at will" (Dikkers 2007, 167). Local fauna of the Pripet basin, a zone known for its distinctive ecology and ancient black alder forests, supposedly include flaming three-foot cockroaches and glowing beavers with the strength of ten men. Belarus's neighbor Ukraine is marketed as the "bridebasket of Europe," playing on the country's historical role as a wheat producer and its current issues with sex-trafficking and mail-order brides.[2] Spouse-farming is a major industry and the exchange rate is 100 USD: 513 *hryvnia*: 1 woman (Dikkers 2007, 181). *The Onion* is certainly not the only media outlet to focus on this aspect of Ukraine's identity. The UK's Channel 4 program *Dawn Porter: Extreme Wife* took on the topic by heading to the country to illuminate the estimated 4,000 mail-order marriages that take place every year, with Ukraine being a prime "sending country."

Short shrift is given to Moldova—"Europe's basement"—which remains "unfinished" due to intransigence on the part of the voting population (Dikkers 2007, 182), thus poking fun at the recent political instability of the country, which has been riven by geographical and ideological divisions since the early days of the Soviet break-up. *The Onion*'s bon mot, however, rings hollow with the so-called "Twitter Revolution" of 2009, which failed to topple the government but provided a clear model for imitation in the Arab Spring of 2011 (see Morozov 2009). In the Caucasus, Armenia is a place where the population digs mass graves for "old times' sake" (darkly pantomiming the trauma of the Ottoman Empire's 1915 massacre of Armenians), Georgia is even more Christian than the "other Georgia" (wryly making a swipe at the religious conservatism of the southern American state), and Azerbaijan, a country where more than half of the nominally Muslim population consumes alcohol (WHO 2004), is a hotbed of Islamic "laxtremism" where sausages are a popular Ramadan treat and the main use of the Quran is as a drink coaster (Dikkers 2007, 205–206).

Travelling across the Caspian to "The Stans," we discover the following: Turkmenistan is home to "world's greatest dictator" (a jibe at Saparmurat Niyazov a.k.a. the "Great Leader of All Turkmen," the late president whose cult of personality rivaled that of North Korea's late Kim Jong-il); proudly independent Tajikistan is poised to "achieve a new era of economic misery that it can truly call its own"; Kyrgyzstan is the bullied wimp of the region that has its lunch money regularly stolen by the Russians, Chinese, and anyone else passing along the Silk Road, cunningly referencing the country as the site of both American and Russian military facilities and a frontline state in the "New Great Game" (Menon 2003). Kazakhstan—the world's ninth largest country—is a "huge waste of space," but, as a result of its oil reserves, "this prehistoric land of horse-eaters may be dictating U.S. foreign policy as early as 2015" (Dikkers 2007, 207–209). The entry on Uzbekistan represents a bit of sublime irony, as no distinguishing information is provided, but the "importance," "significance," and "distinction" of the country and its peoples are constantly stated, wryly mocking Islam Karimov's ham-handed efforts at national identity production since the mid-1990s (see Bell 1999; Adams 2010).

Russia, as a long-time foe of the US, is singled out for vitriolic abuse, a fact noted in the *Chicago Tribune* review of *Our Dumb World* (Darling 2008). The country's terrain consists of "broad plains of scorched irradiated earth, desolate Siberian gulags, and walled-off regions of radioactive, contaminated steppes" (Dikkers 2007, 183), its people a cold, pessimistic lot of racist drunkards, and its national infrastructure a crumbling mass of concrete occupied by feral cats. Russia suffers some of *The Onion*'s sharpest raillery, reflecting submerged fears that linger from decades of ideological conflict combined with a pique that stems from Moscow's dogged resistance to the West's imposition of neoliberal hegemony in the wake of the dissolution of the Soviet Union. Whereas other Newly Independent States are generally treated as non-threatening (at least to the West), *The Onion* preserves a veneer of danger in its conceit of the Russian Federation, thus serving to keep us afraid even as we laugh at the "East."

In its treatment of the former Soviet Union, *Our Dumb World* features a mixture of the real and the trfantastical, seasoned with something just between *Schadenfreude* and contempt that produces a truly "popular geography," i.e., one of and by the people. *The Onion*'s atlas creates an intellectual environment where stereotype reigns supreme and facts lose out to the irresistible allure of truthiness. Similar to Stephen Colbert's political pantomime, *Our Dumb World*'s geography is "from the gut" (see Johnson 2013), carrying just enough verity to seem viable, but never so much as to require people to reassess their deeply held convictions. I now move on to the creation of an unreal but instantly recognizable post-Soviet neverland: Molvanîa.

Out of the Antipodean mind: Molvanîa as a post-Soviet caricature

The first volume in the faux JetLag Travel Guide series, subtitled "Taking You Places You Don't Want To Go," *Molvanîa: A Land Untouched by Modern Dentistry* (2004) is a full-length tourist guide to a fake post-Communist country, one which combines exaggerated but plausible descriptions of the seedier, sadder, and more insalubrious aspects of the former Soviet Bloc. Originally published in Australia as the brainchild of comedy writers Santo Cilauro, Tom Gleisner, and Rob Sitch, the book became a surprise hit and achieved worldwide distribution, thus prompting publication of other titles, including *Phaic Tăn: Sunstroke on a Shoestring* (2004) and *San Sombrèro: A Land of Carnivals, Cocktails and*

Coups (2006). *Molvanîa* has spawned a number of ancillary media projects as well, including a 30-minute video segment about the country which was shown on Australian carrier Qantas's international flights, and two music videos by Molvanîa's own (fake) pop superstar Zladko "Zlad" Vladcik (see Figure 6.1). Most recently, Cilauro and company published another book on the country entitled *Traditional Molvanîan Baby Names: With Meanings, Derivations and Probable Pronunciations* (2011).

Molvanîa provides a seductive, albeit caustic, simulacrum of post-socialist Eastern Europe through its use of genuine photos culled from across the region, including gold-toothed Aeroflot stewardesses, egg-painting peasants in traditional garb, and dour accordionists, as well as a drainage ditch labeled as a thermal spring, a pothole venerated as a holy site, and smog-obscured photos of Stalinist skylines. According to the *New Yorker* review:

> In format and page layout, this inspired send-up of a travel guide looks exactly like the real thing, and it displays an acute feel for all the clichés of the genre, including testimonials that instruct how to have an uncomfortable "authentic" experience, rather than a "bland, westernized" one. (*New Yorker* 2004)

Hotel and restaurant reviews are surprisingly detailed, e.g., that of Romajaci in the capital Lutenblag, which has a menu with "something for everyone—provided

Figure 6.1 Zladko "Zlad" Vladcik (© Working Dog Productions Pty Ltd 2013)

you like pork" (Cilauro, Gleisner, and Sitch 2004, 53). Unlike *The Onion*'s use of real countries, Molvanîa is purposefully fake, as its authors purportedly did not wish to "offend anybody" (BBC 2004). Despite this declaration, former UK minister for Europe Keith Vaz criticized the book, suggesting that *Molvanîa* reflects anti-Eastern European prejudices which are "taking root" in Western Europe (BBC 2004). Paradoxically, Vaz gave the book kudos for unveiling the glaring ignorance of Westerners towards the post-Communist "East."

Molvanîa most resembles Bulgaria in its geographical outlines and physical landscape, though it lacks access to any navigable body of water, possibly bringing to mind the Former Yugoslav Republic of Macedonia (see Figure 6.2). However, the country's history as a former Roman colony evokes shades of Romania or perhaps Croatia (though Bulgaria's recent efforts to highlight Plovdiv's Roman bona fides give puts the Balkan state back in the running). Adding to this confusion, its Soviet heritage (and name) suggests Moldova as the source for the parody. Molvanîa has four geographical regions: the Western Plateau, the Great Central Valley, the Eastern Steppes, and the Molvanîan Alps; the reader is thus presented with a highly precise, yet stunningly prosaic physical geography, reiterating the stereotype that *all* post-socialist European countries are indistinguishable from one another. According to JetLag, Molvanîa is supposedly located "north of Bulgaria and downwind from Chernobyl," while also being the "crossroads of Europe" (Cilauro, Gleisner, and Sitch 2004, 18). Just as impossibly, it shares a border with both Slovakia and Germany, while simultaneously being located in the Baltics

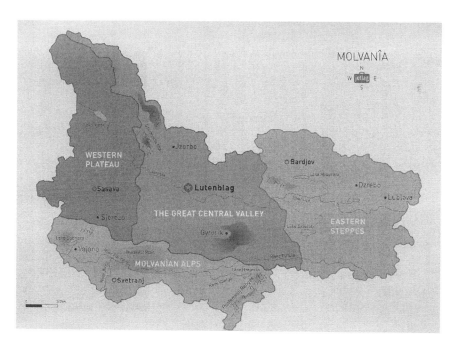

Figure 6.2 Map of Molvanîa (© Working Dog Productions Pty Ltd 2013)

and the Balkans; yet the country is, according to the authors, an "ex-Soviet state" (Cilauro, Gleisner, and Sitch 2004, 17). The artistry of the writers comes across in the ambiguity of space, placing Molvanîa in northeastern Europe on one page and in the southeastern corner of the continent on another, thus hammering home the lack of need for specificity among Western audiences when dealing with the "post-socialist European backyard" (see Imre 2014).

Since the fall of the Iron Curtain, this trend of depicting post-Second World denizens as "European white trash" (Hall 2005) has increased in both intensity and frequency. In fact, this is simply an update of a Cold War-era stereotype of the Soviet Union's citizens, "the unlucky outcasts of Western civilization" (Shlapentokh, Shiraev, and Carroll 2008, 111). Molvanîa, like the post-Soviet countries of *Our Dumb World* and Baron Cohen's "Kazakhstan" (discussed below), is peopled by a permanent underclass of toothless rubes, criminals, drunkards, and social deviants who—if they spoke English—would be at home in the mountains of West Virginia or the estates of South Yorkshire. However, these "Oriental Europeans" (Wallace 2008) do not speak English and are thus instead just "silly foreigners," objects of ridicule that reinforce the notion of Western European superiority and hegemonic dominance over their laggard cousins in the east. Cilauro's (2011) most recent book on the country, a listing of baby names, teases out some of these dismissive clichés in short order: *Nubikgob* ("Often spat at"); *Sidojuglar* ("Filled with self-loathing"); and *Ukokaverny* ("Co-owner of a cave"). This trend is further evidenced by the Orientalist discussion of the Molvanîan language, which is characterized by an abnormal number of silent letters, the use of the triple negative, and four genders: male, female, neutral, and one especially for cheeses (jibes likely to target the funny bone of any diehard monolinguistic Anglophone). Due to Molvanîan's abundance of guttural sounds, the guide includes a health alert warning for non-native speakers of "laryngeal damage that can arise from attempting anything more than few short phrases" (Cilauro, Gleisner, and Sitch 2004, 27).

According to JetLag, Molvanîans are known for being surly and short-tempered, but this tendency is softened by a "Slavic sense of humor" (Cilauro, Gleisner, and Sitch 2004, 20), thus providing some clue to the ethnogenesis of the Molvanîan people. Lampooning the West's obsession with ethnographic studies of Eastern Europe and the tendency within the region to affix heightened importance to one's ethnic affiliation (see Zarycki 2014), *Molvanîa* reports that the population of the country is divided between the Bulgs, who make up over two-thirds of the population, the minority Hungars (29% of the population) who reside in the northern cities, and the Molvs (3%), who can be "found mainly in prison" (Cilauro, Gleisner, and Sitch 2004, 20). Molvanîa no longer has a "Gypsy Problem"—a major issue for many post-socialist European polities (Bancroft 2005)—as the Roma population has been successfully driven abroad in adherence to the dictates of the country's 1987 national anthem, which, incidentally, is sung to the tune of "Flashdance ...What a Feeling" from the eponymously named 1983 motion picture. In a few short pages, Cilauro combines a host of post-1989 stereotypes into a pastiche of anti-Eastern Europeanism, marking the post-socialist realm as intolerant to indigenous diversity, yet desirous of importing culture from abroad.

Shifting to issues of governance, the country's contemporary political culture bears the marks of its despotic rule by Szlonko Busjbusj (known as "Bu-Bu") and subsequent Sovietization following World War II; however, democratization began in Molvanîa much earlier than in the rest of the Bloc, thus resulting in the de facto "post-Soviet warlordization" (see Boghani 21014) of its current political structure (see Figure 6.3). Despite this distinction, the government's failure to allow international biological weapons inspectors into the country has stymied EU membership, thus echoing certain elements of post-Soviet "contagion" reflected in the previous chapter. However, this has not stopped Molvanîa from being awarded provisional status in NATO, a not-so-subtle jeer at Washington, which has been rather lax in its requirements for admitting new post-totalitarian member states to the defense pact since 2001. The country's fake standard is referred to as the "tricolor," but strangely only has two colors—"Communist" red and yellow—as apparently the nation could not afford a third (see Figure 6.4), a particularly cutting comment on the wealth gap between neoliberal post-Soviet states like Estonia and economic "laggards" like Moldova. Emblazoned on the flag is a hammer, sickle, and—Molvanîa's contribution to visual semiotics of socialism—a trowel.

The purported contributions to world culture of Molvanîa, birthplace of the polka and whooping cough, are well-known, but its premier sport—*Plutto*, an Orientalist hybrid between lacrosse and polo played on donkey—has yet to catch on beyond its borders. Conversely, Molvanîan moonshine, a "fierce liquor made from juniper berries and brake fluid" (Cilauro, Gleisner, and Sitch 2004, 106), is turning into a major export product. At the international level, Molvanîa is probably best known for Zladko Vladcik's 2004 Eurovision entry "Elektronik—Supersonik" (Molvanîa n.d.).

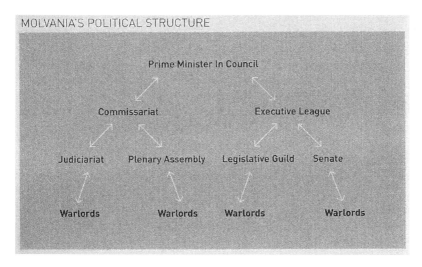

Figure 6.3 The hierarchy of power in Molvanîa (© Working Dog Productions
Pty Ltd 2013)

Figure 6.4 The Molvanîan "trikolor" (Immanuel Giel/GNU GPL)

The song, which is performed by *Molvanîa* co-author Santo Cilauro masquerading as a 1980s synth-rocker version of Borat, replete with a massive "mullet" hairdo and a futuristic silver tracksuit and delivering jaw-dropping lyrics like "I put my butt plug in your socket," was unfortunately disqualified when Vladcik was arrested at Istanbul's Ataturk International Airport and immediately deported. According to the *Molvanîa* web site: "While Eurovision does not normally test for recreational drugs, unfortunately for Vladcik, Turkish Customs do" (Molvanîa n.d.). Despite not winning the contest, the video, which lives on in cyberspace, presents a vivid burlesque of 1980s European synth-pop combined with the *Kitsch und Drang* of Eastern European efforts to embrace all things Western in the wake of the Cold War, while at the same time being moored in a realm defined by "tsars," "dictators," and the Cold War space race.

The political economy of tourism is where the JetLag guide truly outdoes itself in terms of xenophobic invective directed at the post-Second World. Travelers are alerted to fact that many cash machines are operated by mechanical winch handle because of the "country's erratic electrical supply" (Cilauro, Gleisner, and Sitch 2004, 41), thus playing on the myth of a literally "dark" Eurasian backwater evoked in the NBC commercial discussed in the first paragraph of this chapter. A departure tax of 3,000 *strubl*, one of the highest in Europe, is required for all tourists leaving the country; however, "most visitors agree it's well worth the price"(Cilauro, Gleisner, and Sitch 2004, 39). Other oddities associated with the rapid transition from socialism to a market economy include hotel owners who charge their guests to speak English, "service provision" fees at restaurants (i.e., for having a waiter), discounts on room rentals when sharing with an invalid pensioner or a recently released criminal, and extra costs associated with a roof on your hostel room. While such descriptions may be read as simple humor, they tap into common Western stereotypes of Eurasia as underdeveloped yet hyper-capitalist, which—when combined with political, social, and cultural frames and the overarching ideological apparatus of popular media—reinforce the "relentless" (Aligica and Evans 2009) and "systemic" (Sussman 2012) neoliberal propaganda barrage targeting the region.

A guide to Boratistan: Sacha Baron Cohen's final salvo at Kazakhstan

Kazakhstan, a country once described by one of its own diplomats as a "non-descript 'Stan somewhere between China and Dracula" (Wiltenburg 2005), now commands a vibrant, globally recognized national brand. Every year, the Central Asian republic attracts billions in foreign direct investment, welcomes eco-tourists from around the globe, and adds a few more architectural gems to its glittering new capital, Astana. But beginning in 2000, Sacha Baron Cohen, the British come-dian and creator of the fictitious Kazakhstani reporter Borat Sagdiyev, began a long-running parody of the world's ninth-largest country as a medieval backwater and its people as benighted rapists and racists (see Saunders 2006). After several years of modest success on British and American television with *Da Ali G Show*, the character rose to global stardom with the feature-length film *Borat: Cultural Learnings of America for Make Benefit Glorious Nation of Kazakhstan* (2006). The *Borat* motion picture unexpectedly provided Kazakhstan with a precipitous increase in its global profile, though one which came at a hefty price (not least of which was associated with expensive advertisements in US newspapers and journals attempting to undo the public relations damage done by the *Borat* film). Following an ill-conceived attempt at silencing Baron Cohen, the Kazakhstani government changed tactics in 2005 and began to co-opt the parody, making good use of "Boratmania" (Schmid 2010). Under the skillful management of Roman Vassilenko, then press attaché for the country's embassy in the US, Kazakhstan effectively turned a potentially disastrous public relations problem into a nation-branding victory (see Saunders 2008). Despite this, Kazakhstan still remains a victim of geographical imagination and pop-culture geo-graphing gone awry.[3]

Borat's *Touristic Guidings to the Glorious Nation of Kazakhstan* (2007), pub-lished in a single volume with *Touristic Guidings to Minor Nation of U. S. and A.*, stands as a lasting testament to this fact. Meant as a parting shot at the country that turned the British comedian into a household name, Sacha Baron Cohen, writing under the pseudonym of Borat Sagdiyev, and his long-time collaborator Anthony Hines crafted a pejorative atlas of Kazakhstan that not only continued Baron Cohen's pollution of the country's reputation, but actually expanded it in new and often grotesque ways. Whereas *Da Ali G Show*'s impugning of Kazakhstan was often inadvertent—a necessary by-product of Baron Cohen's comedic style wherein he exposed the gross ignorance of Westerners about the "unknown" lands of the post-Soviet East—and the marketing of the *Borat* film laid bare the fact that all the purported Kazakhstan scenes were filmed in Romania, *Touristic Guidings to the Glorious Nation of Kazakhstan* instead takes direct aim at the Central Asian republic. In the book, Baron Cohen employs enough verisimilitude in his bizarre and antagonistic "geography lessons" of the country to be spiteful.

Those familiar with Borat will already be versed in the falsehoods and prevarications of Baron Cohen's effigy of Kazakhstan, a geopolitical imaginary where pubic hair is a main export, dog hunting is a central theme of the national holiday, and horses vote but women are prohibited from doing so. When Baron Cohen began his career as Borat, he knew nothing about the Central Asian republic other than it was anonymous enough to use in his shtick. In his *Touristic Guidings*, it is clear that he knew just enough to be dangerous, as evidenced by the two-page spread on eating horseflesh or the genuine photos of the national sport *kokpar* (polo played with a headless goat carcass). In its more spurious visual representations, the book's images of dirty peasants, ramshackle hovels, and gun-wielding youths play to the most malicious hetero-stereotypes about post-Soviet space and its peoples. Mimicking the transgressive and hyper-masculine oeuvre of the *Jackass* films and MTV series (see Brayton 2007), these representations are combined into a phantasmagoria of porn-star monkeys, swimming pools filled with animal feces, and nude photos of Borat and his "family members."

The opening section of the book provides an *outré* introduction to the Central Asian republic's history and geography. The map of the world exaggerates Kazakhstan's already enormous landmass to include much of Russia's underbelly, as well as northwestern China, and Mongolia, thus making it the "third biggest country in the world" (Hines and Sagdiyev 2007, 8); similarly, the nation's rather small population of 17 million is shrunk by Borat to a paltry seven million. Later in the text, the more detailed national map of Kazakhstan is presented a bit more accurately in terms of the country's size and relative position to its neighbors, although much of China is incorrectly rendered as "Mongolia" and Iran is labeled "Iraq."

Kazakhstan, which in actuality decommissioned its only nuclear power plant in 1999, is shown to have a total of 28 reactors. In the general vicinity of each one are a series of *X*s, each marking a region of "much retardation" and "Strange Ones" (Hines and Sagdiyev 2007, 15), a truly heartless barb given that Kazakhstan continues to grapple with the effects of Soviet nuclear testing in the Semey region, despite assuming the mantle of leadership in creating a nuclear-free world (see EoK 2006). In an ironic twist, about 20 percent of Kazakhstan is marked as the possible location of the country's two apple trees. In fact, Kazakhstan is the source of nearly every one of the world's apple breeds and the "center of origin for many of the planet's major fruit tree crops" (Nabhan 2008). Other highlights include the (defunct 1980s soft drink) PepsiMax Manufacturing Plant, a proposed site for the "biggest hole in the world," and the Great Silk Road (which, incidentally, runs north-to-south in Sagdiyev's bizarre *Weltbild*).

When Baron Cohen and Hines broach the subject of politics, they seem to be taking cues from Stephen Kotkin's intellectually rabid tabloid essay "Trashcanistan" (2002), which lambasted post-Soviet Eurasia as a derelict imaginarium littered with the socio-political detritus of a failed merger of the worldviews of

Karl and Groucho Marx (see Chapter 7). Several pages are devoted to "Premier Nazarbamshev" (a none-too-subtle reference to President Nursultan Nazarbayev), glorifying his vision, wisdom, physical prowess, and leadership. Borat lauds Nazarbamshev's democratization efforts, reporting on the most recent round of voting in the country, "International observers from Zimbabwe, Haiti and Sudans say it was one most fair transparents election they had ever witness [sic]" (Hines and Sagdiyev 2007, 21). *Touristic Guidings* includes a laughable, but unfunny, "Bill of Rights" that requires all citizens over the age of three to bear arms, the right to wrestle one's accusers in a civil trial, and the freedom of worship for all as long as they "recognize that the only true god is the Mighty Hawk Ukhtar" (Hines and Sagdiyev 2007, 23), an oblique nod to the Kazakhs' veneration of the golden eagle (see Privratsky 2013).

True to his previously established modus operandi, Baron Cohen takes particular aim at American popular culture, dragging Kazakhstan down in the process. Echoing Sussman's (2010) assertion that there has been steady escalation of promotional culture in post-Soviet Eurasia whereby a neoliberal "ordering of society" places the market first and popular culture is glutted with Western-style reality TV with a transitological twist, the country's most popular television shows are reported to be *Gypsy Bingo* (which involves betting on which Roma will successfully cross a minefield) and *Kazakhstan's Next Top Prostitute*. Borat's geographical primer perpetuates the trite myth that Eastern Europeans are obsessed with (bad) 1980s pop music, as the largest stadium in the country is named after schlock-rocker Huey Lewis.

In a scathing allusion to Kazakhstan's history as a sending country in inter-country adoptions, Borat invites "western moviestars" Madonna and Angelina Jolie to consider adopting his grandchild for the price of $100, stating "I sure this much more cheap than you payings for your Africans [sic]" (Hines and Sagdiyev 2007, 93). Continuing Baron Cohen's well-hewn mythologizing of Kazakhstan as a hotbed of anti-Semitism (most notoriously via his "Throw the Jew Down the Well" YouTube hit), *Touristic Guidings* includes a section on the "817 kilometer-high" statue of "Melvin the Redeemer," a monument to Mel (Melvin) Gibson in recognition of his declaration that "The Jews are responsible for all the wars in the world" (qtd. in Lubrano 2006) during an tequila-fueled tirade in 2006.

In terms of political economy, the purported depletion of the Aral Sea due to over-production of PepsiMax and the use of the company's soda cans as a form of legal currency in Kazakhstan belie the generally positive status of the country, as does the assertion that the country's leading export is pubic hair, clipped from hirsute peasant women. Likewise, Borat's claim that the Kazakh National Museum can be obtained by doing a "swap of a Cadillac" with Premier Nazarbamshev presents an unfair subreption of the Kazakhstani economy and the country's attitude towards its cultural and historical heritage (see Saunders 2007). Despite Kazakhstan's buoyed geo-economic status and Astana's prodigious expenditures on nation-branding, from advertisements in *The New York Times* to hosting of international sporting events to promotion of interfaith conferences, Baron Cohen and Hines find it easy to fall back on post-Soviet tropes of corruption, backwardness, and blind faith in Western capitalism, expecting very little from their audience. Unlike The Onion's *Our Dumb World*, which stresses the quiddity of each post-socialist nation, or *Molvanîa: A Land Untouched by Modern Dentistry*, which collapses the whole of the post-Second World into a single, homogenous agglomerate, Borat's *Touristic Guidings to the Glorious Nation of Kazakhstan* takes a middle path, using the unique attributes of a real nation against it, while simultaneously selecting stereotypes drawn from around the region to create a pornographic hybrid, though—importantly—one which is politically, economically, and cultural neutered. Whereas Cold War hetero-stereotypes of Kazakhstan focused on the country as fountainhead of the "evil empire's" (Reagan 1983) space program and 1990s-era depictions of the nation positioned it as a cesspool of nuclear and biological threats (Burrows 1994), Baron Cohen's Kazakhstan is—in the end—a joke, though not necessarily a funny one.

Post-Soviet space as a comedic prop: An analysis of the faux geographies of the former USSR

As evidenced above, the vast expanse of post-Soviet Eurasia has come to serve as a ready-made space for cultural producers seeking a backdrop of post-transitional misery in the current era of globalization. In order to more fully flesh out the role of the region in "grounding" dispositional humor, i.e., the "advantaged party" (West) disparages and lords its "triumph over the disadvantaged party" (the post-Soviet East) (Wilson 1999, 173), I will discuss just a few of the dozens of films, novels, and other media products that have similarly employed a defeated and defunct post-Soviet mirror-world as a vehicle for spoof, satire, and/or parody.

We will begin with "Krakozhia," the fictional Slavic country constructed in Steven Spielberg's motion picture *The Terminal* (2004). The plot of the film centers on a good-hearted traveler, Viktor Navorski (Tom Hanks), who becomes a "citizen of nowhere" when he is stranded in JFK International Airport after his passport is invalidated as the result of a coup in his homeland, "the tiniest country in the region . . . [which] has been involved in a civil war throughout the late '80s and '90s as it has tried to transition from Communist rule." The "Krakozhian" language spoken by Tom Hanks is gibberish with a few Slavic-like utterances, but, according to the actor, is phonetically and structurally based on Bulgarian (Hanks 2004). Regarding the language and the positioning of Krakozhia, Hanks stated in an interview on NPR: "Well, we started with a Slavic beginning to it and a Cyrillic-alphabet language; and Krakozhia means that it comes from one of those nations with an "-ia" at the end of it: Estonia, Latvia, Yugoslavia" (Hanks 2004).

Geopolitically, Krakozhia is thus a pop-culture *réchauffé* of the so-called TAKO nations of post-Soviet space, i.e., Transnistria, Abkhazia, (Nagorno-) Karabakh, and (South) Ossetia (see King 2008). These states each declared their independence from their respective post-Soviet "parent" countries (Georgia, Azerbaijan, and Moldova), but have yet to achieve international recognition, or to paraphrase Nivorski's apprehending border control agent in *The Terminal*: "Technically, these places do not exist" (a rather germinal piece of analysis given that this accurately describes most of what Hollywood projects about post-Soviet Eurasia). Describing Navorski as "part-genius, part-idiot, at once the hero and victim of globalism," Hoberman (2007) suggests that Spielberg's choice to create an imaginary country for his character Navorski shows that—at least geopolitically-speaking—the region has been declawed and is thus appropriate for parody, unlike other parts of the world where coups and counter-coups represent genuine threats to global stability (notably, the Middle East and South Asia).[4]

While the previously discussed Molvanîan artifice is relatively bereft of political content (though quite acicular in cultural terms), the American author Gary Shteyngart's bestselling and well-reviewed *Absurdistan: A Novel* (2006) contributes to the belligerent anti-branding of post-Soviet space (see Saunders 2012) in a self-cognizant and frequently normative fashion.[5] Much of the novel's action takes place in the post-Soviet Caspian republic of Absurdsvanï. Ethnically divided between the Sevo and Svanï, the so-called "cretins of the Caucasus" (Shteyngart 2007, 125), this post-Sovietesque world-in-miniature is a jolting bricolage of new wealth and crushing poverty, thus serving as a paragon of what Parag Khanna calls the "new Second World" country, one that is "*both* first- and third-world at the same time" (Khanna 2008, xxv). The Absurdi capital, described as "a miniature Cairo after it had crashed into a rocky mountain" (Shteyngart 2007, 134), is awash in investment from American multinationals, particularly oil conglomerates and designer clothiers, while the country ranks "slightly below Bangladesh" on the UN Human Development Index (Shteyngart 2007, 138). Trapped in a vicious cycle of rapacious leaders bent on establishing dynastic rule, Russian and American malfeasance, and civil and ethnic conflict, Absurdsvanï is a stand-in for the post-Soviet republics of Georgia, Armenia, and Azerbaijan, while not actually being any of them. Bringing home the point, the generic term "Absurdistan" has been in use for some time to make reference to the border republics of the Commonwealth of Independent States, symbolizing a "recognition of just how tragically comic life had become" in these post-Communist states (Buchanan 2002, 3). Unlike Steven Spielberg, Santo Cilauro, and Sacha Baron Cohen, Shteyngart's keen "awareness of the former Soviet Union" (Lee 2008, 32) provides a rich simulacrum despite its churning animosity towards the post-Second World.

Similar to *Absurdistan* in its critique of both post-Soviet banana republics and the American companies that love them (Dick Cheney's Halliburton, or "Golly Burton" as it is known to the Absurdis, is as much of a target for Shteyngart as is any regime in the region) is G. B. Trudeau's *Tee Time in Berzerkistan: A Doonesbury Book* (2009). The bound collection of comic strips recounts the attempts by President-for-Life Trff Bmzklfrpz (pronounced "Ptklm") to improve the status of "Greater Berzerkistan" in Washington with the help of highly paid lobbyists. Intentionally or not, Trudeau's book is a sublime, graphic rendering of Ken Silverstein's *Turkmeniscam: How Washington Lobbyists Fought to Flack for a Stalinist Dictatorship* (2008), which uncovered K Street's complicity in gussying up Saparmurat Niyazov's unsavory regime.[6] *Tee Time* opens with a former terrorist (now Berzerkistan's foreign minister) chatting with lobbyists about how to overcome the association of his country with genocide, torture, drug trafficking, and tyranny. The lobbyists lead with the obvious: "The goal of the campaign would be to rebrand your pariah state as valued a partner in the global war on terror so that you'd qualify for a military aid package of, say, $25 billion" (Trudeau 2009, 10). Next they advise the minister on how to "manage" the ethnic cleansing issue: "gentrification" associated with housing issues. Dealing with the president, however, proves more difficult, as he refuses to abandon his Hitlerian haircut-cum-mustache because it "scares the homosexuals and gypsies" (Trudeau 2009, 23), thus thwarting the necessary make-over prescribed by K Street's geopolitical imagineers. While Trudeau directly targets DC insiders, Central Asia gets caught in the crossfire as the region is made out to be something far worse than it actually is.

The same can be said of the playful and less biting "Kreplachistan," a creation of the master of contemporary cultural iconography, *Saturday Night Live* veteran Mike Myers (Kreplachistan, like Krakozhia, has generated a minor cottage industry of "fake country" souvenirs, including coffee mugs and t-shirts available for sale over the Internet). The country, which is referenced in the *Austin Powers* films, is associated with the bumbling villain Dr. Evil's (Myers) plans for world domination. Playing on both the "anarchist/madman" fears explored in previous chapter and presaging the era of post-Soviet "goofiness" of Molvanîa and the like, in *Austin Powers: International Man of Mystery* (1997), Dr. Evil steals a nuclear warhead being transferred to the United Nations from the "breakaway Russian republic" in order to use it ignite the Earth's core. The "Kreplachistan Situation" builds on the "loose nuke paradigm," which has come to be closely associated with Russia and Kazakhstan in the post-Soviet period, and has served as a plot device for films including those mentioned in the previous chapter. Kreplachistan makes a vague reference to the autonomous republic of Karakalpakistan, a constituent part of Uzbekistan, and *kreplach*, an Eastern European Jewish meat-filled dumpling. While Myers' antics lack the prejudicial overtones of contemporary geopolitical cinema, his choice of a fictional 'Stan furthers the Orientalist narrative in Western media treatments of the southern tier of post-Soviet space.

While a beautiful and culturally sensitive representation of the post-Soviet condition, *Everything is Illuminated* (2005), a tragicomedy based on Jonathan Safran Foer's 2002 quasi-autobiographical novel, the film still renders Ukraine as a vast gray and undifferentiated mass populated by an older generation of disaffected cranks and a younger coterie of laughable neoliberalized rubes. The affectual elements of the film (Ukraine as an anonymizing space, a post-Soviet imagescape) support larger geopolitical codes and visions associated with the Western antagonist's "discovery" of his (Jewish) roots, which, in the end, are as self-negating as they are fulfilling. However, for the purposes of this chapter, we should focus on the persona of musician Eugene Hütz,[7] i.e., Alex, Foer's (Elijah Wood) tour guide and Western pop-culture enthusiast, a sort of Borat *in situ*. His dog's name is Sammy Davis Junior Jr., he is a great fan of Michael Jackson (particularly the track "Billie Jean"), his mangled English provokes laughter based on its strangely formal grammar (he invariably uses the verb "repose" in lieu of "sleep" and the adjective "proximal" for the preposition "near"), and he—like Borat—seems obsessed with "making sex" (especially with American "girls"). While the film turns tragically poignant at the end, the viewer is left with the impression that the transition of the post-Soviet world had created a schism wherein the young have voluntarily abandoned their history and identity for trickle-down pop culture, while the older generation *Homo Sovietici* are so mired in the past they can never hope to escape it.

A sort of gendered mirroring of Hütz's Alex is guest commentator Olya Povlatsky (Kate McKinnon), a recurring character on NBC's long-running late-night comedy show *Saturday Night Live*. She regularly appears in the program's parodic news segment, which makes timely references to contemporary events in the United States, as well as overseas. Olya appears in "traditional" peasant garb, and functions as the alternatively naïve and wise *muzhik*, commenting on events in the region, including the Chelyabinsk meteor (16 February 2013), the 2014 Winter Olympiad (25 January 2014), and Russian intervention in eastern Ukraine (3 May 2014). Regarding the latter, she critiqued Putin's actions, lamenting the annexation of Crimea: "Great job, Putin! Make Russia even bigger because Russia's not big enough already." On the choice of Sochi for the 2014 winter games, she pondered the question: "How could the Olympics pick Russia? What was the other option—Haiti or middle of ocean?" When she is not impugning her own country, she details her attempts at dating (often thwarted by ravenous wolves invading the discotheque) and her love of Western popular culture, including the family sitcom *Full House* (1987–1995)—which supposedly just premiered in Russia—and the most recent season of *Game of Thrones* (2011–). When questioned about how she can view such recent airings of highly sought-after media, she responds "I have HBO GO password. I am not animal!" Through her (faux) editorial comments, Olya reintegrates *Homo post-Sovieticus*'s mythologized obsession with Western pop culture with the popular conception of Russians as the world's foremost population of intellectual property bandits (see US Congress 2013).

Lastly, we must turn our gaze to the "goofy" immigrant trope. With the collapse of the Soviet Union, immigration from Russia and Ukraine to the West, particularly the US and Great Britain, has been steady. As explored in Chapter 5, this mass movement of post-Soviet citizens has been presented as a threat to law-abiding Anglo-Saxon society through films such as *Little Odessa* (1994), *Eastern Promises* (2007), and *RocknRolla* (2008). However, there is another side to the post-Soviet/Western immigration story: the rube or greenhorn (see Borenstein 2008). Emerging as a common trope due to Sacha Baron Cohen's long-running Borat character, the awkward, unpolished, and often-oversexed post-Soviet immigrant has become a staple of British and American comedy.

In the wake of Borat, the new personification of the post-Soviet everyman is "Oleg" (Jonathan Kite) from the American sitcom *2 Broke Girls* (2011–). Oleg, a regular on the CBS program, is a Ukrainian immigrant who works as a short-order cook in a retro-hip Williamsburg, Brooklyn diner. While Borat was always positioned as a visitor to the West, Oleg has put down roots in his new country. When he is not preparing eggs and bacon for hung-over hipsters, he is hitting on the restaurant's two attractive waitresses Max (Kat Dennings) and Caroline (Beth Behrs) with lines like "Once you go Ukraine, you will scream with sex pain" and "You may refer to me as Sir Oleg of the Amazing Party in My Pants." Like Borat before him, Oleg personifies the over-sexed post-Soviet male, who ironically cannot seem to find a willing partner, despite the equally stereotyped sexual voraciousness of post-socialist women.[8] According Kite, "There's a looseness than comes with his sexuality and his Eastern European culture. There's a lot of sex in that part of the world. It's not taboo. He's just from a different place" (qtd. in Richenthal 2011).[9] His attempts to pick up his coworkers (or anyone else) always end in failure until he meets Sophie Kachinsky (Jennifer Coolidge), a Polish immigrant whom Max and Caroline originally took for a madam based on her attire, attitude, and wealth (in fact, she runs a successful cleaning service, thus reflecting a different but equally embedded stereotype of immigrant post-socialist Slavic women). Sophie and Oleg eventually enter into a torrid affair, but one characterized by Sophie's constant emotional abuse of Oleg, whose menial job makes the long-term chances of the match impossible. Oleg is thoroughly feminized by the relationship, becoming a "sex toy" for the domineering and brutish Sophie, who has "made it" in (neoliberal) Brooklyn, televisually reifying the relative influence of their states' respective position in post-1989 world politics. In its treatment of Eastern Bloc immigrants, *2 Broke Girls* reminds us of the collapse of the socialist system and the disruption it has caused to these societies, not least of which being the transformation of the "dangerous" post-Soviet man into a goofy, weak, and feminized version of his former self.

In the current era of deterritorialized transnational media flows, geopolitical imagineers like Sacha Baron Cohen, Santo Cilauro, and the editors of *The Onion* are on the frontlines of the battle to control national images in the twenty-first century. By marking the post-Second World through the use of parodic mental maps that present the former Soviet republics as—in turn—absurd, uncouth, untutored, and offensive, more than just geographical imagination is affected: geopolitical codes of behavior are influenced as well. Boratian, Molvanîc, and Onionesque geographies allow the West to maintain its borders with the post-socialist "East" long after the fall of the Iron Curtain. Ironically, this occurs as these peoples, states, and lands are becoming "Europe/European" through various formal (EU, NATO, etc.) and informal (economic, social, cultural, etc.) processes of inclusion, thus underscoring what has been labeled the "economic power of Western industries of the imagination" (Goldsworthy 1998, x). Whether on a conscious or subconscious level, Western cultural producers engage in mass-mediated objectification of the region and its constituent parts as "actors" on the world stage, theatrically assuming their choreographed, subaltern roles in the theater that is international politics. Not incidentally, these practices tell us more about how the culture-producing society conceives itself than about how it sees foreign peoples. In the next chapter, we transition from people to places in an effort to address other important aspects of this trend.

Notes

1 Albanian independence only pre-dated the Great War by two years, and were it not for Woodrow Wilson's support, the Balkan state would have likely disappeared from the map in 1919, so it too may be considered "new." The Russian Federation does not fit neatly among either the "new" states or the preexisting ones, given the radical changes wrought by the dissolution of the USSR.
2 While Ukraine is not the only country with such associations, it is an important aspect of the national image. For analysis on Russia's popular-geopolitical representation as a sender of potential brides to the West, see Williams (2012).
3 In the decade since the film, Kazakhstan has had some difficulty sloughing off the "Boratistan" patina. Most notoriously, the Borat-inspired spoof "O Kazakhstan" was played at a high-profile sporting event in Kuwait rather than the country's actual national anthem. Less visible, but nonetheless important, is a steady use of the Borat referent when the country is mentioned (even in a positive light).
4 While satire is driven by a desire for positive change, parody does not necessarily need to contribute to political discourse (Gray, Jones, and Thompson 2009). In effect, the post-Second World is now "safe," unlike the Muslim world, which poses—in the words of former US President George W. Bush—an "existential threat" to contemporary (Western) civilization. In this dichotomy, we might also place sub-Saharan Africa in the same camp as the former Eastern Bloc, given that political upheavals in the region after the end of the Cold War are little more than geopolitical footnotes and produce no real impact on ongoing global struggles.
5 Shteyngart was born in Leningrad, but relocated with his family to the United States in 1979. His first novel, *The Russian Debutante's Handbook* (2003), was written after a visit to Prague, Czech Republic, and is partially set in Prava, a fictional European city.
6 Silverstein originally recounted his controversial strategy of posing as a shill for Ashgabat and the startling results in the *Harper's* July 2007 article "Their Men in Washington: Undercover with D.C.'s Lobbyists for Hire."
7 Hütz expanded his role in "exposing" the geopolitical imaginary of his birth country in the subsequent documentary *The Pied Piper of Hützovina* (2007). Filmed as a sort of musical road-trip film (Hütz fronts the Manhattan-based Gypsy punk band Gogol Bordello) combined with a desperate attempt on the part of the filmmaker (Pavla Fleischer) to reignite her romantic relationship with Hütz, the film ends up showcasing the barren geographies of Ukraine, from its anonymous Stalinist apartment blocks to its squalid Gypsy villages, rather than the country's vibrant capital and its deep forests and rolling fields (see Chapter 7).
8 Consider the Austin Powers' character "Ivana Humpalot," a secret agent who is sent to kill the bumbling British spy, but fails to do so when she falls for him, claiming he is too sexy. Humpalot ultimately sleeps with Austin in a Bondesque parody of *The Thomas Crown Affair* (1968).
9 According to the condom manufacturer Durex's annual Sexual Wellbeing Survey (2012), the countries of the former Eastern Bloc do have sex more often than most nations (Russia and Poland both ranked in the top tier, with 80 percent and 76 percent of respective respondents reporting that they had sex weekly, trailing only Greece and Brazil). Ukraine was not specifically polled.

References

Adams, Laura L. 2010. *The Spectacular State: Culture and National Identity in Uzbekistan.* Durham, NC: Duke University Press.

Aligica, Paul D., and Anthony J. Evans. 2009. *The Neoliberal Revolution in Eastern Europe: Economic Ideas in the Transition from Communism.* Cheltenham, UK and Northampton, MA: Edward Elgar Publishing.

Armeyskov, Sergey. 2013. "#Klyukvification: Representation of Russia(ns) in Western Popular Culture." *Russian Universe*, available at http://russianuniverse.org/2013/10/31/russian-stereotypes-2/ [last accessed 14 July 2015].

Bancroft, Angus. 2005. *Roma and Gypsy-Travellers in Europe: Modernity, Race, Space and Exclusion*. Aldershot, UK: Ashgate.

Bardan, Alice, and Anikó Imre. 2012. "Vampire Branding: Romania's Dark Destinations." In *Branding Post-Communist Nations: Marketizing National Identities in the "New" Europe*, edited by Nadia Kaneva, 168–192. New York and London: Routledge.

BBC. 2004. "Molvanîa Spoof Mocks Travel Books." *BBC News*, available at http://news.bbc.co.uk/2/hi/europe/3592753.stm [last accessed 27 July 2012].

Bell, James. 1999. "Redefining National Identity in Uzbekistan: Symbolic Tensions in Tashkent's Official Public Landscape" *Cultural Geographies* no. 6 (2):183–213.

Berglund, Sten. 2001. *Challenges to Democracy: Eastern Europe Ten Years after the Collapse of Communism*. Cheltenham, UK and Northampton, MA: Edward Elgar Publishing.

Boghani, Priyanka. 21014. "What Comes Next in Ukraine?" *Frontline*, available at www.pbs.org/wgbh/pages/frontline/foreign-affairs-defense/battle-for-ukraine/what-comes-next-in-ukraine/ [last accessed 13 July 2015].

Borenstein, Eliot. 2008. "Our Borats, Our Selves: Yokels and Cosmopolitans on the Global Stage." *Slavic Review* no. 66 (1):1–7.

Brassett, James. 2015. "British Comedy, Global Resistance: Russell Brand, Charlie Brooker and Stewart Lee." *European Journal of International Relations* online (prior to print):1–24.

Brayton, Sean 2007. "MTV's "Jackass": Transgression, Abjection and the Economy of White Masculinity." *Journal of Gender Studies* no. 16 (1):57–72.

Buchanan, Donna A. 2002. "Soccer, Popular Music and National Consciousness in Post-State-Socialist Bulgaria, 1994–1996." *British Journal of Ethnomusicology* no. 11 (2):1–27.

Burrows, William E. 1994. "Nuclear Chaos." *Popular Science* no. 245 (2):54–59, 76.

Cilauro, Santo, Tom Gleisner, and Rob Sitch. 2004. *Molvanîa: A Land Untouched by Modern Dentistry*. Woodstock, NY and New York: The Overlook Press.

——. 2011. *Traditional Molvanîan Baby Names: With Meanings, Derivations and Probable Pronunciations*. Melbourne: Hardie Grant Books.

Colletta, Lisa. 2009. "Political Satire and Postmodern Irony in the Age of Stephen Colbert and Jon Stewart." *Journal of Popular Culture* no. 42 (5):856–874.

Condee, Nancy. 2006. "Learnings of Borat for Make Benefit Cultural Studies." *Pittsburgh Post-Gazette*, 12 November, available at www.post-gazette.com/opinion/Op-Ed/2006/11/12/Next-Page-Learnings-of-Borat-for-make-benefit-cultural-studies/stories/200611120163 [last accessed 2 January 2007].

Daniels, Stephen. 2011. "Geographical Imagination." *Transactions of the Institute of British Geographers* no. 36:182–187.

Darling, Cary. 2008. "Onion's 'Our Dumb World' Could Pass for What It's Mocking." *Chicago Tribune*, available at www.chicagotribune.com/chi-0103reader_jan03-story.html [last accessed 28 June 2013].

Debord, Guy. 1983 [1967]. *Society of the Spectacle*. Detroit: Black & Red.

Dikkers, Scott. 2007. *Our Dumb World: The Onion's Atlas of Planet Earth, Seventy-Third Edition*. New York: Little, Brown and Company.

Dodds, Klaus. 2005. *Global Geopolitics: A Critical Introduction*. Harlow, UK: Prentice Hall.

Durex. 2012. "Sexual Wellbeing Global Survey." *Durex*, available at www.durex.com/en-US/SexualWellbeingSurvey/ [last accessed 30 December 2012].

EoK. 2006. *Kazakhstan's Nuclear Disarmament: A Global Model for a Safer World.* Washington DC: Embassy of the Republic of Kazakhstan to the United States of America and the Nuclear Threat Initiative.

Gille, Zsuzsa. 2000. "Cognitive Cartography in a European Wasteland: Multinational Capital and Greens Vie for Village Allegiance." In *Global Ethnography: Forces, Connections, and Imaginations in a Postmodern World*, edited by Michael Burawoy, 240–267. Berkeley and Los Angeles: University of California Press.

Goldsworthy, Vesna. 1998. *Inventing Ruritania: The Imperialism of the Imagination.* Bury St. Edmunds, UK: Edmundsbury Press.

Gray, Jonathan, Jeffrey P. Jones, and Ethan Thompson. 2009. "The State of Satire, the Satire of State." In *Satire TV: Politics and Comedy in the Post-Network Era*, edited by Jonathan Gray, Jeffrey P. Jones, and Ethan Thompson, 3–36. New York: New York University Press.

Grayson, Kyle, Matt Davies, and Simon Philpott. 2009. "Pop Goes IR? Researching the Popular Culture–World Politics Continuum." *Politics* no. 29 (3):155–163.

Groth, Paul Erling, and Todd W. Bressi. 1997. *Understanding Ordinary Landscapes.* New Haven, CT: Yale University Press.

Hall, Andrew. 2005. "Images of Hungarians and Romanians in Modern American Media and Popular Culture." *Nationalism, History, and Memory in Eastern Europe*, available at http://homepage.mac.com/khallbobo/RichardHall/pubs/huroimages060207tk6.html [last accessed 28 July 2012].

Hanks, Tom. 2004. "Interview: Tom Hanks Discusses His new Film 'The Terminal'." *National Public Radio: All Things Considered*, 18 June, available at www.npr.org/templates/story/story.php?storyId=1964704 [last accessed 3 March 2010].

Hines, Anthony, and Borat Sagdiyev. 2007. *Borat: Touristic Guidings to Glorious Nation of Kazakhstan.* New York: Flying Dolphin Press.

Hoberman, J. 2007. "Laugh, Cry, Believe: Spielbergization and its Discontents." *Virginia Quarterly Review* no. 83 (1):119–135.

Imre, Anikó. 2009. *Identity Games: Globalization and the Transformation of Media Cultures in the New Europe.* Cambridge, MA: The MIT Press.

——. 2014. "Postcolonial Media Studies in Postsocialist Europe." *Boundary 2* no. 41 (1):113–134.

Johnson, David Kyle. 2013. "Colbert, Truthiness, and Thinking from the Gut." In *Stephen Colbert and Philosophy: I Am Philosophy (And So Can You!)*, edited by Aaron Schiller, no pp. Chicago: Open Court Publishing.

Johnson, Steve. 2011. "Blooming Onion." *Chicago Tribune*, available at http://articles.chicagotribune.com/2011-01-10/entertainment/ct-live-0111-onion-tv-20110110_1_steve-hannah-onion-news-network-national-lampoon [last accessed 8 December 2012].

Kaneva, Nadia. 2007. "Meet the 'New' Europeans: EU Accession and the Branding of Bulgaria." *Advertising & Society Review* no. 8 (4):15–24.

Khanna, Parag. 2008. *The Second World: Empires and Influence in the New Global Order.* New York: Harcourt, Brace and Company.

King, Charles. 2008. *The Ghost of Freedom: A History of the Caucasus.* Oxford: Oxford University Press.

Kotkin, Stephen. 2002. "Trashcanistan." *New Republic* no. 226 (14):26–38.

Landry, Charles. 2012. *The Art of City Making.* London and New York: Routledge.

Lee, Steven S. 2008. "Borat, Multiculturalism, *Mnogonatsional 'nost'*." *Slavic Review* no. 67 (1):19–34.

Lenart, Silvo, and Harry R. Targ. 1992. "Framing the Enemy." *Peace & Change* no. 17 (3):341–362.

Lipovetsky, Mark, and Daniil Leiderman. 2008. "Angel, Avenger, or Trickster? The 'Second-World Man' as the Other and the Self." In *Russia and Its Other(s) on Film: Screening Intercultural Dialogue*, edited by Stephen Hutchings, 199–219. Basingstoke, UK: Palgrave Macmillan.

Lubrano, Alfred. 2006. "Unconventional Wisdom: Booze Loosened Mel's Jaw, But His Rant Was from Bile." *The Philadelphia Inquirer*, 12 August.

Menon, Rajan. 2003. "The New Great Game in Central Asia." *Survival* no. 45 (2):187–204.

Molvanîa. n.d. "Molvanîa Disqualified from Eurovision . . . Again." *Molvanîa*, available at www.molvania.com/molvania/eurovision.html [last accessed 8 January 2012].

Morozov, Evgeny. 2009. "Moldova's Twitter Revolution." *Foreign Policy*, available at http://neteffect.foreignpolicy.com/posts/2009/04/07/moldovas_twitter_revolution [last accessed 29 July 2012].

Muller, Benjamin J. 2008. "Securing the Political Imagination: Popular Culture, the Security *Dispotif* and the Biometric State." *Security Dialogue* no. 39 (2–3):199–220.

Naarden, Bruno, and Joep Leerssen. 2007. "Russians." In *Imagology: The Cultural Construction of National Characters—A Critical Survey*, edited by Joep Leerssen and Manfred Beller, 226–230. Amsterdam and New York: Rodopi.

Nabhan, Gary Paul. 2008. "The Fatherland of Apples." *Orion*, May/June, available at https://orionmagazine.org/article/the-fatherland-of-apples/ [last accessed 4 March 2013].

Nestingen, Andrew. 2013. *The Cinema of Aki Kaurismäki: Contrarian Stories*. London and New York: Wallflower Press.

New Yorker. 2004. "Molvania (Book)." *New Yorker* no. 80 (29):105–10.

Privratsky, Bruce. 2013. *Muslim Turkistan: Kazak Religion and Collective Memory*. London and New York: Routledge.

Purcell, Darren, Melissa Scott Brown, and Mahmut Gokmen. 2010. "Achmed the Dead Terrorist and Humor in Popular Geopolitics." *Geoforum* no. 75:373–385.

Reagan, Ronald. 1983. Speech to the National Association of Evangelicals. Orlando, Florida (8 March).

Richenthal, Matt. 2011. "2 Broke Girls Interview: Jonathan Kite on Pride in Oleg." *TV Fanatic*, available at www.tvfanatic.com/2011/11/2-broke-girls-interview-jonathan-kite-on-pride-in-oleg/#ixzz21pwp4XOz [last accessed 4 November 2013].

Ridanpää, Juha. 2009. "Geopolitics of Humour: The Muhammad Cartoon Crisis and the *Kaltio* Comic Strip Episode in Finland." *Geopolitics* no. 14 (4):729–749.

Saunders, Robert A. 2006. "Cultural Learnings: Welcome to Boratistan." *Transitions*, available at www.tol.org/client/article/17850-welcome-to-boratistan.html [last accessed 16 November 2006].

———. 2007. "In Defence of *Kazakshilik*: Kazakhstan's War on Sacha Baron Cohen." *Identities: Global Studies in Culture and Power* no. 14 (3):225–255.

———. 2008. "Buying into Brand Borat: Kazakhstan's Cautious Embrace of Its Unwanted 'Son'." *Slavic Review* no. 67 (1):63–80.

———. 2012. "Brand Interrupted: The Impact of Alternative Narrators on Nation Branding in the Former Second World." In *Branding Post-Communist Nations: Marketizing National Identities in the "New" Europe*, edited by Nadia Kaneva, 49–78. New York and London: Routledge.

Schmid, Susanne. 2010. "Taking Embarrassment to Extremes: Borat and Cultural Anxiety." In *Facing the East in the West: Images of Eastern Europe in British Literature, Film and Culture*, edited by Barbara Korte, Eva Ulrike Pirker, and Sissy Helff, 259–274. Amsterdam: Rodopi.

Semeneko, Irina, Vladimir Lapkin, and Vladimir Pantin. 2007. "Russia's Image in the West (Formulation of the Problem)." *Social Sciences* no. 38 (3):79–92.

Shlapentokh, Dmitry, Eric Shiraev, and Eero Carroll. 2008. *The Soviet Union: Internal and External Perspectives on Soviet Society.* New York: Palgrave Macmillan.

Shteyngart, Gary. 2003. *The Russian Debutante's Handbook.* New York: Riverhead.

——. 2007. *Absurdistan: A Novel.* New York: Random House.

Silverstein, Ken. 2007. "Their Men in Washington: Undercover with D.C.'s Lobbyists for Hire." *Harper's Magazine* no. 315 (1886):53–61.

——. 2008. *Turkmeniscam: How Washington Lobbyists Fought to Flack for a Stalinist Dictatorship.* New York: Random House.

Sperling, Valerie. 2009. *Altered States: The Globalization of Accountability.* Cambridge: Cambridge University Press.

Sussman, Gerald. 2010. *Branding Democracy: U.S. Regime Change in Post-Soviet Eastern Europe.* New York: Peter Lang Publishing.

——. 2012. "Systemic Propaganda and State Branding in Post-Soviet Eastern Europe." In *Branding Post-Communist Nations: Marketizing National Identities in the "New" Europe,* edited by Nadia Kaneva, 23–48. New York and London: Routledge.

Trudeau, Garretson Beekman (G. B.). 2009. *Tee Time in Berzerkistan: A Doonesbury Book.* Riverside, NJ: Andrews McMeel Publishing.

US Congress. 2013. *House Hearing, 109th Congress: Intellectual Property Theft in China and Russia.* Washington, DC: US Government Printing Office.

Wallace, Dickie. 2008. "Hyperrealizing 'Borat' with the Map of the European Other." *Slavic Review* no. 67 (1):36–49.

WHO. 2004. Azerbaijan. In *Global Status Report on Alcohol,* edited by World Health Organization. Geneva.

Williams, Kimberly A. 2012. *Imagining Russia: Making Feminist Sense of American Nationalism in U.S.–Russian Relations.* Albany, NY: SUNY Press.

Wilson, Wayne. 1999. *The Psychopath in Film.* Lanham, MD: University Press of America.

Wiltenburg, Mary. 2005. "Backstory: The Most Unwanted Man in Kazakhstan." *Christian Science Monitor,* available at www.csmonitor.com/2005/1130/p20s01-altv.html [last accessed 9 December 2007].

Zarycki, Tomasz. 2014. *Ideologies of Eastness in Central and Eastern Europe.* Abingdon, UK and New York: Routledge.

7 Mapping Trashcanistan

The post-Soviet badlands in popular culture, news media, and academe

In his 2002 essay "Trashcanistan," which appeared in the leftwing tabloid magazine *New Republic*, Princeton professor Stephen Kotkin notoriously described the entire post-Soviet realm as a "dreadful checkerboard of parasitic states and statelets, government-led extortion rackets and gangs in power, mass refugee camps, and shadow economies," which, despite its nuclear and chemical weapons, was "ignorable" (2002, 26–27). To Kotkin, Russia—the "world's superstore for doomsday weapons"—is a "basket case" where WMDs sit comfortable alongside "shit" (i.e., fertilizer) as some of the country's most important export products (2002, 26). Looking beyond the Russian Federation, Kotkin views the geopolitics and domestic economic development of the other post-Soviet republics as evidence of the pernicious policies of national self-determination, which led to regimes based on avaricious corruption, unrepentant criminality, and ethnic antagonisms. Intended as a review essay of a number of books that appeared in the first decade after the dissolution of the USSR, "Trashcanistan" instead functions as a geopolitical smear campaign, lambasting the region as a whole, while individually targeting certain states for unabridged opprobrium. In his analysis of Central Asia scholar Martha Brill Olcott's *Central Asia's New States* (1996), Kotkin slyly pans her analysis of the region by making a comparison of Brill Olcott's treatise to the collective geography of the "Stans", stating that the text "covers a lot of ground, most of it barren" (Kotkin 2002, 27). Traversing the Caspian, the Princeton don characterizes Yo'av Karny's *Highlanders: A Journey to the Caucasus in Quest of Memory* (2001) as a pornographic wallowing in the "squalor" and bloodlust of the indigenous peoples of the region. While Kotkin finds much he likes about Charles King's *The Moldovans* (2000), he informs his reader that Moldova evokes all the worst aspects of his geopolitically imagined Trashcanistan, from a "permanently impoverished and ineradicably corrupt" Gagauzia to "hot-blooded" Romanian nationalists and murderous Slavic apparatchiks (Kotkin 2002, 31). Turning to the second most populous post-Soviet republic, Kotkin's review of the collected volume *Ukraine: The Search for a National Identity* (Wolchik 2000) represents the country as ideologically adrift, hopelessly corrupt, and hamstrung by its "dodgy" history.

If this is how a leading American academic with extensive expertise in the region frames the post-Soviet realm, what can we expect from cable-news pundits, sport reporters, travel writers, filmmakers, TV series producers, graphic novelists, and videogame developers? This chapter examines that question, offering a critique of how the popular-geographical imagination of space and place, a practice that is wholly dependent on manipulation of the Anglophone West's *Weltbild* of the region, has contributed to a continuation of the enemy frame in populist discourses associated with Eurasian states since 1991. While Kotkin's "Trashcanistan" is perhaps the most glaring example of this trend vis-à-vis post-Soviet Eurasia, there is a surfeit of other similar treatments of the (formerly Communist) East. This chapter attempts to assess the negative spectacularization of the Russian/Eurasian realm via news coverage/analysis/punditry. This is done with the aim of exposing the underlying geographical assumptions about the region, and especially how clichés materialize in popular culture via mediated landscapes and symbolic geographies, thus proving the contention that "visual communication is powerful because it binds the viewer in a communicative relation where agency is hidden and meaning is ambivalent" (Stocchetti and Kukkonen 2011, 4). In order to accomplish this melding of ideology and text with visuals, I draw extensively on a variety of other thinkers in the field of cognitive mapping, geographical visualization, semiotic landscapes, and geopolitical aesthetics (Hopkins 1994; Jameson 1995; Shapiro 2008; Cosgrove 2008).

Following Dixon and Zonn, I examine a number of popular-cultural forms as visual representations of "broader-scale discourses that continually construct and deconstruct the world as we know it" (2005, 292), paying special attention to how such renderings of "reality" shape real-world political assumptions.[1] This includes not only reinforcing the "appropriateness" of US hegemony of the globe, but also validating (rather than producing a sense of guilt about) the sorry state of many parts of the former Soviet Union. Recognizing that the very concept of "landscape" is a human construct enabled in the mind's eye,[2] the body of this chapter focuses on the post-Soviet realm's pop-cultural rendering as a zone of snow, grayness, dilapidation, and radiation. Following this critical analysis of the major geovisual themes of Western/Anglophone popular culture, the focus shifts to an interrogation of the real-world effects of such representation, utilizing François Debrix's (2004) helpful concept of tabloid realism to assess the framing of Eurasian space since 1991. As Debrix states,

> The desire to condition the public to certain beliefs and attitudes through the dissemination of a popular but often paranoid political discourse is not novel . . . Of particular interest is the relationship between these popular media outlets in the construction of American identity both within the borders (who the American citizen is, what the nation is made of) and outside them (who America's enemies are, who we are at war with). (2004, 155)

In addition to the theoretical scaffolding provided by Debrix, this chapter also employs Roland Bleiker's recent work on aesthetics (2012) and assemblage (2015) and Gregory's (1994), Dijkink's (1996), and Ó Tuathail's (2000) foundational expositions on geopolitical codes and visions to argue how a wide variety of everyday attitudes towards a given polity can sustain flawed assumptions and influence geopolitical codes framed by popular-culture visions of the post-Soviet East.

Projecting place and space: Popular cartography, affective landscapes, and (geo)politicized aesthetics

Popular geopolitics is a discipline that is obsessed with ocularcentric representations of physical and human geography, as well as the embedded politics in the depiction of place and space. Driven by myths, mindsets, and visions of the unreal as "real," such politicized imaginaries are dependent on symbols, archetypes, and repetition (Dijkink 1996, 1). As discussed in Chapter 3, images of certain nations exist in the mind's eye of many people purely as affect, with no knowledge base whatsoever (Kunczik 1997). This is certainly the case for Russia in much of the Anglophone world. Stereotypes of Russia range widely, but certainly the trite triad of vodka, bears, and extreme cold pervades the generic "understanding" of the vast realm that is encompassed within Russia's borders. As the Russian historian Anna Pavlovskaya (2013) argues, these stereotypes about the country are surprisingly stable and have been in existence for centuries, typically crafted by early travelers to the region and sustained through countless works of popular culture. In the two previous chapters, we have focused principally on representations of "actors" and "agents" in the popular geopolitical imagination, i.e., revanchist Russian generals, bumbling Kazakhstani reporters, etc. However, this chapter examines how popular culture represents the physical realm, which is often just as important as how human populations are depicted, perhaps even more so given the importance of aesthetics and affect in shaping geopolitical visions (see, for instance, Dijkink 1996; Salter 2011; Robinson 2012).

At this point, it is prudent to distinguish between the interrelated concepts of affect, emotion, and aesthetics. Following geographer Nigel Thrift (2008), *affect* represents a way of thinking, knowing, and seeing the world, a sort of parallel intelligence that dynamically influences the social contours of interaction; in short, affect is a matrix that informs and orders experience (Thien 2005). *Emotion*, distinguished here from affect (but not feeling), is an often transitory "positive or negative evaluative state with neurological, neuromuscular, and sometimes cognitive manifestations" (Thye, Lawler, and Yoon 2008, 43). Emotions include love, joy, happiness, serenity, confidence, pride, fear, anxiety, embarrassment, envy, disgust, anger, sadness, shame, guilt, repentance, loneliness, awe, and surprise (see Turner and Stets 2005; Riis and Woodhead 2010; Pile 2010). It is important to remember, however, that emotions are not simply "inner states but also relational stances" (Riis and Woodhead 2010, 5), thus serving as gauges that structure encounters, interactions, and other forms of social exchange (Anderson and Smith 2001; Bondi 2005; Thrift 2008). Given the parameters of these distinct but inextricably linked concepts, it is then vital to understand the ways in which the "world is mediated by feeling" (Thien 2005, 451), politically, socially, and culturally. From a representational standpoint, affect is closely tied to aesthetics and, for our purposes, the technologically enhanced aesthetics of mass media, which engage "sensorial commodities creating virtual spaces which expose the player to a variety of affects" (Shaw and Warf 2009, 1333). In order to further distinguish affect from aesthetics, this chapter to will refer to affect as the non-representational awareness, embodiment, and other "moments of intensity" (O'Sullivan 2001, 126) produced by aesthetics or the visual-audio elements of design which are focused on enhancing the mediatic experience.

A variety of scholars from fields as diverse as history, sociology, literature, and art have demonstrated the importance of aesthetics of physical space and the natural world in national imagination (see, for instance, Inglis 1977; Olwig 2002; Cusack and Bhreathnach-Lynch 2003; Ryan 2011; Wright 2014); however, the majority of this work has focused on the homeland, *Heimat*, or *patria*. Nevertheless, a few scholars, including some critical geographers, have chosen to analyze the ways in which non-human elements associated with the "land" (maps, landscapes, scenery, photographs, iconic structures, etc.) influence everyday geopolitical thinking about foreign countries (and by extension, the people who inhabit them). Consequently, there has been a realization that the relationship between geography and "seeing" is particular important in everyday views on international relations (see Jay 1988; Rose 2001). Geographer Denis E. Cosgrove provides an authoritative analysis of the centrality of the visual to geographical understanding:

[V]ision is more than the ability to see and the bodily sense of sight. Vision's meaning incorporates imagination: the ability to create images in the mind's eye, which exceed the various ways those registered on the retina of the physical eye by light from the external world. Vision has a creative capacity that can transcend both space and time: it can denote foreseeing as well as seeing. (2008, 8)

Jonathan Swift once noted, "vision is the art of seeing the invisible" (qtd. in Ogao 2002, 13); without such visually-charged, imagined truths (Leerssen 2007), it is impossible to conceive of the world beyond our own parochial and quotidian realm(s). Consequently, such "false seeings" are ineluctably imbricated in how one's world is constructed; focusing more on how things ought to be *and* are taught to be, rather than on what they really are, thus brings into being "phantasms" of place and space that suffer from nationalist moralizing (Rogoff 2000, 3). As Cosgrove (2008) points out, geographical vision is highly prone to ideology. Whether we are talking about the backdrop for an action-thriller, *mise-en-scène* footage for a news segment, or a place-based video montage that precedes the biographical profile of a (foreign) Olympic athlete, there are judgments about what imagery to include—or more specifically, what scenery will frame the narrative and situate the viewer along a specific ideological matrix (Pisters 2003). For purposes of clarity, "a frame operates to select and highlight some features of reality and obscure others in a way that tells a consistent story about problems, their causes, moral implications, and remedies" (Entman 1996, 77–78). Such landscape framing exudes politicized orientations, engages ideological matrixes, and operationalizes geopolitical codes which have been primed by decades—if not centuries—of socializing groundwork perpetuated through militarism, statist education, political inculcation, and various us/them pathologies.

In *Critical Geopolitics*, Gearóid Ó Tuathail delineates the ways in which "geo-power" emerges through the visualization of place and space, and how the imperial eye/I structures *what* we see *when* we see from afar. As he states: "Geography is not only taught in classrooms, but is also projected at the citizen-spectator in films, newspapers, advertisement, postcards and travel brochures" (2000, 162). Cinema, sequential art, videogames, travelogues, novels, and television programs all require settings where action can take place. Likewise, reports of foreign events, tourism advertisements, and international sport reporting all demand that imagery be used to situate the audience in a place (if only for a fleeting moment of imagination). Selecting the content for these representations of place and space is deeply bound up with the geopolitical understanding of the cultural producer, but is also dependent on the targeted audience's baseline understanding of geographical space. As Bertrand Westphal states: "A *place* is only a place because of the ways which we, individually and collectively, organize space in such a way to mark topos as special, to set it apart from the spaces surrounding it and infusing it" (2011, x).

In his preface to Frederic Jameson's *The Geopolitical Aesthetic* (1995), Colin McCabe notes we must always engage in cognitive mapping based on limited information, yet we have the capacity to produce maps in our minds as long as the provided information "overlap(s) at certain crucial points with other grids of interpretation" (Jameson 1995, xv); this process then allows for political and economic analysis. In his work on the mapping of cinematic places, Jeff Hopkins cautions that the "cinematic landscape is not . . . a neutral place entertainment or an objective documentation or mirror of the 'real,' but an ideologically charged cultural creation whereby meanings of place and society are made, legitimized, contested, and obscured" (1994, 47). Stuart Aitken and Leo E. Zonn remind us that "American films set in 'exotic' locations usually say more about capitalism and Hollywood than they do about the cultural poetics within which they are filmed" (1994, 14); instead, "place becomes spectacle, a signifier of the film's subject, a metaphor for the state of mind of the protagonist" (1994, 17). Looking at this relationship from another angle, Jacques Rancière argues, "The real must be fictionalized to be thought" (2004, 38). Failure to aesthetically produce an intelligible "place" will result in a disconnect that will undermine the final cultural product. For instance, if a comic book artist wishes to situate their character's adventure in a Central American country, it is likely that they will employ easily identifiable geographical markers (alongside a caption noting the location). A "Central American" frame thus dictates that certain scenic tropes be employed, be they beaches, rainforests, *barrios pobres*, or Mesoamerican pyramids. Such use of symbolic geography is relatively unquestioned in most forms of popular-cultural production, though it is a source of critical analysis in various fields of academic scholarship; however, scholars in the field of popular geopolitics have made the first attempt to yoke these representations to political culture and real-world manifestations of Ó Tuathail's "geo-power."

Strongly influenced by the works of Yi-Fu Tuan (1977, 1979) and Edward Said (1979), scholars of popular geopolitics have increasingly included depictions of the physical world in their analyses. One subset of this research focuses on popular cartographies, with excellent analysis of both World War II geographies (Cosgrove and della Dora 2005; Zhrauliova 2014) and those associated with the Cold War conflict (Sharp 2000; Dodds 2003; Sage 2008). More recently, the focus has turned to the aesthetics of the War on Terror, and popular-culture projections of landscapes of danger associated with jihadi violence (see Dodds 2007; Šisler 2008; Schopp and Hill 2009; Höglund 2014). The importance of contrasting "safe" geographies of home (spaces of dwelling, attachment, and rootedness) with "dangerous" geographies abroad (zones of unintelligibility, mystery, and threats) is key to understanding the popular-geopolitical representation of the Other. Through the use of emotion-laden landscapes (Berberich, Campbell, and Hudson 2013), affective maps (Craine and Aitken 2011), and visual rhetoric (Bogost 2010) in popular culture, feelings are activated—particularly the emotions of fear and loathing—and with this manipulation also comes the additional power of persuasion (Robinson 2012). Hollywood, television executives, comic book artists, videogame developers, and the news/sport media are all active participants in what has been deemed the "fear-industrial complex" (Stossel and Jaquez 2007), which includes trepidations about everything from child-kidnapping and shark attacks to WMDs and pandemics.

Eurasia's "danger zones": An analysis of four semiotic landscapes

Post-Soviet Eurasia presents an interesting case study for understanding the power of geographical vision, particularly given the embeddedness of stereotypical landscapes associated with Russia/USSR. Due to its climate, Russia has long been visually depicted as an infinite and mostly unexplored icebound wasteland, one which history has seen fit to validate through the decimation of invading armies (most notably Napoleon's Grand Armée in 1812 and the Axis forces during World War II). The representation of the Eurasian landscape as perennially covered in snow remains dominant in popular culture to this day. In the coming pages, I discuss how this imagery flows through a variety of media, from film/television/graphic novels to news/sport reporting, and how such representation manifests in geopolitical "knowledge" on the part of citizens in the Anglophone West via politicized linkages between danger, barrenness, and cold in the post-Soviet realm. While the frame of Eurasia (and more specifically Russia) as a frigid desert is centuries-old, alternative representations have come into vogue in the past quarter century, each of which carries with it the notion of "decay."

The second geopolitical landscape I explore is that of the urban warzone. A variety of media have depicted Moscow, the largest city in the Eurasian realm, as the site of unmitigated violence. Seemingly lawless, the capital is shown as a metropolitan arena where gangsters and terrorists are free to act as they will, and with the surfeit of weapons left over from the Cold War, the carnage these villains can inflict is limited only by the imagination. As I will argue, Moscow is utilized in ways that distinguish it from other major world capitals (e.g., London, Paris, Rome), arguably due to residual ideological frames that place the city beyond affective associations for Western viewing audiences.

The third set of symbolic geographies discussed is that of post-Soviet ruination, i.e., imagery of debris associated with the USSR's lost greatness. Such "ruin porn" is a staple of Olympic coverage of Russian, Ukrainian, and other post-Soviet athletes (even ones born after the dissolution of the USSR).[3] However, this frame is also common in geopolitical thrillers, tabloid news coverage of the region, travelogues, videogame settings, and satirical novels, among other media formats.

Finally, stemming from the meltdown of Chernobyl Nuclear Power Plant on 26 April 1986, the geopolitical frame of Eurasia as a "zone" (see Dyer 2012) of radiological/biochemical contamination has come into vogue, resulting in the graphic rendering of parts (or even all) of the region as a no-man's-land. In this frame, desolation is man-made, not natural, and is geopolitically linked to Soviet scientific and technological prowess gone awry. Videogames, horror films, and science fiction all make use of such landscapes (sometimes combining these with other tropes mentioned above), ultimately resulting in pop-culture "axes of intelligibility" (Grayson 2014) whereby the reader/viewer possesses a surprising ability to associate the post-Soviet East with certain attributes, although they may be at a loss to locate Russia, Ukraine, or Turkmenistan on a map of the world (see Goldsworthy 1998).

Siberian nightmares, frozen wastelands, and icy dangers

From its earliest depictions in Western popular culture, the Russian landscape has been linked to a seemingly perpetual winter. Imagery of snow, ice, taiga, and tundra are commonplace in everyday depictions of the country. The resonance of such bathetic clichés is so strong that many first-time visitors to the country are often unprepared for the oppressive summertime heat that reaches even into the country's "northern capital," St. Petersburg. While premiering in theatres while there was a still a Soviet Union, *The Hunt for Red October* (1990) is emblematic of the hyperborean representation of Russia. The film opens with a slow panning shot of snow-covered taiga, eventually opening up to show the Polyarny inlet in Murmansk Oblast, the site of the Soviet Navy's primary submarine station. Standing on the tower, the saturnine Captain Marko Ramius (Sean Connery) and his First Officer Vasili Borodin (Sam Neill) exchange a precious few words of (bad) Russian in the frigid air before setting out into the open ocean just as Ramius comments on the weather as "cold" and "hard." Surrounded as they are by the projected barrenness of the Russian Far North, one can easily take these words as reflective of the country itself. In fact, the only other images of the USSR to be shown in the film are inside acier, constructivist Moscow buildings; even here we are reminded of cold, as we see a Politburo member entering his offices, methodically shedding layer after layer of heavy coats and outerwear.

The Hunt for Red October builds on hackneyed stereotypes of the Russian realm as a ghostly, white wasteland, limitless in its icy embrace. Key scenes in *Doctor Zhivago* (1965), *Nicholas and Alexandra* (1971), and *Rocky IV* (1985) all depict such open stretches of endless taiga, a "bleak, suspenseful, and threatening" space (Dodds 2011, 90). This representative theme would be continued and magnified in the years after Russia "opened up" to the West. These assemblages of images are capable of producing (and likely intended to produce) affective responses that include revulsion, shivering, or attempting to get warm.[4] A variety of post-1991 Western films showcase frozen geographies in an effort to contextualize "Russia" in their various narratives, from the Bond film *Goldeneye* (1995)[5] to the adaptation of comic book *Hellboy* (2004).[6] In all of these treatments of Russia as cold, hard, and forbidding, there is a reminder that Westerners will find little welcome therein.[7] Immutable Russia—as ever—is impenetrable and hostile, in terms of both its geography and the effect of that geography on its people, a subtle evocation of the "swallowing up" of the Westerner, from the real-world invaders of Napoleon and Adolf Hitler to the latter-day mediatic interlopers. Two poignant examples include FedEx systems engineer Chuck Noland (Tom Hanks) in *Cast Away* (2000), who is "lost" for years after his plane from Moscow to Malaysia crashes, and the Western characters in the crime-thriller *Transsiberian* (2008), a morality tale wherein a American woman's sexual dalliance enmeshes her in organized crime and drug-smuggling, threatening her life as well as that of her husband.[8] To this list of pop-culture artefacts, we might also add Marcel Theroux's novel *Far North* (2010),[9] which depicts a post-climate-change world where Siberia is one of the few inhabitable places on the globe; however, despite the continued persistence of flora and fauna in Russia's north, the geography is only matched in its hostility towards its (American) settlers by the ferocity and violence of the indigenous peoples of the region (i.e., Russians, displaced Central Asians, and Tungus). In the words of the narrator Makepeace Hatfield, a climate refugee from the US who came to the region with her family and other members of a religious sect:

> I believe Siberia was suggested to them as a joke. People thought of the place as a land of ice, a desert of rocks and snow, with the wind blasting it ten months a year from the Urals to the Pacific Ocean. (Theroux 2010, 51)

According to Hatfield, Moscow let in the Americans to offset the "hungry-eyed Chinese" with their territorial designs on Russia's eastern hinterlands, thus (fictionally) fulfilling geopolitician Dmitry Trenin's (2002) prescription for offsetting Russia's demographic weakness in Siberia with immigrant populations with no irredentist aspirations.

The NBC faux-reality television series entitled simply *Siberia* (2013) represents the most interesting (and controversial) manifestation of the arctic badlands as a (popular) symbolic geography of post-Soviet Eurasia. Ironically presented as a "real" reality TV program, this science fiction-cum-horror series uses Siberia as a backdrop for a *Survivor*-like competition. Transported to a remote area of northern Eurasia with nothing but the clothes on their backs, 16 contestants—most of whom hail from First World countries (US, Iceland, Canada, and Taiwan)—are pitted against one another to be the last to (voluntarily) remain in the Eurasian wilderness. Almost immediately, the situation takes a deadly turn, with one participant dying. Over time, the group realizes they have been abandoned by the producers, who may or may not have perished in an unexplained attack on a neighboring campsite. Soon the group is faced with starvation and hypothermia as the green realm they arrived in quickly turns to a (familiar) white. Through a series of events, they learn they are situated very near to the location of the Tunguska Event of 1908, the largest near-Earth asteroid impact in recorded history. This metaphysical anomaly has purportedly created conditions that preternaturally threaten the lives of the survivors, who face myriad dangers including attacks by carnivorous predators, the local Tungus people, and Russian paramilitaries (dressed like Spetsnaz operators).[10]

Upon its premiere, the series drew sharp criticism from Russia, particularly for its rather obvious (mis)use of the Canadian terrain (as the setting for "Siberia") and First Nation peoples (labeled "Evenki," despite their stark differences from the diverse indigenous populations of Siberia). The juxtaposition of (western) Canada *qua* Siberia only serves to underscore the ideological nature of this form of Russian-themed cultural production. The two countries share much in their geographical and demographic makeup; however, whereas Canada is represented from multiple angles in Western pop-culture (a sophisticated and "European" Montreal, a downhome and maritime Nova Scotia, urbanized and commonplace Toronto, and a cosmopolitan and trendy British Columbia), the geography of Siberia is rendered as the synecdoche for all of European Russia, despite the latter's continental climate, which more closely resembles that of New York State than Nunavut. Indubitably, the shibboleth that distinguishes Russia (as "immutable" *Sibir'*) from the US's (known, civilized, and non-threatening) "neighbor to the north" is rooted/routed in a host of geopolitical factors (Pabst 2006), which inevitably bleed through into filmic and other forms of pop-culture representation. Such neo-Cold War overtones ultimately provoked a backlash, with the *Siberian Times* excoriating the show's "keeping with the US stereotype of Siberia, mixed with a cocktail of Hollywood horrors and then frozen in a Cold War time warp" (Stewart 2013). The critique went on to directly connect the issue of attracting tourism and foreign direct investment to calumnious popular-cultural production. In reinforcing the septentrional frame alongside unpredictable threats that the "wild expanses" of Russia present to Western visitors, *Siberia*—by adding a secondary layer of supernatural horror, including mysterious creatures and mutation-based experimentation—effectively posits the post-Soviet realm as an alien zone where the uncanny reigns.

Urban warzones, everyday battlefields, and criminal neverwheres

With Russia's rapid transition to market-based capitalism, there was an undeniable rise in violent crime in the country. A similar situation affected many of the other post-Soviet republics as well, particularly Ukraine and certain Central Asian states. As discussed earlier in the text, the KGB agent was quickly displaced by the *mafioso* as the primary villain in Western popular-culture treatments of the post-Soviet realm. However, in this section, I move beyond the personification of Russians as mobsters (or more commonly, the hybrid *mafioso*-FSB agent), reflecting instead on how mediatized "worlds of affect" (Shaw and Warf 2009) associated with the former USSR sculpt an evolving geopolitical imaginary of Eurasia as a lawless cacotopia, triggering affective responses that range from gnawing apprehension to abject fear.

Scriptwriters of Western action films have repeatedly used their creative capacities to set all or part of their narrative imagescapes within the boundaries of the old USSR, with central Moscow being a particularly popular site. While it is common to see large swathes of urban space destroyed in Hollywood blockbusters, the cities or urban landmarks targeted for annihilation tend to be extremely familiar to the viewing audience, through either actual experience or regular representation via popular culture. A shortlist would include the ruination of the White House in films such as *Independence Day* (1996), *2012* (2009), and *Olympus Has Fallen* (2013) (see Hall 2013), the decimation of New York City in *Armageddon* (1998), *Cloverfield* (2008), and *The Avengers* (2012) (see Acuna 2012), or scenarios wherein Californian cities meet their doom in movies like *Terminator 2: Judgment Day* (1991), *Battle: Los Angeles* (2011), and *San Andreas* (2015) (see Grad 2015). However, in the case of non-US cities, there is a tendency among filmmakers to place a premium on *why* a particular city is targeted in the filmic narrative. Whether we are talking about London, Paris, or Rome, there is something special about the setting, and its perdition redounds throughout the plot.

Obliterating portions of Moscow, however, seems to be outside of this metaphoric structuration. There are no ramifications for unleashing machine gunfire in the middle of the city, blowing up buildings, killing pedestrians on the street, crashing scores of cars, etc. In *Mission: Impossible—Ghost Protocol* (2011), the Kremlin is razed by terrorist bombs planted under Red Square; however, other than serving as a plot device leading to the main character Ethan Hunt (Tom Cruise) being arrested by Russia's Foreign Intelligence Service (SVR), the act is of minor importance in the overall scheme and no attention is paid to collateral damage. In the aforementioned *A Good Day to Die Hard* (2013), the viewer is treated to a sustained heavy munitions attack on a Moscow skyscraper by a Mil Mi-24 Hind gunship and a car chase through the always traffic-jammed streets of the Russian capital involving a Ural Typhoon MRAP infantry vehicle which flattens everything in its path. In no instance does either act of mass destruction seem to attract the attention of local authorities. Moscow is (blessedly, at least for the action-addicted audience) without any controls on mass violence, destruction, or mayhem.

The same cannot be stated about *The Bourne Supremacy* (2004), as the film involves an epic car chase through Moscow in which the protagonist Jason Bourne (Matt Damon) is pursued by the mercenary FSB agent Kirill (Karl Urban). While visible, the Moscow police serve as (anti-Western) agents of "law enforcement" in their attempts to interdict Bourne (and thus permit Kirill to assassinate him for his oligarch boss). Despite the actual presence of law enforcement in the Bourne film, the depredation and death wrought by the plot device of a car chase remains peripheral. Moscow is evidently there to be destroyed.[11] Other recent films set in Moscow also treat the capital as an expendable space, where rampant crime, bloodshed, and car crashes are the norm, including *The Saint* (1997), *Hitman* (2007), *Iron Man 2* (2010), and *Jack Ryan: Shadow Recruit* (2014). The notable (existential) exception to this axiom is *The Darkest Hour* (2011), in which an unseen alien peril attacks Earth via its power sources, ruthlessly dispatching any and all humans who interrupt the harvest of resources. Rather than serving as a set piece to be blown up, Moscow *as a space* figures prominently in beating back the alien interlopers (like the Napoleonic and fascist invaders before them). However, such an anomaly is quite easily explained given that Russo-Kazakh filmmaker Timur Bekmambetov produced the film. Bekmambetov was the driving force behind a variety of other cinematic projects which represented post-1991 Russia as a complex and multifaceted realm, from the Russian blockbusters *Night watch* (2004) and *Day watch* (2006) to the documentary *Happy People: A Year in the Taiga* (2010).

Moving beyond the Russian capital, a number of other Western films set in post-Soviet space expound upon the notions of criminality and everyday battlefields. John McTiernan's remake of *Rollerball* (2002) and *Babylon A.D.* (2008) (both discussed in Chapter 5) are remarkably similar in their dystopian depictions of Russia/Kazakhstan as zones of wanton normlessness and the debasement of the human experience, veritable popular cultural analogs-cum-hyperbole to Kotkin's "Trashcanistan" frame. In the former, Kazakhstan is constructed as a unbounded and barren land of detritus, where only bloodthirsty criminals flourish while the "worker" is destined for a short and brutish life, punctuated only by the carnal pleasures of watching blood sport. In a pivotal (if bizarre and even laughable) scene,[12] the futility of "escaping" Kazakhstan is reluctantly realized by the protagonists, who are utterly unable to get out of this Central Asian republic controlled by nefarious enterprises. In *Babylon A.D.*, geography is a major factor in the narrative; in fact, degrees of longitude and latitude accompany many of the changes in scene, as do snippets of cinematic geovisualization which situate the action at specific points on the planet. The film jumps impossibly around Eurasia, including to Vladivostok, where an ancient Soviet submarine transports illegal migrants to western Canada. The train sequence is particularly evocative, beginning in Kazakhstan where post-Soviet masses, including inexplicably *niqab*-clad women, scramble aboard a maglev before it passes directly over a nuclear bomb site at the edge of an anonymous Siberian city. To all intents and purposes, post-Soviet Eurasia is one giant failed state (to be subsequently contrasted with a sleek and Tokyo-esque "America," glittering in its gauche opulence).

Providing an interactive imaginary to the broadcasted mediascape that is film, a variety of videogame creators have also set their narratives within terrorism-prone post-Soviet pandemoniums, including the fake countries of Novistrana in *Republic: The Revolution* (2003), Adjikistan in *SOCOM: U.S. Navy SEALs Combined Assault* (2006), and Arstotzka in *Papers, Please* (2013). However, other game developers have been bold enough to situate the action in "real" places, with Russia being one of the most popular choices, e.g., *Hitman 2: Silent Assassin*

(2002), *Delta Force: Xtreme* (2005), *Call of Duty 4: Modern Warfare* (2007), and *GoldenEye 007* (2007), as well as a variety of Tom Clancy-based games including *Politika* (1997), *Rainbow Six: Rogue Spear* (1999), *EndWar* (2008), and *Ghost Recon: Future Soldier* (2012). As Vít Šisler (2008) has argued, videogames' popular geopolitical exploitation of stereotypes and clichés has an even more acute effect on the game-player than other forms of popular culture given the "immersion factor" involved in the ludic environment, which inevitably creates virtual realities for the gamer through moments of emotional intensity (Shaw and Warf 2009). Unlike cinema, where the viewer is a passive participant in the narrative, digitized game worlds allow the spectant (agent) to engage with the spected (subject). Since the narrative arc unfolds based on contingencies related to the player's actions in the game world, players who lack real-world interactions with the post-Soviet East are more likely to assume highly subjective prejudices associated with the enemies and (hostile) geographies they encounter (see Ash and Gallacher 2011). As Russia and/or a generic post-Soviet realm (alongside the Middle East and Latin America) serve as regular settings for terrorist attacks, loose nukes, and seething iniquity, such popular-cultural artefacts have an important role to play in shaping geographical imagination in a world where more than one in ten Americans spends more than 22 hours per week immersed in online gaming (Siegal 2014).

The gray zone—A failed utopia filled with derelict ruins and anonymizing edifices

When asked to describe urban space in Russia, or the former USSR more generally, most Westerners are likely to comment on the country's grayness, a side effect of the rapid industrialization of the 1930s as well as further building projects undertaken decades later. Foreigners who do travel to Russia are quite likely to spend their time in St. Petersburg, strolling amid the colorful architecture, historic mansions, and the picturesque boulevards paralleling the iconic canal system that earned the city the sobriquet of "Venice of the North," even while continuing to trade in cinereous stereotypes about the country. Socialist urbanism seems to pervade popular associations more than almost any other association, a fact that owes much to sustained representations of Russian space via popular culture. Consequently, it is important to understand the affect produced by such imagery, whether it be in the form of annoyance, derision, or disgust.

In the James Bond film *Goldeneye* (1995),[13] a cornucopia of Cold War relics, from satellites to statues, litter the landscape of the film, which—like much of Western cinema—portrays Russia as graveyard locked in a dull gray winter. The optics of this motion picture are particularly fecund for geopolitical analysis: the prologue, set prior to the end of the USSR, begins at a Soviet chemical weapons facility, while the film's opening credits show a disintegrating host of Sovietesque icons (socialist art, crumbling statues, a lowering of the flag of the USSR, broken hammers and sickles, etc.). These settings remind the viewer of the recent past before the film goes on to sculpt a present where Russia could potentially strike at the heart of the West once again (effectively presenting post-Soviet geography itself as a sort of villain). In one scene, a map of Russia is shown, but—of course—still delimiting the older borders of the USSR (and displayed in characteristic crimson).[14] In a pivotal scene, James Bond (Pierce Brosnan) poignantly confronts his Russo-British adversary Alec Trevelyan (Sean Bean) in a graveyard of Soviet flotsam, a geovisual metaphor for the "death of the Soviet experiment" and the stillborn nature of the "new Russia," a country that is purportedly incapable of getting over its history.[15] This visual-scripting of the former Second World (with post-Soviet Russia as its gravitational center) as a colorless dystopian zone, bereft of either the historical gravitas of Europe or the crass vitality of North America, is nothing new. As Klaus Dodds (2003) has demonstrated, despite the decades separating *From Russia with Love* (1963) and *The World Is Not Enough* (1999), the Bondian narrative of "Russia" has remained remarkably stable. In fact, one of the earliest popular-cultural treatments of post-revolutionary Russia, *Tintin in the Land of the Soviets* (1930) constructed an eerily similar picture even as the Soviets were attempting to build their (doomed) socialist utopia (see Dunnett 2009).

Beyond film, a variety of other media platforms have made use of the gray expanse of the post-Soviet realm as their chosen setting. The WildStorm comic book *The Winter Men* (2005–2008) is a case in point. The series revolves around an attempt by the Soviet state to create a superhuman, focusing on a Russian policeman whose life exposes the vagaries of reforms and the ensuing chaos wrought by the transition from totalitarianism to market-capitalism in the 1990s and beyond. A positive review of the series lauded John Paul Leon's artwork as "perfect for bringing the cold, stark, and dingy quality of the backdrop to life" (*Fourth Rail* 2008). While the series deals with a variety of issues in contemporary Russia, the settings are universally bleak and oppressive, (geo-)graphically reproducing the Cold War-era, Tom Clancy-esque gestalt of "rundown gray buildings, bad cars, grim-faced people and lots of vodka" (Curran 2014). WildStorm also released *The Programme* (2008), which pits super-powered Americans and (post-)Soviets against one another. Created by Cold War-era experimentations based on Nazi dark science, these godlike creatures were intended for use in World War III, but instead begin to clash only after the end of the Soviet–US conflict. Writer Peter Milligan and artist C. P. Smith's Eurasian settings are highly predictable: subterranean gray zones underneath Red Square, a Siberian gulag, and a war-torn Central Asian hinterland (which is ultimately targeted for a tactical nuclear strike by the US).[16]

Such geographically belligerent and purposefully exoticized treatments of the former USSR occur in other comic series/graphic novels as well, from Marvel Comics' *Winter Guard* series to Mark Millar's reboot of Superman as an "alien immigrant" to Soviet Ukraine (rather than Kansas) in the critically acclaimed *Superman: Red Son* (2003). Sal Abbinanti's *Atomika* (2005–2011) and Christian Gossett's *The Red Star* (2003) similarly present Eurasian space as a hyper-urbanized constructivist nightmare-scape; however, both series add elements of Russian mythology and sorcery combined with high technology to the mix, thus deepening the techno-Orientalization of Russia (see Roh, Huang, and Niu 2015) by suggesting it as a realm where magic reigns (akin to India, China, or other mysterious Eastern lands). Regardless of the presence of the occult or not, all these graphic renderings of the region have something in common: colorlessness combined with anomie. Depicting a realm antithetical to humor, fun, and pleasure, they provide an interesting paradox to the laughter-inducing scenes discussed in the previous chapter;[17] however, it is important remember that the (Western) mirth associated with Borat, the Absurdis, and President for Life Trff Bmzklfrpz is a triumphalist laughter, not one of comradeship or childish joy.

Complicating this generalization is Tony Hawks' best-selling book (2002) and film adaptation (2012) *Playing the Moldovans at Tennis*. While the travelogue spends a great deal of time constructing Moldova (and especially the breakaway republic of Transnistria) as Europe's "Mos Eisley," i.e., a "wretched hive of scum and villainy" where almost anything can be traded, bought, or sold, the author is a sensitive observer of the people of the country. Hawks provides a passionate defense of their plight as denizens of Europe's poorest country, while conveying the congeniality of the Moldovan people. The book, though funny at times, does not produce the triumphalist laughter of other pop-culture treatments of the region; instead, humor often results from Hawks' cultural errors and personal failures to understand basic facts about how things are done in the former USSR. Before setting off on a dare to play the entire country's Olympic football team in individual tennis matches, Hawks admitted to a friend that he only knew two things about Moldova: "The wine is good and the brandy is better."

Perhaps the most dramatic manifestation of the semiotic "ruin porn" (see Vultee 2013; Pyzik 2013; Rann 2014) of the former USSR manifests in sport media, particularly premier international competitions such as the Olympic Games.[18] Coming as it did on the heels of the dissolution of the USSR, the 1992 Summer Olympics in Barcelona, Spain, established a powerful tradition for American sport journalism, as reporters and producers sought to convey the challenges of individual athletes competing for the combined Commonwealth of Independent States Olympic team, known as the Unified Team/Équipe Unifiée.[19] Considered by some historians as the final victory of the Soviet experiment, the Unified Team took the greatest number of medals at the games (112 in total, including 45 gold medals). Despite a commanding performance by its members, Western coverage of the team proved invariably pessimistic, framing each and every competitor in terms yoked to the demise of the Soviet Union. Faced with no shortage of negative nouns and adjectives for division, regret, and uncertainty, the *New York Times* essay on the "splintered future" of the team (and the geopolitical formation that birthed it) used terms such as "tragedy," "disintegration," "catastrophe," "loss," and "defunct," while simultaneously poking fun at the choice of beige as the color for the combined team (Erlanger 1992). In terms of visual presentations of post-Soviet sport infrastructure, NBC—the de facto broadcaster in the US for the Olympics since 1988—engaged in what can only be described as pitiless reveling in the "collapse" of the USSR (as opposed to a "love of the game" frame [see Blain 2012] which would have served NBC's audience just as well), even as the former Soviet athletes stood tall and proud above American sportsmen and women on the podia in Barcelona.

False seeings of dilapidated training facilities and maudlin recounting of personal tragedies associated with privatization and transition peppered, even dominated, the requisite backstory provided on each Russian, Ukrainian, or Kazakhstani competitor. Four years later when the US hosted the games in Atlanta, Georgia, the phenomenon only increased in its effect, with many such vignettes suggesting that the only way for a post-Soviet athlete to have a fighting chance in the Games was to relocate to the US for training (with the American state of Connecticut being branded the location of choice for this sporting diaspora, abandoned by the State). Bright and spacious American arenas were juxtaposed against gray, molding, and crumbling Soviet-era facilities in Kyiv, Almaty, and elsewhere. Such media coverage fulminated with victory, perpetuating the myth that the US (and its Western European allies) had "won" the Cold War and now were "responsible" for providing succor to the defeated, i.e., the men and women of the former Soviet Union (who were not typically shown on screen). Despite these conditions, the only identifiable moments of empathy came when an individual post-Soviet athlete was "rescued" by their relocation to the West in order to train, a common US reporting tactic that allows for the use of personalization to make the Other one of "us."[20] In many ways, this form of coverage continued what Dodds (2005) has identified as the integral role that popular geopolitics plays in the world of sport, specifically surrounding first-order ideological conflicts like the Cold War. Interestingly, there was little or no analogous coverage of the "problems" associated with the former Eastern Bloc sport powerhouses like the Romanians, Poles, and Hungarians (though the issue of Eastern German doping and its after-effects continued to be a popular theme in analysis of the combined German Olympic team's successes and failures). This represents a visible shift from Cold War-era coverage, in which all Eastern Bloc states were targeted for journalistic derision. Following their inclusion in the "West," these countries could be safely ignored, unlike Russia and other post-Soviet nations.

The most recent Winter Olympic Games, held in the southern Russian city of Sochi, only served to amplify the defamatory nature of American sport coverage, though one which shifted away from the "gray zone" geopolitical aesthetic of past games. Abetted by Snapchat, Imgur, Flickr, and countless other photo-sharing platforms, as well as Twitter, Facebook, and various social networking sites, US- and UK-based sport media organizations utilized citizen-journalists in the Black Sea resort to present a failed Olympic Games well before the torch arrived in Sochi (as well as dire predictions about the hostility that would meet LGBT athletes and visitors). Rather than reprising a stale narrative about broken-down gymnasiums, melting ice rinks, and bankrupt sporting teams (an improbable storyline for Russia, which at the time was riding high on its petroleum wealth), sportscasters instead presented a novel semiotic landscape, replete with "criminally" (Weiss 2014) unfinished infrastructure of an impossibly grandiose nature (a sort of ruins-in-waiting narrative), bizarre side-by-side toilets (Petchesky 2014), Cossack militiamen sadistically whipping Pussy Riot members (Walker 2014), yoghurt blockades (Kaplan 2014), wolves in elevators (Daily Mail 2014), and a panoply of other "epic fails" on the part of Russia's Olympic Committee.[21]

Despite a highly successful and, by most accounts, breathtaking opening cere-mony, American media coverage obsessed over a technical glitch that prevented the red or "American" (North and South) ring of the Olympic symbol from light-ing up, suggesting a purposeful geopolitical slight against the athletes from the New World, but particularly those from the United States. Bringing the media bias into clear focus, *The Moscow Times*—an English-language daily newspaper con-demned by the Russian press as virulently "anti-Putin" (Luhn 2015)—criticized Western media coverage of Sochi as "venomous . . . vicious, [and] often hypo-critical and extremely biased," aimed at "sabotaging" the Olympics (Lozansky 2014). As will be explored in the next chapter, Sochi represented the lengths to which Moscow was willing to go to brand Russia via geopolitical "imagineering" (Müller 2014, 629), while also reinforcing the country's exposure to the tender mercies of deeply ensconced pop-culture referents, highlighting the very tensions this book seeks to address.

In a partial defense of filmmakers, comic book writers and artists, sport jour-nalists, and other cultural producers, there are a host of market-based reasons for treating Russia and the post-Soviet republics in this way. Audiences expect to see such scenes when a narrative is situated in "Russia"; yet, we must ques-tion how much of this expectation is in fact the result of long-standing popu-lar cultural coding of Russianness. A poignant counter-example to this trend is the National Geographic series *Wild Russia* (originally produced by the German NDR Naturfilm/Studio Hamburg Doclights), which shows a colorful side of Russia where nature rules and Stalinist architecture is absent. The six-episode series explores Russia's "wildest" regions: Siberia, Kamchatka, the Arctic basin, the high Caucasus, the Pacific-Rim Primorye region, and the Ural Mountains. However, such "divergent" geographical excavations of Russia are few and far between in Western popular culture, given that, as Dijkink reminds us, "Each human being yearns for a kind of world order, a sensible pattern of people, things, and behavior in the world" (1996, 15). For representations of Russia to diverge from the "expected" would be a psychic trauma for many raised on the myth of the Soviet bogey, thus challenging the fundamental ideological landscapes and political imaginaries that reify the "West" (see Gregory 1994).

Ground Zero—An irradiated anti-world of mutants and monsters

Of the various thematic aesthetics, the portrayal of the post-Soviet realm as an apocalyptic zone of radiation, mutation, and environmental degradation is perhaps the most affective, aggressive, and politically charged. Operating in a represen-tational matrix of "good and evil," the West assumes the moral high ground vis-à-vis the former USSR, which—according to the geovisual narrative—tam-pered with forces beyond its control, resulting in the complete disruption of the "natural order" of things. With the disaster at Chernobyl and the subsequent aban-donment of Pripyat (see Figure 7.1) serving as the wellspring for such politicized geo-imaginaries, the Anglophone West is able assuage its own collective guilt for the nuclear horrors of Los Alamos, Hiroshima, Nagasaki, Mururoa, Maralinga, and Ekker (see Saunders 2014).[22] Beyond *Schadenfreude*, we might also consider other forms of affect associated with the radiated/mutated "East," from being placed in a state of anxiousness to actual panic. Given the propensity of the USSR to remake nature to serve the needs of the proletariat, perhaps such triumphalist criticism is not totally misplaced (a position with which many in the environmentalist movements of Estonia, Latvia, and the Russian Federation would agree). Yet, the ways in which Anglophone pop-culture producers (as well as Russian and Ukrainian videogame developers seeking ever larger markets for their own products) have sculpted post-Soviet space as a sort of "toxic Disneyland" (Pyzik 2013) is particularly belligerent in its tone and scope.

While there is no need to reprise the previous analyses of *Chernobyl Diaries* (2012) and *A Good Day to Die Hard* (2013), both films featured "Pripyat"[23] as an important site, with the former using the space as a backdrop for cannibalistic mutants to terrorize and then feast on Western visitors, whereas the latter employing Pripyat as a setting for unearthing the depths of Russia's pre- and post-1991 corruption (the film's villains were uranium smugglers operating in the mid-1980s when the accident at Chernobyl interrupted their lucrative business). In both films, the level of radioactivity is greatly exaggerated, reaffirming the widely held perceptions of Western audiences about the ecological threat posed by the purported failure of Soviet science.[24] However, the irony of the Chernobyl Exclusion Zone (CEZ) is that it has actually emerged as a "earthly paradise for wildlife" (Byshniou

Figure 7.1 Iconic ferris wheel in Pripyat, Ukraine (Author)

2006, 44), including local flora and fauna, as well as newly introduced species like the herds of Przewalski's horses (*Equus Przewalskii*). In fact, it was the flourishing of nature in the CEZ that led Israeli-American director Oren Peli to make *Chernobyl Diaries*. As the review of the film in *Sight & Sound* states:

> Peli apparently found his inspiration in film and photographs of the area's eerie buildings and infrastructure, as well as reports of a resurgence in local wildlife—lynx, wolves, boars, wild horses and elks—in the 18–mile exclusion zone surrounding it. (Hammond 2012, 53)

However, despite Peli's reverence for the return of the natural world, the film ultimately conveys the "exploitative, paranoid convictions" (Hammond 2012, 53) about Ukraine and, through its link to "Soviet" scientists, the whole of the former USSR. Similarly, the blurring of the Russian and Ukrainian borders in *A Good Day to Die Hard* conflates past and present, real and unreal, leaving the viewer thinking that an irradiated (Ukrainian) wasteland lies just at the outskirts of Moscow. Extending the "contaminated East" trope, a pivotal scene in the aforementioned *Babylon A.D.* (2008) involves the main characters traversing Siberia via a maglev train, which crosses directly over a nuclear-decimated cityscape. As the train approaches the city, its passengers are warned of the danger as prophylactic measures are enacted on the train to shield them from the effects of some long-past atomic blast. Par for the course, Russia is a site of the presumed nuclear war which girds *Babylon A.D.*'s dystopian world-space (as opposed to the seemingly untouched West, symbolized by a pristine and over-developed "New York" shown later in the film).

While these films are emblematic of how "dangerous space" has been constructed in the Western popular-culture imaginary since 1991, they are actually mild in their collective renderings when contrasted with other media, particularly videogames which have set their action in post-apocalyptic zones linked to the former USSR. Some of these products are produced within the former Soviet Union, but have become highly successful in North America and Europe, including the first-person shooter games *S.T.A.L.K.E.R.: Shadow of Chernobyl* (2007)[25] and *Metro 2033* (2010)/*Metro: Last Light* (2013).[26] The former, created by GSC Game World, based in Kyiv, Ukraine, is set in and around the CEZ, which is now affected by a second nuclear event, causing large-scale human, animal, and plant mutations of otherworldly proportions. The player's avatar is a "stalker," an illegal scavenger within the zone. The *Metro* series is based on Dmitry Glukhovsky's 2005 post-apocalyptic novel *Metro 2033*, and involves subterranean gameplay in a post-nuclear holocaust Moscow where aliens, mutants, neo-fascists, and neo-Bolsheviks bloodily vie for power. Like the *S.T.A.L.K.E.R.* series, these games were developed in Ukraine, but enjoy global distribution. A number of other videogames, many originating in the Anglophone West, also make use of Eurasian space for apocalyptic scenarios, including *Dark Sector* (2008), with its action set in the fictional post-Soviet republic of Lasria where Cold War-era gulag experimentations have wrought a biological menace that created zombie-like creatures.

Such aesthetic worlds, where the action is fast, killing is easy, and dangers abound, lay bare the ease with which Eurasia serves as a handy geovisual sandbox for Anglophone cultural production.

Putting popular-culture codes and visions to work: Tabloid realism and the post-Soviet frame

One might well ask: Why is all this important? The answer relates to international relations, and particularly how democratic publics conceive, interpret, and act on their geographical imaginations, as this process is "central" to the formulation and conduct of foreign policy in the twenty-first century (Debrix 2008, 9). As Nicholas Kiersey and Iver Neumann note, "Popular culture shapes how constituencies understand the world. Since public worldviews are one of the factors constraining what politicians can do and at what cost, the popular cultural artefacts that contribute to shaping them are indirectly important to political outcomes" (2015, 79). By framing post-Soviet space as a realm of uncanny danger producing affects of revulsion, policymakers are able to manipulate fear to achieve a variety of geopolitical outcomes, from NATO expansion to economic sanctions to popular support for foreign opposition movements. Through popular culture and mass media, citizenries formulate their views of the outside world and establish their political attitudes towards other polities, categorizing them as friends, enemies, and generic (typically unimportant) Others.

> Of necessity, "imaginative geographies" have recourse to all sorts of popular and tabloid media and styles of presentations in order to facilitate the commonsensical cultural deployment of those political geographies that ultimately seek to "locate, oppose, and cast out". (Gregory qtd. in Debrix 2004, 43)

Given the repetitiveness of geographical staging of the former USSR as a frozen, crime-ridden, gray, and/or irradiated wasteland in popular culture, it is then no surprise how easily post-Soviet countries are negatively rendered in tabloid media, from evening news broadcasts about protests in Moscow to foreign policy treatises on Central Asia. As Iver Neumann states, "Imagination and exclusion are two sides of the same coin" (1999, 37); for the post-Soviet East, this could not be more true.

Since the Bolshevik Revolution, the Anglophone affective mapping of northern Eurasia has been one based on fear, loathing, and abjection.[27] As discussed earlier in the text, the notion of the "mad" Russian—drunk, scheming, and violent—pervades the popular consciousness, later enhanced by the Cold War-era branding of the "utopian, self-serving deceitful, and paranoid" Soviet (Shaw 2006, 90), producing an accretive gestalt that has been recently alloyed with the "goofy" *Homo post-Sovieticus* frame. But beyond this geopolitical individuation of national identity, there is something deeper: a (sub)conscious fear of the enormity of Russian space. For Britons, Americans, and Canadians, the sheer size of the country is often intimidating, and as such there is a seemingly natural reticence to engage with Eurasian space on a neutral footing. Instead, if we accept that popular culture provides us with "maps of meaning" or "ways to navigate the world" (da Costa 2004, 191) then we can begin to conceive of how great the Anglophone West's collective insecurities are, namely that Eurasian space is something to be shunned.

This pop-culture-fueled orientation manifests in tabloid media coverage of most news events emanating from post-Soviet space, while also presenting in more academic geopolitical analyses of the region. In keeping with Kotkin's Trashcanistan modus, statesman and political pundit Zbigniew Brzezinski represents the vanguard. While he is a throaty defender of Ukraine as the "true heartland" of European values, his characterization of the rest of the former Soviet republics borders on the malicious. In his 1997 book *The Grand Chessboard*, he predicted a complete collapse of Central Asia into a morass of feuding ethnicities and Islamist satrapies (20 years on, nothing of his "shatterbelt" thesis has come to pass). Moreover, the political commentator rails against Russia at every turn, attempting to provoke the American public and politicians into a new war footing with the Federation, an orientation that has made him quite popular on the pundit circle since Russia's annexation of Crimea. Robert D. Kaplan is no kinder in his travelogue-cum-geopolitical exposé *Eastward to Tartary: Travels in the Balkans, the Middle East, and the Caucasus* (2000), as the popular geographer re-Orientalizes the previously "European" states of Azerbaijan and Georgia, sketching out a darkened landscape where warlords are presidents, chaos reigns, and danger lurks in every corner. This text builds on Kaplan's (in)famous essay "The Coming Anarchy" (1994), a paragon of geopolitical fear-mongering in which the South Caucasus played a key role. Even the "new generation" of geopoliticians suffer from the anti-Eurasian frame. Parag Khanna's *The Second World: Empires and Influence in the New Global Order* (2008) sculpts post-Soviet Eurasia as an anarchic mix of luxuriant wealth and abject poverty, an *Absurdistan*-like meeting place of Fifth Avenue opulence and Lagos-slum hardship, where greed has resulted in Moscow "leasing" out its vast domain to the highest bidder (usually Chinese).

While these are just three examples of the tabloidization of geopolitical media that has characterized the post-1991 milieu, they are fairly representative of the ways in which Eurasian space is represented. As Debrix argues:

> In the tabloid media, reality must be described and truth must be revealed in a flashy, surprising, gripping, shocking, often moralizing, and sometimes anxiety-spreading manner. The reality of tabloid realism is a sensational one. But the tabloid narrative must also be made accessible to a large amount of people. It must use images and languages that can be readily understood and easily recognized by the vast majority of "Americans." (2004, 152)

For this so-called "average American," raised on a steady diet of action films where Russia goes "boom," NBC's Olympic Russia-bashing, and Chernobyl-based videogames, there is a rather easy co-constitution of norms of statecraft with the outcomes of stagecraft (see Dodds 2015). When Russia, Ukraine, or another post-Soviet republic is featured on the nightly news (or referenced in the monologue of a late-night talk show host, *The Daily Show*, or some other "fake news" outlet), well-hewn geopolitical codes and visions are at the ready. By tapping into a reservoir of mythical aesthetics and affective worlds rooted in forbidding icy taiga, soulless gray buildings, crime-infested capitals, and burnt-out nuclear reactors, citizens in the Anglophone West "know" how to the receive the ensuing information, whether it be about the latest member of Pussy Riot to be jailed, an international dispute over natural gas, a border skirmish, or an assassinated opposition politician.

Hitherto, the focus of the book has been mostly on how those outside the Eurasian realm represent its content via semiotic landscapes, the politics of aesthetics, emotion-laden staging of space, imaginative geo-graphing, and ideological matrixes linked to visual and textual clues. In the coming chapter, we turn to the 15 post-Soviet republics' efforts at representing their own people, places, and spaces (as well as products, histories, ideas, and "spirits"). In some cases, these efforts will be direct responses to the popular culture discussed up until now. In other instances, such branding will function independently from such mediatic projections, while in a few cases there will actually be synthesis (if not symbiosis) of branding and pop-culture. As the world becomes increasingly connected by new media, there is something like a pop-culture commonwealth emerging that makes it impossible to ignore Hollywood (and its various adjuncts) when seeking to create, manipulate, or burnish the national image. As a major player in the field notes, "Almost every nation and culture on earth is now sharing the elbow-room in a single information space. No conversation is private any longer, no media is domestic, and the audience is always global" (Anholt 2007, 52). In the following pages, I will interrogate how nation branders operate in this protean realm which is barraged by foreign images, affective mapping, and geopoliticized representation.

Notes

1 Adapting Paletz and Schmid's (1992) assessment of the reality between the real and the imagined, I contend that "reality" is constituted by three realms: the objective world (actual); the symbolic world (presented); and the subjective world (perceived).
2 Appadurai's (1996) groundbreaking reification of other "scapes" (ethno-, media-, techno-, etc.) only serves to underscore how fissiparous our accepted notions of reality actually are, including the notion that landscapes are, at their root, acts of geopoliticized imagination.
3 Ruin porn, also known as ruin photography, is a relatively novel art form wherein the residuum of "once-great" acts of human ingenuity that have fallen into disrepair (or have been completely given over to the forces of time and effects of nature) are visually represented. Post-industrial urban spaces tend to be the most common source material for these "highly politicized" exercises in the aestheticization of abandonment (Mullins 2012).
4 Reflecting how deeply embedded such filmic stereotypes of Russia are, the Cinetrain international documentary film, in an effort to get beyond such "geopolitically frozen" clichés, sent 24 young filmmakers from 15 countries across Russia, stopping at St. Petersburg, Murmansk, Moscow, and Tomsk in an effort to film "short documentary

pieces about Russian stereotypes—snow, ice, vodka, colossal landscapes, beautiful women, and Lada cars" (Kvasha 2013).

5 Nearly everything in Russia is depicted in varying shades of gray and icy white, especially those scenes in the preposterously named city of "Severnaya" ("Northern"), a secret weapons site "somewhere" in Siberia.

6 Harking back to pre-Soviet Russia, Hellboy—a demon conjured by and rescued from the (Nazi) Thule Society—takes on the Russian mystic Rasputin, who hopes to bring about a dark paradise on Earth by opening a portal to a Lovecraftian dimension filled with unspeakable horrors. Much of the film is set in post-Soviet space, including a site in Moldova where a blood-sacrifice is made to revivify the "mad monk" many decades after his death (in a bit of geographical legerdemain, the "Burgó Pass" or Pasul Tihuţa made famous by Bram Stoker [1897], is relocated from Romania to the post-Soviet state of Moldova).

7 Even cinematic representations of pre-Soviet Russia from the post-1991 period engage in such renderings of Eurasian space, including the farcical climax of *The League of Extraordinary Gentlemen* (2003), which is set in a late-nineteenth-century techno-Orientalist phantasmagoria on the Amur River. Like *Hellboy*, this film is an adaptation of a popular comic book series, and provides further testament to how such tropes flow across a variety of mass media.

8 Extending the Russian stereotype, one of the film's main antagonists, Carlos (Eduardo Noriega), is using matryoshka dolls—the "most globally recognized symbol of traditional Russian national identity" (Williams 2012, p. 1)—to transport drugs from the Russian Far East to Moscow. When he tries to force himself on the female protagonist Jessie (Emily Mortimer) in the snowy wilderness near an Orthodox church, she kills him with a fencepost, leading to a series of ever more dangerous (and snow-laden) encounters with Russian security personnel.

9 In addition to his fictional accounts of post-Soviet space, Theroux has also worked on *The State of Russia* series for British broadcaster More4, detailing a number of problems in the country including demographic challenges, the AIDS epidemic, and the plight of the displaced Meskhetian Turks.

10 Even from within the group, threats abound as the company's sole Eastern European, Miljan (Miljan Milosevic), a DJ from Podgorica, Montenegro, emerges as a paranoiac with potentially murderous tendencies, thus casually conflating all "Eastern" dangers into one undifferentiated whole.

11 I do not mean to suggest that geopolitical thrillers should be bereft of car chases, explosions, and gun battles in the street. Certainly, the Bourne series has made a habit of trashing cities across Europe and even farther afield (Morocco, India, etc.). However, in the case of Russia's urban space, the Western cinema-goer is rarely subjected to any other imagescapes than those involving rampant destruction. Moreover, the miniscule penetration of Russian films in the US and UK marketplaces does little to balance the "Hollywood" representation of Moscow as a warzone.

12 Several minutes of the "escape scene" are shot, somewhat ridiculously, through night vision goggles, an effect that nearly every review of the film lampooned. The resulting cinematography seems almost juvenile, and only enhanced the ire of the film's many negative critiques. To put the critical reception in context, Rollerball received a 3 percent rating on the popular motion picture review website *Rotten Tomatoes*; see www.rottentomatoes.com/m/rollerball/.

13 The plot revolves around a conniving general, Arkady Ourumov (Gottfried John), who is positioned as the "Next Iron Man of Russia," thus linking the narrative to the revanchist trope discussed in Chapter 5.

14 This is a curious inversion of the politically staged cognitive mapping in *X-Men: First Class* (2011), which shows a map of the USSR in 1962 anachronistically labelled "Russia."

15 The conceit of the film is that a trusted member of MI5 is in fact a Russian fanatic who hid his Cossack identity to infiltrate the service and avenge his kin slaughtered

by Stalin following the (Yalta Agreement-sanctioned) repatriation of Soviet nationals by British forces after WWII. Bringing together the pre- and post-Cold War concerns about Russia, as well as notions of "infiltration" by agents of the "dangerous East," *Goldeneye* employs Russian villainy in a rather complex way, and one which was likely lost on many viewers.

16 In a bit of sublime irony, one of the (post-)Soviet super-beings, the gigantic mute Stalingrad, laments the lackluster feeling that destroying Las Vegas gives him. His comrade-in-arms, Revolution (who resembles a Sovietesque Audrey Hepburn), tries to convince the team to head to California for their next act of ideologically programmed annihilation, stating: "We should move on to Los Angeles. Burn Hollywood. Hollywood is worth destroying" (Milligan and Smith 2008, 12). Rather than targeting the political or economic centers of the US (i.e., Washington and New York), these rampaging "Reds" go after consumerist and image-making meccas.

17 Looking at issues of gender and sexuality, Western popular culture tends to draw a sharp line between sexualized subjects *within* post-Soviet space and *Homo post-Sovietici outside* of it. For those still in the "badlands," sex is off the table as life is too dreary; for those who travel to the "West," sex is everywhere and thus these subjects assume hyper-sexualized states of being. Such representations interestingly replicate Soviet-era media depictions of "too much sex" in the West and "no sex in the Soviet Union" (RT 2015). For a more in-depth analysis of the gender dynamics of the sexualized female post-Soviet subject, see Kimberly A. Williams' *Imagining Russia: Making Feminist Sense of American Nationalism in U.S.–Russian Relations* (2012).

18 Rebecca Litchfield's haunting photo journal *Soviet Ghosts: The Soviet Union Abandoned: A Communist Empire in Decay* (2014) brings the Western desire to gaze at its own "victory" into sharp focus, particularly given the text's promotional material, which suggests that the author risked life and limb to expose what the local "authorities" did not want you to see (Piepenbring 2014), i.e., the shame associated with the "loss" of the Cold War and the ultimate failure of the Soviet experiment.

19 The Baltic republics of Estonia, Latvia, and Lithuania each sent separate contingents, reflecting their common position that each of these countries was never legally part of the USSR (in fact, none of these opted for any post-1991 affiliation with the Russian Federation or CIS). Lithuania's participation in the games was notable for its basketball team's sponsorship from the American psychedelic jam band The Grateful Dead (see Siegel 2012). Georgia, though not an official member of the CIS at the time, participated as part of the Unified Team.

20 In his analysis of media biases that actually matter, W. Lance Bennett identifies personalization as the "tendency to downplay the big social, economic, or political picture in favor of the human trials, tragedies, and triumphs that sit at the surface of events" (2002, 45).

21 Notably, the last of these was a prank perpetrated by late-night comedian Jimmy Kimmel; however, at the time of its airing, it was plausible within a larger arc of events positioning Sochi as a complete and utter disaster.

22 In the category of such pop-culture responses to the 1986 events at the Chernobyl Nuclear Power Plant, one should consider not only visual representations of space, but also aural ones, including the pop songs David Bowie's "Time Will Crawl" (1987), Tim Dennehy and Christy Moore's "Farewell to Pripyat" (1999), and the metal band Neurosis's "The Horror of Chernobyl" (2011). A number of other lesser-known bands have also written tracks on Chernobyl/Pripyat, including CiLiCe, Municipal Waste, Huns & Dr Beeker, Ash, Zavod, and Cursed Sails.

23 The *Die Hard* scenes for Pripyat were filmed in Kiskun, Hungary, while *Chernobyl Diaries* was filmed in an abandoned Soviet Air Force base in Kiskunlacháza, Hungary.

24 By way of example, I would offer a personal anecdote. On a visit to the site of the Chernobyl disaster, the highest recorded radiation measures I experienced were less than one-half of the level of radiation I exposed myself to by taking the transatlantic

flight to get the country. Yet in every instance where I spoke about my trip to the defunct nuclear reactor and its immediate environs, I was queried by my (typically well-traveled) co-nationals with concerns for my safety. While thoroughly unscientific, I attribute this paranoia to the impact of popular-culture treatments of the region post-1986.

25 Two subsequent releases include *S.T.A.L.K.E.R.: Clear Sky* (2008) and *S.T.A.L.K.E.R.: Call of Pripyat* (2010). The series borrows several concepts from Soviet popular-cultural traditions, including Boris and Arkady Strugatsky's sci-fi novella *Roadside Picnic* (*Piknik na obochine*, 1971) and Andrei Tarkovsky's film *Stalker* (1979) (see Sokolova 2012).

26 Theroux's aforementioned *Far North* also centers on "stalking" for relics in an irradiated Siberian city, Polyn 66, with a sly reference to Chernobyl, as the author notes that before entering the deadly "Zone" in a forced search for booty: "[T]he mood in the camp that night was sour as wormwood" (2010, 192). "Wormwood" is the English translation of Chernobyl.

27 In Great Britain, this phenomenon can be seen to begin from an even earlier date, given the jingoism associated with the Great Game and Russo-British rivalries in the eastern Mediterranean (see Chapters 1 and 3).

References

Acuna, Kirsten. 2012. "Check Out 15 Movies In Which New York City Gets Destroyed." *Business Insider*, available at www.businessinsider.com/15-movies-where-new-york-city-gets-destroyed-2012-4?op=1 [last accessed 2 July 2015].

Aitken, Stuart, and Leo E. Zonn. 1994. "Re-Presenting the Place Pastiche." In *Place, Power, Situation, and Spectacle: A Geography of Film*, edited by Stuart Aitken and Leo E. Zonn, 3–25. Lanham, MD: Rowman & Littlefield.

Anderson, Kay, and Susan J. Smith. 2001. "Editorial: Emotional Geographies." *Transactions of the Institute of British Geographers* no. 26:7–10.

Anholt, Simon. 2007. *Competitive Identity: The New Brand Management for Nations, Cities and Regions*. Houndsmills, UK: Palgrave Macmillan.

Appadurai, Arjun. 1996. *Modernity at Large: Cultural Dimensions of Globalization*. Minneapolis: University of Minnesota Press.

Ash, James, and Lesley Anne Gallacher. 2011. "Cultural Geography and Videogames." *Geography Compass* no. 5 (6):351–368.

Bennett, Lance W. 2002. *News: The Politics of Illusion*. New York: Longman.

Berberich, Christine, Neil Campbell, and Robert Hudson. 2013. "Affective Landscapes: An Introduction." *Cultural Politics* no. 9 (3):313–322.

Blain, Neil. 2012. "Beyond 'Media Culture': Sport as Dispersed Symbolic Activity." In *Sport, Media, Culture: Global and Local Dimensions*, edited by Alina Bernstein and Neil Blain, 227–254. London: Routledge.

Bleiker, Roland. 2012. *Aesthetics and World Politics*. New York: Palgrave Macmillan.

——. 2015. "Visual Assemblages: From Causality to Conditions of Possibility." In *Reassembling International Theory: Assemblage Thinking and International Relations*, edited by Michele Acuto and Simon Curtis, 75–81. Houndsmills, UK: Palgrave Macmillan.

Bogost, Ian. 2010. *Persuasive Games: The Expressive Power of Videogames*. Cambridge, MA: The MIT Press.

Bondi, Liz. 2005. "Making Connections and Thinking through Emotion: Between Geography and Psychotherapy." *Transactions of the Institute of British Geographers* no. 30:433–448.

Brill Olcott, Martha. 1996. *Central Asia's New States: Independence, Foreign Policy, and Regional Security*. Washington, DC: United States Institute of Peace Press.

Brzezinski, Zbigniew. 1997. *The Grand Chessboard: American Primacy and Its Geostrategic Imperatives*. New York: Basic Books.

Byshniou, Ihar. 2006. "Forest Diary." *Index on Censorship* no. 35 (2):44–53.

Cosgrove, Denis E. 2008. *Geography and Vision: Seeing, Imagining and Representing the World*. London: I. B. Tauris.

Cosgrove, Denis E., and Veronica della Dora. 2005. "Mapping Global War: Los Angeles, the Pacific, and Charles Owens's Pictorial Cartography." *Annals of the Association of American Geographers* no. 95 (2):373–390.

Craine, James, and Stuart C. Aitken. 2011. "The Emotional Life of Maps and Other Visual Geographies." In *Rethinking Maps: New Frontiers in Cartographic Theory*, edited by Martin Dodge, Rob Kitchin, and Chris Perkins, 149–167. London: Routledge.

Curran, Bill. 2014. "Visit to Russia Erases Stereotypes." *Buffalo News*, available at www.buffalonews.com/opinion/my-view/bill-curran-visit-to-russia-erases-stereotypes-20141106 [last accessed 2 July 2015].

Cusack, Tricia, and Sighle Bhreathnach-Lynch. 2003. *Art, Nation and Gender: Ethnic Landscapes, Myths and Mother-figures*. Farnham, UK: Ashgate.

da Costa, Maria. 2004. "Cinematic Cities: Researching Films as Geographic Texts." In *Cultural Geography in Practice*, edited by Alison Blunt, Pyrs Gruffudd, Jon May, Miles Ogborn, and David Pinder, 191–201. Oxford: Oxford University Press.

Daily Mail. 2014. "Comedian Who Cried Wolf." *Daily Mail*, available at www.dailymail.co.uk/news/article-2564317/Comedian-cried-wolf-Jimmy-Kimmel-teams-Olympic-luger-hoax-video-showing-animal-prowling-Sochi-hotel.html [last accessed 2 July 2015].

Debrix, François. 2004. "Tabloid Realism and the Revival of American Security Culture." In *11 September and Its Aftermath: The Geopolitics of Terror*, edited by Stanley D. Brunn, 151–190. London and Portland, OR: Frank Cass.

——. 2008. *Tabloid Terror: War, Culture, and Geopolitics*. London and New York: Routledge.

Dijkink, Gertjan. 1996. *National Identity and Geopolitical Visions: Maps of Pride and Pain*. New York and London: Routledge.

Dixon, Deborah, and Leo Zonn. 2005. "Confronting the Geopolitical Aesthetic: Fredric Jameson, *The Perfumed Nightmare* and the Perilous Place of Third Cinema." *Geopolitics* no. 10 (2):290–315.

Dodds, Klaus. 2003. "License to Stereotype: Popular Geopolitics, James Bond and the Spectre of Balkanism." *Geopolitics* no. 8 (2):125–156.

——. 2005. *Global Geopolitics: A Critical Introduction*. Harlow, UK: Prentice Hall.

——. 2007. "Steve Bell's Eye: Cartoons, Geopolitics and the Visualization of the 'War on Terror'." *Security Dialogue* no. 38:157–177.

——. 2011. "Gender, Geopolitics, and Surveillance in *The Bourne Ultimatum*." *Geographical Review* no. 101 (1):88–105.

——. 2015. "Popular Geopolitics and the 'War on Terror'." In *Popular Culture and World Politics: Theories, Methods, Pedagogies*, edited by Federica Caso and Caitlin Hamilton, 51–61. Bristol, UK: E-International Relations.

Dunnett, Oliver. 2009. "Identity and Geopolitics in Hergé's *Adventures of Tintin*." *Social & Cultural Geography* no. 10 (5):583–599.

Dyer, Geoff. 2012. *Zona: A Book About a Film About a Journey to a Room*. New York: Knopf Doubleday.

Entman, Robert M. 1996. "Reporting Environmental Policy Debate: The Real Media Biases." *Harvard International Journal of Press/Politics* no. 1 (3):77–92.

Erlanger, Steven. 1992. "Unified Team Faces Splintered Future." *The New York Times*, available at www.nytimes.com/1992/07/19/sports/olympics-unified-team-faces-splintered-future.html [last accessed 3 July 2015].

Fourth Rail. 2008. *"The Winter Men #1* Review." *Fourth Rail*, available at www.thefourthrail.com/reviews/critiques/080805/wintermen1.shtml [last accessed 4 July 2010].

Glukhovsky, Dmitry. 2005. *Metro 2033*. Moscow: Eksmo.

Goldsworthy, Vesna. 1998. *Inventing Ruritania: The Imperialism of the Imagination*. Bury St. Edmunds, UK: Edmundsbury Press.

Grad, Shelby. 2015. "Ranking Movies That Destroy Los Angeles: Did 'San Andreas' Do It Best?" *L.A. Times*, available at www.latimes.com/local/california/la-me-hollywood-disasters-20150529-story.html [last accessed 2 July 2015].

Grayson, Kyle. 2014. "Encounters Through the Cooking Glass: Geopolitics and Aesthetic Subjects in *Breaking Bad*." Paper presented at Interdisciplinarity of Popular Geopolitics: Popular Culture and the Making of Space and Place seminar, University of Leeds (22 October).

Gregory, Derek. 1994. *Geographical Imaginations*. Cambridge, MA: Blackwell.

Hall, Jacob. 2013. "A Brief History of Cinematic Assaults on the White House." *Screen Crush*, available at http://screencrush.com/cinematic-white-house-attacks/ [last accessed 2 July 2015].

Hammond, Wally. 2012. "The Chernobyl Diaries." *Sight & Sound* no. 22 (8):53–53.

Hawks, Tony. 2002. *Playing the Moldovans at Tennis*. New York: St. Martin's Griffin.

Höglund, Johan. 2014. *The American Imperial Gothic: Popular Culture, Empire, Violence*. Farnham, UK and Burlington, VT: Ashgate Publishing.

Hopkins, Jeff. 1994. "A Mapping of Cinematic Place: Icons, Ideology, and the Power of (Mis)representation." In *Place, Power, Situation, and Spectacle: A Geography of Film*, edited by Stuart Aitken and Leo E. Zonn, 47–66. Lanham, MD: Rowman & Littlefield.

Inglis, Fred. 1977. "Nation and Community: A Landscape and Its Morality." *The Sociological Review* no. 25 (3):489–514.

Jameson, Frederic. 1995. *The Geopolitical Aesthetic: Cinema and Space in the World System*. Bloomington and Indianapolis: Indiana University Press.

Jay, Martin. 1988. "Scopic Regimes of Modernity." In *Vision and Visuality*, edited by Hal Foster, 3–23. Seattle, WA: Bay Press.

Kaplan, Robert D. 1994. "The Coming Anarchy." *Atlantic Monthly* no. 273 (2):44–77.

——. 2000. *Eastward to Tartary: Travels in the Balkans, the Middle East, and the Caucasus*. New York: Random House.

Kaplan, Thomas. 2014. "Russia Blocks Yogurt Bound for U.S. Athletes." *The New York Times*, available at www.nytimes.com/2014/02/06/nyregion/russia-blocking-a-yogurt-shipment-from-reaching-us-olympians.html?_r=0 [last accessed 2 July 2015].

Karny, Yo'av. 2001. *Highlanders: A Journey to the Caucasus in Quest of Memory*. New York: Farrar, Straus and Giroux.

Khanna, Parag. 2008. *The Second World: Empires and Influence in the New Global Order*. New York: Harcourt, Brace and Company.

Kiersey, Nicholas J., and Iver B. Neumann. 2015. "Worlds of Our Making in Science Fiction and International Relations." In *Popular Culture and World Politics: Theories, Methods, Pedagogies*, edited by Federica Caso and Caitlin Hamilton, 74–82. Bristol, UK: E-International Relations.

King, Charles. 2000. *The Moldovans: Romania, Russia, and the Politics of Culture*. Stanford, CA: Hoover Press.

Kotkin, Stephen. 2002. "Trashcanistan." *New Republic* no. 226 (14):26–38.

Kunczik, Michael. 1997. *Images of Nations in International Public Relations*. Mahwah, NJ: Lawrence Erlbaum Associates.

Kvasha, Semyon. 2013. "Foreign Filmmakers Catch Russian Stereotypes." *Russia Beyond the Headlines*, available at http://rbth.com/articles/2013/01/15/foreign_cinematographers_will_film_russia_stereotypes_21899.html [last accessed 20 December 2014].

Leerssen, Joep. 2007. "Imagology: History and Method." In *Imagology: The Cultural Construction of National Characters—A Critical Survey*, edited by Joep Leerssen and Manfred Beller, 17–32. Amsterdam and New York: Rodopi.

Litchfield, Rebecca. 2014. *Soviet Ghosts: The Soviet Union Abandoned: A Communist Empire in Decay*. London: Carpet Bombing Culture.

Lozansky, Edward. 2014. "Don't Rain on Putin's Olympic Parade." *The Moscow Times*, available at www.themoscowtimes.com/opinion/article/dont-rain-on-putins-olympic-parade/493807.html [last accessed 4 July 2015].

Luhn, Alec. 2015. "Hackers Target Russian Newspaper Site Accused of Being Anti-Putin." *The Guardian*, available at www.theguardian.com/world/2015/feb/05/russia-moscow-times-cyber-attack [last accessed 4 July 2015].

Milligan, Peter, and C. P. Smith. 2008. "The Stars and Stripes." *The Programme* no. 7:1–30.

Müller, Martin. 2014. "After Sochi 2014: Costs and Impacts of Russia's Olympic Games." *Eurasian Geography and Economics* no. 55 (6):628–655.

Mullins, Paul. 2012. "The Politics and Archaeology of 'Ruin Porn'." *Archaelogy and Material Culture*, available at http://paulmullins.wordpress.com/2012/08/19/the-politics-and-archaeology-of-ruin-porn/ [last accessed 4 August 2015].

Neumann, Iver B. 1999. *Uses of the Other. The "East" in European Identity Formation*. Minneapolis: University of Minnesota Press.

O'Sullivan, John. 2001. "The Aesthetics of Affect: Thinking Art beyond Representation." *Angelaki: Journal of the Theoretical Humanities* no. 6:125–136.

Ó Tuathail, Gearóid. 2000. *Critical Geopolitics*. Abingdon, UK: Taylor & Francis.

Ogao, Patrick J. 2002. *Exploratory Visualization of Temporal Geospatial Data Using Animation*. Enschede, Netherlands: International Institute for Aerospace Survey and Earth Sciences.

Olwig, Kenneth. 2002. *Landscape, Nature, and the Body Politic: From Britain's Renaissance to America's New World*. Madison: University of Wisconsin Press.

Pabst, Naomi. 2006. "'Mama, I'm Walking to Canada:' Black Geopolitics and Invisible Empires." In *Globalization and Race: Transformations in the Cultural Production of Blackness*, edited by Kamari Maxine Clarke and Deborah A. Thomas, 112–131. Durham, NC: Duke University Press.

Paletz, David L., and Alex P. Schmid. 1992. *Terrorism and the Media*. Newbury Park, CA: SAGE.

Pavlovskaya, Anna. 2013. "Western Stereotypes of Russia: Historical Tradition and Current Situation." *Elektronnii zhurnal Rossia Zapad: dialog kul'tur* no. 4, available at www.regionalstudies.ru/journal/homejornal/rubric/2012-11-02-22-03-27/312-anna-pavlovskaya-qwestern-stereotypes.html [last accessed 1 July 2015].

Petchesky, Barry. 2014. "A User's Guide To The Bizarre Toilets Of Sochi." *Deadspin*, available at http://deadspin.com/a-users-guide-to-the-bizarre-toilets-of-sochi-1516518904 [last accessed 3 July 2015].

Piepenbring, Dan. 2014. "Soviet Ghosts." *Paris Review*, available at www.theparisreview.org/blog/category/look/ [last accessed 4 July 2015].

Pile, Steve. 2010. "Emotions and Affect in Recent Human Geography." *Transactions of the Institute of British Geographers* no. 35:5–20.

Pisters, Patricia. 2003. *The Matrix of Visual Culture: Working with Deleuze in Film Theory*. Stanford, CA: Stanford University Press.

Pyzik, Agata. 2013. "Toxic Ruins: The Political and Economic Cost of 'Ruin Porn'." *Australian Design Review*, available at www.australiandesignreview.com/features/29607-toxic-ruins-the-political-economic-cost-of-ruin [last accessed 3 July 2015].

Rancière, Jacques. 2004. *The Politics of Aesthetics*. London and New York: Continuum.

Rann, Jamie. 2014. "Beauty and the East: Allure and Exploitation in Post-Soviet Ruin Photography." *The Calvert Journal*, available at http://calvertjournal.com/features/show/2950/russian-ruins-photography#.VZbC3EYfPNg [last accessed 3 July 2015].

Riis, Ole, and Linda Woodhead. 2010. *A Sociology of Religious Emotion*. Oxford and New York: Oxford University Press.

Robinson, Nick. 2012. "Videogames, Persuasion and the War on Terror: Escaping or Embedding the Military-Entertainment Complex?" *Political Studies* no. 60 (3):504–522.

Rogoff, Irina. 2000. *Terra Infirma: Geography's Visual Culture*. London: Routledge.

Roh, David S., Betsy Huang, and Greta A. Niu. 2015. *Techno-Orientalism: Imagining Asia in Speculative Fiction, History, and Media*. New Brunswick, NJ: Rutgers.

Rose, Gillian. 2001. *Visual Methodologies: An Introduction to the Interpretation of Visual Materials*. Thousand Oaks, CA: SAGE.

RT. 2015. "On This Day: Russia in a Click (17 July)." *RT Russiapedia*, available at http://russiapedia.rt.com/on-this-day/july-17/ [last accessed 30 December 2015].

Ryan, Terre. 2011. *This Ecstatic Nation: The American Landscape and the Aesthetics of Patriotism*. Amherst: University of Massachusetts Press.

Sage, Daniel. 2008. "Framing Space: A Popular Geopolitics of American Manifest Destiny in Outer Space." *Geopolitics* no. 13:27–53.

Said, Edward. 1979. *Orientalism*. New York: Vintage Books.

Salter, Mark B. 2011. "The Geographical Imaginations of Video Games: Diplomacy, Civilization, *America's Army* and *Grand Theft Auto IV*." *Geopolitics* no. 16 (2):359–388.

Saunders, Robert A. 2014. "The Short Life and Slow Death of Captain Euro: Popular Geopolitics and the Pitfalls of the Generic European Superhero." *Aether: The Journal of Media Geography* no. 12:1–22.

Schopp, Andrew, and Matthew B. Hill. 2009. *The War on Terror and American Popular Culture: September 11 and Beyond*. Madison and Teaneck, NJ: Fairleigh Dickinson Press.

Shapiro, Michael J. 2008. *Cinematic Geopolitics*. Abingdon, UK: Taylor & Francis.

Sharp, Joanne P. 2000. *Condensing the Cold War: Reader's Digest and American Identity*. Minneapolis: University of Minnesota Press.

Shaw, Ian G. R., and Barney Warf. 2009. "Worlds of Affect: Virtual Geographies of Videogames." *Environment and Planning A* no. 41:1332–1343.

Shaw, Tony. 2006. *British Cinema and the Cold War: The State, Propaganda and Consensus*. London and New York: I. B. Tauris.

Siegal, Jacob. 2014. "This is How Much Time the Average Gamer Spends Playing Games Every Week." *BGR*, available at http://bgr.com/2014/05/14/time-spent-playing-video-games/ [last accessed 4 July 2015].

Siegel, Alan. 2012. "Remembering the Joyous, Tie-Dyed All-Stars of the 1992 Lithuanian Basketball Team." *Deadspin*, available at http://deadspin.com/5931282/remembering-the-joyous-tie-dyed-all-stars-of-the-1992-lithuanian-basketball-team [last accessed 3 July 2015].

Šisler, Vít. 2008. "Digital Arabs: Representation in Video Games." *European Journal of Cultural Studies* no. 11 (2):203–220.

Sokolova, Natalia. 2012. "Co-opting Transmedia Consumers: User Content as Entertainment or 'Free Labour'? The Cases of *S.T.A.L.K.E.R.* and *Metro 2033*." *Europe-Asia Studies* no. 64 (8):1565–1583.

Stewart, Will. 2013. "New NBC Show *Siberia* is Scorned by Russians for Stereotyping and Being 'Frozen in a Cold War Time Warp'." *Daily Mail*, available at www.dailymail. co.uk/news/article-2353073/NBC-Siberia-scorned-Russians-frozen-Cold-War-time-warp.html#ixzz3IDG5HX3I [last accessed 3 November 2014].

Stocchetti, Matteo, and Karin Kukkonen. 2011. *Images in Use: Towards the Critical Analysis of Visual Communication*. Amsterdam and Philadelphia, PA: John Benjamin Publishing Co.

Stossel, John, and Natalie D. Jaquez. 2007. "The 'Fear Industrial Complex'." *ABC News*, available at http://abcnews.go.com/2020/story?id=2898636 [last accessed 3 November 2014].

Theroux, Marcel. 2010. *Far North: A Novel*. New York: Picador.

Thien, Deborah. 2005. "After or Beyond Feeling?: A Consideration of Affect and Emotion in Geography." *Area* no. 37 (4):450–454.

Thrift, Nigel. 2008. *Non-Representational Theory: Space/Politics/Affect*. London and New York: Routledge.

Thye, Shane R., Edward J. Lawler, and Jeongkoo Yoon. 2008. "Social Exchange and the Maintenance of Order in Status-Stratified Systems." In *Social Structure and Emotion*, edited by Dawn T. Robinson and Jody Clay-Warner, 37–63. San Diego, CA: Academic Press.

Trenin, Dmitri. 2002. *The End of Eurasia: Russia on the Border between Geopolitics and Globalization*. Washington, DC: Carnegie Endowment for Global Peace.

Tuan, Yi-Fu. 1977. *Space And Place: The Perspective of Experience*. Minneapolis: University of Minnesota Press.

——. 1979. *Landscapes of Fear*. New York: Pantheon Books.

Turner, Jonathan H., and Jan E. Stets. 2005. *The Sociology of Emotions*. New York: Cambridge University Press.

Vultee, Fred. 2013. "Finding Porn in the Ruin." *Journal of Mass Media Ethics* no. 28 (2):142–145.

Walker, Shaun. 2014. "Pussy Riot Attacked with Whips by Cossack Militia at Sochi Olympics." *The Guardian*, available at www.theguardian.com/music/2014/feb/19/pussy-riot-attacked-whips-cossack-milita-sochi-winter-olympics [last accessed 4 July 2015].

Weiss, Michael. 2014. "Putin's Criminal Olympics." *The Daily Beast*, available at www. thedailybeast.com/articles/2014/01/27/putin-s-olympic-shame.html [last accessed 2 July 2015].

Westphal, Bertrand. 2011. *Geocriticism: Real and Fictional Spaces*. New York: Palgrave Macmillan.

Williams, Kimberly A. 2012. *Imagining Russia: Making Feminist Sense of American Nationalism in U.S.–Russian Relations*. Albany, NY: SUNY Press.

Wolchik, Sharon L. 2000. *Ukraine: The Search for a National Identity*. Lanham, MD: Rowman & Littlefield.

Wright, Julia. 2014. *Representing the National Landscape in Irish Romanticism*. Syracuse, NY: Syracuse University Press.

Zhurauliova, Tatsiana. 2014. "Imagining Global War: Popular Cartography During World War II." *Artbound*, available at www.kcet.org/arts/artbound/counties/los-angeles/popular-cartography-during-world-war-ii.html [last accessed 3 November 2014].

8 Branded!

Marketing Eurasia's new nations to the (Western) world

On 1 May 2006, the *Financial Times* newspaper reported that the Kremlin had retained the services of public relations firm Ketchum to improve Russia's image during the country's stint as head of the Group of Eight (G8). With the Russian Federation joining a list of corporate clients that included Kodak, FedEx, and IBM, it became clear that nation branding had truly arrived in post-Soviet Eurasia. It can be argued that Moscow was somewhat late to the table as the Baltic States were then in their second decade of robust nation branding efforts, some of which were already being heralded as surprise successes. However, unlike Tallinn, Riga, and Vilnius, Moscow could not simply enter into the global supermarket of nations as a new product. With its storied past, vast economic interests, and complex geopolitical entanglements, the Russian Federation needed rebranding more than branding. In fact, President Putin's long-running spokesperson Dmitry Peskov advocated the need for a PR makeover in rather stark terms, noting that a recent dispute over natural gas with Ukraine had made it clear that Russia was doing rather poorly on the communications front despite "very successful" outcomes in its role as leader of the G8 (qtd. in Buckley 2006). A decade on, Russia—via Gazprom's heavy-handed *neftpolitik*—still looks the bully when it comes to Ukraine, and it seems that no amount of nation branding will attenuate its image as the big, bad brother of Eurasia. In the words of nation branding maven Simon Anholt, "Everybody has an opinion about Russia, and it is likely to be a poor one" (2006, 186). Regardless, the Kremlin continues its efforts to present the image of a stable, responsible, and prosperous federation to the outside world, often through the use of various tactics including state branding.

Yet Russia is not alone among the post-Soviet republics in this regard. Irrespective of how democratic, capitalist, or globalized they actually are, each and every one of the Newly Independent States is engaged in some form of nation branding effort, flogging their credentials as the model entity in the marketplace of nation brands. For some of these countries this requires crafting brand new images, while other states are rebuilding older brands from the rubble of the pre-socialist period. As Anholt points out:

> One of the most damaging effects of Communism was the way in which it destroyed the national identity and the nation brands of the countries within

the Soviet Union. By stopping the export of their national products and preventing people from traveling abroad . . . the Soviet regime effectively deleted old, distinctive European brands. (2007a, 118).

Another critic is even harsher: "Countries that languished under Soviet rule had uniformly dour brands reflected in unsmiling citizens, poor infrastructure, ugly new buildings and crappy consumer goods" (Cromwell 2013). The ramifications of Soviet rule are made more acute by the perceived "loss" of the Cold War by the USSR, and the ensuing ideological baggage associated with the East/West hermeneutic, with the "victorious West" being situated as the "locus of superior knowledge, technology, and ideals" (Kaneva 2012, 8). These issues are further complicated by the decomposition/recomposition of these states and the transitological paradigm, which often interrupts attempts at presenting economic stability and an openness to the neoliberal world system.

Moreover, nation branding is not easy. Opinions change slowly and the ghosts of the past are always "haunting" states' national images and influencing international relations (Auchter 2014). For Russia, in particular, there is no "clean slate" scenario on the table (Simons 2013, 15). In a recent report made by the Valdai Club, an organization of international experts focused on Russia and its role in the world:

> International public opinion is not unlike the mind of a seven-year-old child— highly superficial, inert and influenced by the baggage of each country's history and relations with other countries. It doesn't see the full picture and only snatches images from 24[-hour] news channels and Facebook. Global perceptions of a nation can shift noticeably in some 40 or 50 years, provided that a country engages in the painstaking and never-ending work of shaping its image. (Andreev 2014)

This parsimonious yet trenchant appraisal of the challenges that face any country, when combined with the additional issues facing post-Soviet nations, brings into focus the challenges of becoming a brand state in Eurasia. Yet, this does not even broach the discussion about the cascading hurdles produced by popular-cultural production of the type explored in the previous several chapters, nor does it take into account the domestic popular culture and its undeniable impact on national identity in transitional states like those in the region (see, for instance, Borenstein 2008; Beumers 2012; Miazhevich 2012; Havens and Lustyik 2013; Čvoro 2014). The branded imagination of the Western—and more specifically Anglophone— world makes any undertaking associated with national image a problem of both ideology and praxis (see Kaneva 2012), further complicating what scholars have come to label as commercial nationalism (Volčič and Andrejevic 2015). While Russia grapples with its past and squaring its regional ambitions with its desire to be seen as "normal," the Baltics endeavor to create bright futures and forget their way out of their own geopolitical conundrums; as Kazakhstan battles Borat to guarantee foreign investment and a privileged place in the international community, Azerbaijan revels in Eurotrash culture and auto-exoticism, hoping for residual geo-economic benefits. Taken together, these countervailing trends in commoditizing the national identity present a complex, variegated, and possibly prescient vision of how the nation is being (re)made in the twenty-first century and how individual states are attempting to escape the procrustean confines of their respective "world regions."

Rebranding Russia: Paroxysms, paradoxes, and Putin

Despite the dramatic changes that have occurred since 1991, the Soviet mantle still sits heavily upon Russia's national image, curiously paralleling the ways in which the peculiarities of the autocratic Empire of All the Russias weighed on the Soviet Union from the 1920s until its demise. For many in the West, Russia—as the USSR before it—continues to make true the oft-repeated quip: "It is a riddle, wrapped in a mystery, inside an enigma" (Churchill 1939). For the reformist head of the new Russia, Boris Yeltsin, the first years of independence were meant to be a time for the radical rebranding of the Russian state and a sloughing off of the lingering trappings of the Stalin (1927–1953) and Brezhnev (1964–1982) eras. Yeltsin strongly desired to rewrite representations of the country as "backward," "isolated," and possessing a general "iciness" in its relations with the outside world. Yeltsin's seemingly modest, but ultimately unattainable, goal of positioning Russia as a "normal, civilized country" (Dunlop 2000) involved rapprochement with its erstwhile enemies (China, Turkey, Japan, China, Great Britain, Canada, and the US), integration into what Gorbachev had earlier labeled the "common European home" (see Lukyanov 2014), and a shift from superpower patronage to standard bilateral relations with countries in the Middle East, Latin America, Sub-Saharan Africa, and Southeast Asia. Through the application of "shock therapy," Yeltsin and his economic advisors hoped to turn Russia into a (responsible) capitalist model, replete with ample consumer goods, more foreign direct investment than it could handle, thriving light and information industry sectors, and a modern and globalized banking and financial system. Instead, the ultimate debacle looked more like the "economic genocide" prophesied by Yeltsin's then-vice president and future rival Aleksandr Rutskoi in 1992. Life expectancy dropped, whole industries disappeared, and people's savings evaporated almost overnight.

On the cultural front, the newborn Russia sought recognition for its rich cultural and architectural heritage (see Figure 8.1), combined with the dynamism of a country emerging from decades of restriction on artistic expression. However, deprived of the generous subsides of the Soviet era, Russia's production of culture waned in the 1990s. Opera, ballet, and other forms of high culture were forced to adapt to the market, while Russian cinema quickly wilted in the onslaught of Hollywood imports (notably, the first genuine Russian-made blockbuster would not come until the new millennium in the form of the aforementioned Bekmambetov film *Night Watch*). Likewise, the country looked to throw open the doors to tourism and invite the world to view its national treasures and historic sites, which had been rather difficult to access under the auspices of the state-run, bureaucratic nightmare of Intourist. While Russian tourism did receive a boost in the first decade of independence, these advances were often overshadowed by Russians going abroad to experience other places. However well intentioned, Yeltsin's administration did not make good on all—or even most—of its efforts to rebrand the Russian Federation; instead, journalists, politicians, and academic experts typically describe Russia as a "collapsed and criminal state" rather than what it actually was, a middle-income country struggling to find its "place in the world" and overcome its totalitarian past (Shleifer and Treisman 2005, 151).

By the end of the 1990s, the Kremlin had lost all control of its international image, and the country's national brand on the home front floundered and flailed as neo-Eurasianist revanchists, Soviet nostalgics, pro-Western intellectuals, and radical Russian nationalists battled for the right to define the "New Russia" (Salmin 2000). As Vladimir Lebedenko has argued, Yeltsin-era efforts to define a national image for Russia never moved beyond sloganeering, as those in key

Figure 8.1 St. Basil's Cathedral, Moscow, Russia (CIA Factbook)

positions failed to concern themselves with the issue: "Left to its own devices, this image just 'drifted'—by virtue of the tradition of the personification of state authority peculiar to Russia—along with the contradictory image of Boris Yeltsin" (2008, 108). On the international front, Yeltsin's initial flirtation with Euro-Atlanticism had crashed and burned by 1998, with Russia realizing it had given much and received very little in return. This was no trivial concern when it came to Russia's *Fremdbild* west of its borders. The eastward expansion of NATO, the war in the Balkans, and round condemnation of the situation in Chechnya prompted a complete rethink towards integration with the West, just as the Euro-Atlantic community began to grow suspicious about the commitment of the Kremlin to abandoning its domination over the former USSR, given Moscow's unsubtle meddling in the affairs of Georgia, Tajikistan, Moldova, and its other former colonies. In a high profile essay in *Foreign Affairs*, Democratic Senator Sam Nunn from the US state of Georgia, referred to Russia as a "recovering great power," (Nunn and Stulberg 2000, 46), subtly suggesting that, like an alcoholic or a drug addict, one never stops being a great power, as it is an addiction that cannot be cured, only treated.

Deindustrialization, rampant inflation, and successive global financial crises had shrunken Russia's influence so greatly that it could do little to dictate the terms of its clumsy embrace of globalization. Reflecting the country's harsh representation in film and other media throughout the 1990s, for much of the world, Russia's prime exports were loose nukes, gangsters, and mail-order brides (see Chapters 3 and 4). Russia's cultural industries, bereft of the state support that sustained the field for decades, suffered an ideological, financial, structural, and social systemic failure (see Tchouikina 2010). As far as tourism went, other than a handful of East European pensioners, few thronged to see the collapsing Sovietesque sanatoria or sports complexes. The image of Russia as a gray, foreboding hinterland, populated by miserable losers of the Cold War afflicted by the grotesqueries of the politico-economic transition, seemed to only gain greater currency after the fall of the Soviet Union. While, by 2000, grains of fact backed up these prejudicial perceptions, the overall picture of Yeltsin's Russia as a derelict behemoth rotting beside the superhighway to global prosperity was largely the result of negative portrayals in the world's popular media combined with a lack of cogent strategy for countering these depictions, rather than any accurate representation of reality.

Reflecting the realization that Russia's image in the West was "much worse" than the reality, Moscow began to view its desultory, fulsome nation brand as a serious threat to its security and economic well-being at the turn of the millennium (Feklyunina 2008). At the forefront of this trend were leaders in industry, who made it clear that Russia's anti-image was harming exports, quashing FDI, and making international partnerships difficult. Not coincidentally, the year 2000 saw the introduction of the Concept of the Foreign Policy of the Russian Federation which declared the state's commitment to ensure "a positive perception of Russia in the world and popularize the Russian language and the cultures of Russia's peoples and ethnic groups in foreign states" (qtd. in Semeneko, Lapkin, and Pantin 2007, 79),[1] thus enshrining national-image projection in the fabric of foreign policy. Through an active management of the country's international profile under the new leadership of Vladimir Putin (including the aforementioned hiring of the global PR firm Ketchum to represent Russia), serious attempts to improve the country's nation brand began in earnest.

From sport to architectural heritage to cultural outreach, the Kremlin made enormous strides in elevating the global opinion of the Russian state during the 2000s. Eschewing the politically charged term "propaganda," Russian brand marketers instead referred to their activities as public relations (PR) or, more uniquely in the Russian sphere, as the application of "humanitarian technologies" to the question of national image.[2] Whereas Yeltsin had turned his back on using the tools of the state to manage Russia's image abroad, Putin and his team of political technologists enthusiastically embraced the tools of PR to ensure a positive perception of Russia in the world, even when foreign policy complications, human rights violations, and the ebbing of democracy seemed to get in the way.[3] As one critic of Russian efforts at (re)branding opines:

> The communist era created powerful, visually sensational graphics and obscenely empty slogans but behind the shiny hoardings were dark secrets and deeds. In the post-Cold War world it seems that Russia is trying to promote openness by being all message, no marketing. (Burt 2007)

However, this critique only examines part of the efforts undertaken since the turn of the century. In his analysis of Russian rebranding efforts over two decades, IR scholar Greg Simons (2011) argues that a number of factors weigh heavily on any efforts, including the embeddedness of negative images, historical associations with the USSR and tsarist Russia, and over-simplification of identity. Valentina Feklyunina (2008) adds to this the problem of lack of coordination on the government–civil society level, combined with a seemingly perpetual post-Soviet identity crisis. Recognizing this, Moscow's nation branders began efforts at redefining key aspects of the country's image abroad in the second decade of the twenty-first century. With ample media platforms at its disposal and the linguistic-cultural reservoir associated with the Russian language, the Kremlin has made significant strides in branding the country beyond its borders, thought this process has not been without its problems. In fact, the current decade has seen Russia's brand rise and fall like a rollercoaster.

The highly changeable nature of Russia's national image is rooted in two areas: first, the ingrained nature of "Russia" in the international mind's eye is something that does not change very quickly; and second, it has been tethered to Vladimir Putin's own brand since about 2000. During Putin's rise to international prominence and (sometimes-begrudging respect) in the West, he has steadily become the image-bearer of a new (or re-tooled) Russia, thus presenting a new, vigorous national gestalt that provided the country with the raw materials to begin to articulate its own identity as well as project itself as a "cultural archetype" for others to emulate (Lebedenko 2004, 76). Undoubtedly, "Brand Putin" is trending in tandem with "Brand Russia" (Nagornykh and Safronov 2015).

Putin, as captain of both brands, commands 85 percent approval ratings and an entire industry has grown up around his visage (see Cassiday and Johnson 2010; Goscilo 2013; Foxhall 2013; Rainsford 2015). Even in the West, Putin remains a dominant figure in the popular geopolitical imagination, despite the fact that he is increasingly associated with Hollywood super-villains (see Saunders forthcoming). Putin, despite all his faults, is universally viewed as a confident and powerful leader of his country. In popular culture, journalistic analysis, and IR scholarship, the Russian president is often counterpoised with Western heads of state to highlight their weaknesses. He is the anti-Yeltsin: a tee-totaling, athletic, hyper-masculine, and stone-faced man to be reckoned with. Consequently, Russian national pride has witnessed measurable increase since Putin's ascension, translating into a stronger brand abroad despite geopolitical controversies (Leone and Gorshkov 2004). However, yoking Putin to Russia presents a double-edged sword, given his "strong man" associations, bellicosity in military intervention in Georgia in 2008,[4] and well-publicized disdain for many foreign leaders. Yet Putin is not the only way that Russia is finding ways to alter its national image.

On the level of elite person-to-person diplomacy, Russia has done well by integrating top-level policymakers and academics into a wide array of non-profit organizations focused on embedding Russia in the West. The most well-known of these is the Valdai International Discussion Club (http://valdaiclub.com/). Established in 2011, the working group has focused its efforts on presenting a positive face for Russia, while also endeavoring to create a dialog that allows Russia to present its case to the international scholarly community and political elites. Demonstrating the importance of nation branding in Putin's Russia, the Valdai Club's Tenth Anniversary Meeting featured Simon Anholt (2013) as the plenary speaker. In his address, which touched on the polarizing and complex nature of Russia's brand, he noted: "There are some things people very much admire about Russia, and other things they strongly dislike." Anholt ended his remarks by arguing that rather than trying to make the world stand in awe of the country, "it needs to find ways of making people feel glad that Russia exists." Also influential is the Russian International Affairs Council (http://russiancouncil.ru/en/), which aims at building links between policymakers, civil society, and academe in an "effort to find foreign policy solutions to complex conflict issues." There have also been a number of smaller non-profit organizations with often murky links to political technologists working to promote Russia's image abroad, including the Conservative Friends of Russia and the Positive Russia Foundation, among others (see Foxhall 2015).

Beyond such intellectually inclined outreach, Putin has actively supported a transnational renewal of Russian language and culture as a mechanism for building up the Russian brand, while also providing it with content that is not exclusively drawn from the Russian Federation. According to *Ethnologue* (2009), Russian—spoken in some 33 countries by approximately 144 million native speakers—is the eighth most spoken native language in the world. Estimates of the total number of speakers range upwards of 275 million (Babich 2007), thus ranking Russian as the fifth most spoken language worldwide, trailing only

English, Chinese, Spanish, and Hindi. Russian is also one the most commonly used languages in cyberspace, consistently ranking in the top ten among Internet languages (IWS 2011), with nearly 60 million Russian speakers accessing the web per month and steady growth in new users expected over the next five to ten years (Minenko 2012). Given the size and scope of virtual and real-world Russophonia, it is not surprising that this geolinguistic space has emerged as fertile ground for rebranding Russia (see Saunders 2014a).

Certainly, the most successful attempts at utilizing the Russian language as a branding tool come from the Russkiy Mir Foundation, described as "the most concerted effort to date at conceptualizing a notion of 'Russianness' that transcends ethnic bloodlines and geographical boundaries" (Gorham 2011, 30). The ethos of Russkiy Mir is decidedly international in its scope and focused on the "promotion of Russian culture and Russian language as a global language" with the aim of "bringing the world together," which Hudson (2013) describes as a form of civilizational meta-politics. Heartily backed by the Kremlin and a new cadre of political technologists in the Putin administration, Russkiy Mir has set up shop not only in post-Soviet space and major world capitals (London, Washington, Rome, and Beijing), but in some less likely locales as well, including Guayaquil, Ecuador and Bangkok, Thailand. While it is clear that Russkiy Mir is focused on extending Russian's geolinguistic presence in new parts of the globe, the primary focus remains the traditional spaces of Russophonia, with more than one-third of all centers located in the former Soviet Union, and roughly the same number located in former Soviet satellites. According the organization's web site:

> The Russkiy Mir Foundation has undertaken an international cultural project to develop Russian Centers in partnership with educational organizations around the world. The Foundation's Russian Centers are created with the aim of popularizing Russian language and culture as a crucial element of world civilization, supporting Russian language study programs abroad, developing cross-culture dialog and strengthening understanding between cultures and peoples. (RMF 2015)

Russkiy Mir's 80 centers around the world facilitate Russian-language learning through print and digital publications, distance learning facilities, arts programs, cultural events, and support services for educators. Such efforts seemed to be aimed at shifting Russian from its Soviet-era identity as the language of "interethnic communication" within the socialist realm to an idiom that commands the status of a genuine "global language," partially if not totally depoliticized and delinked from the Russian Federation.

While Russkiy Mir's work is closely modeled on preexisting cultural-linguistic paradiplomacy employed by Germany (Goethe-Institut), China (Confucius Institute), and the UK (British Council), its partner, Rossotrudnichestvo— or the Federal Agency for the Affairs of the CIS States, Compatriots Living Abroad, and International Humanitarian Cooperation—is somewhat different. Rossotrudnichestvo's mission is explicitly political, proudly dedicated to integration efforts across the CIS, "formation of a positive image of Russia abroad," and protecting the "linguistic and cultural needs" of Russia's 20 million compatriots living outside the country (Rossotrudnichestvo 2013). Established in 2008, this autonomous federal government agency, which is under the jurisdiction of the Ministry of Foreign Affairs, operates 59 centers of science and culture, with plans to have a presence in 100 countries by 2020 (ITAR-TASS 2012). Organizations like Rossotrudnichestvo, the International Council of Russian Compatriots, and the Rodina Association provide support to Russophones living outside of Russia, but increasingly, they also expect something in return from "compatriots," i.e., active social and professional activities intended to preserve the Russian language and Russian culture in their states of residence (see Gorham 2011).

Taken together, the activities of Russkiy Mir and Rossotrudnichestvo represent two planks of a worldwide campaign by Russia to expand its influence through the vehicle of the Russian language, an effort made clear by President of the International Association of Russian Language and Literature Teachers and Chair of the Russkiy Mir Foundation Board of Trustees Lyudmila Verbitskaya when she stated: "The well-coordinated work of the Russkiy Mir Foundation, foreign Russian specialists, language experts of Russia and Rossotrudnichestvo should facilitate the realization of our dream—the planet should start speaking Russian" (RMF 2012). While the Russian language undoubtedly benefits from the funding of the Russian government abroad, it is the explicit linkage of the Kremlin's political aspirations that may threaten Russian's ability to shift from being the Soviet *lingua franca* to being a true global language. According to language historian Michael Gorham:

> It is at once logical and ironic that language should serve as one of the primary means of strengthening Russia's image, especially regionally. Logical, because language as a vehicle of cultural expression and affiliation enjoyed more cultural capital than most things "Russian" at the time . . . Ironic since the Russian language . . . had served as one of the very first and most vulnerable targets by means of which the newly independent states declared their independence from Soviet (and Russian-language) dominance in the late- and post-perestroika period. (2014, 160)

By promoting "common humanitarian values" across the "multinational Eurasian region" while attempting to disabuse its CIS partners of their belief in its imperial ambitions (Feklyunina 2008), Russia's aim has been to renew bonds broken by the dissolution of the Soviet Union; however, it is clear that post-colonialism continues to shape not only non-Russian attitudes towards such policies, but the orientation of the (former) colonizers as well (Hudson 2013). President Putin declared in 2007, "As the common heritage of many peoples, the Russian language will never become the language of hatred or enmity, xenophobia or isolationism" (qtd. in Babich 2007). Yet, despite Russian's purported inoculation against anything untoward, the close linkage between language and power is problematic primarily because of Russia's tendency to treat the geographical sphere in which the

Russian language remains the dominant medium of communication as the true political border of the Russian state (Bogomolov and Lytvynenko 2012). Such behavior, if sustained, could prove a poison pill for the realization of "global Russian" (see Ryazanova-Clarke 2014).

In addition to elite forums and cultural-linguistic outreach, new media has formed a key part of Russia's nation branding strategy. At Ketchum's urging, Russia launched a host of Internet-based initiatives to present a modern, ready-for-business environment to the outside world. One of these was the *Modern Russia* Twitter feed, which was meant to provide regular updates on the "economic, political and social modernization of Russia" (Simons 2011, 334). However, @modernrussia (http://twitter.com/modernrussia) has since been delinked from Russia, and as of July 2015 was a get-rich-quick advertisement with only two followers, a rather pregnant metaphor given that Putin fired Ketchum in March of 2015. Speaking of the split, Peskov stated: "We decided not to renew the contract because of the anti-Russian hysteria, the information war that is going on" (qtd. in Kottasova 2015). While this effort produced little fruit, other new media strategies have worked better, including one-term president Dmitry Medvedev's regular use of blogs, Twitter, and other forms of social networking to engage both the Russian people and the wider international community in a discussion about Russian economic reforms and greater openness to investment and international trade (see Yagodin 2012), as well as efforts to market Russia's "Silicon Valley" in Skolkovo and positive associations with Russia's membership in the BRIC countries (see Simons 2013). Print-cum-digital advertisements in *The Telegraph* and other major news media outlets have also played an important role.

Perhaps the most successful tool for rebranding Russia has been RT (formerly known as Russia Today), which has reversed the trend of Russian media being "widely used as a source of negative information" about the country (Semeneko, Lapkin, and Pantin 2007, 82). RT broadcasts in English, Arabic, German, French, and Spanish, in over 100 countries and reaching a population of 700 million worldwide (RT 2015). RT is a highly effective tool for Moscow as it seeks to promote a positive view of Russia on the world stage. The broadcaster is state-funded and closely affiliated with the Kremlin, generally adhering to a pro-government stance on most issues. While Western media analysts are quick to label RT as a latter-day *Pravda* or *Soviet Life* (see, for instance, Shafer 2007; Orttung, Nelson, and Livshen 2015; Zwick 2012), this critique is a hollow one, as RT's coverage of international events is essentially comparable in terms of its pro-Russian bias to CNN International, which toes the Washington line, and the BBC, which, despite its many claims to the contrary, is decidedly pro-British in terms of its international coverage. However, when it comes to "internal issues" in certain countries (namely the US, UK, and, more recently, Ukraine[5]), RT's coverage often diverges from the norms of the BBC, CNN International, and other broadcasters like Euronews, DW, and Al Jazeera.

RT is a sleek, highly professional, cosmopolitan affair, replete with dynamic graphics and reporters drawn from around the globe. Unlike CNN and BBC, both of which carry an enormous amount of political baggage in the developing world as a result of the geopolitical associations of the networks with their respective governments, RT is comparatively unburdened by such historical blemishes on its character. Alongside Al Jazeera English, RT has quickly emerged as an international media platform with a strong focus on the Global South, proudly reporting stories from a perspective which is critical of Western hegemony and suspicious of American and British motives in Africa, Asia, and Latin America (see Pomerantsev 2014). This type of framing is powerful for Russia as the country seeks to rehabilitate its brand in parts of the world where the Soviet Union was once respected and emulated. Furthermore, by challenging lingering Cold War stereotypes of the country, which can be found in nearly any CNN or BBC broadcast about Russia, RT is able to promote Russia as a genuinely "new" country in terms of its foreign policy goals, economic relationships, and geopolitical profile.

When it comes to coverage of Russia itself, RT is even more effective in revising embedded prejudices associated with the country. Through its documentaries on Russia's diverse regions and peoples, special programs on the business climate, and regular coverage of technological innovation in the country, RT is able to advertise qualities not normally associated with the country, which is still often seen through the dark and distorting lens of Soviet Communism. This trend became particularly evident in coverage of terrorist attacks in Russia from the mid-2000s onwards. Prior to the advent of RT, foreign journalists residing in Russia tended to be the only source for Western media in the wake of terrorist bombings in the country (which from 1999 until 2005 were tragically frequent). However, since the network first aired in December 2005, RT is now the first (and generally the only) source for video footage, context, and on-the-ground reporting when terrorists strike the country. This has enabled Russia to radically alter the discourse associated with its "war on terror" against Chechen and other North Caucasian insurgents, as well as to speak truth to power when it comes to Hollywood-style presentations of the country as a lawless neverland. However, with the abating of terrorist attacks within the Russian Federation, this mechanism has lost much of its previous utility.

Being able to influence the way foreign media covers Russia is extremely important, and RT is thus seen as generating a positive return on investment for the network's owners (who, we need not be reminded, are closely linked to the Kremlin's political technologists, that small army of intellectual elites charged with shaping perceptions of Russia's government at home and abroad). However, it is even more important for RT to be watched as a stand-alone broadcast platform by key decision-makers outside the country, especially Western European countries with economic ties to Russia. It is here where RT is doing extremely well. RT has a broadcasting reach of over 84 percent in Great Britain and is available to more than 120 million people in Europe (RT 2015). Perhaps most surprisingly, RT is the second most popular foreign broadcaster (after BBC) among US viewers, with a daily audience more than six times larger than that of Al Jazeera English (Rizvi 2015).

Where it cannot reach its intended audience by direct broadcast via its satellite network, RT compensates by giving its content away for free on the Internet and through a live streaming video application for iPhone and iPad (something that Al Jazeera charges for and which BBC and CNN do not yet offer everywhere). RT has become so effective at linking new media—particularly YouTube (see Shuster and Mcdonald-Gibson 2015)—to satellite television broadcasting that certain scholars have begun to argue that the network is actually setting benchmarks and starting trends that are remaking the global information ecosystem (see Strukov and Zvereva 2014). Taken together, RT's efforts to provide a Russian perspective is emerging as a major plank in the country's wide-ranging effort to rebrand itself for the twenty-first century.

More traditional mechanisms have also been used to burnish the country's image, particularly in the areas of investment, sport, and culture, though also touching on some subjects related to politics. *Russia Beyond The Headlines* (RBTH)—produced by *Rossiyskaya Gazeta*, the government's official newspaper—is a weekly insert that runs in most of the world's most influential newspapers, including *The Wall Street Journal*, *The New York Times*, and the *Daily Telegraph*. In addition to traditional print runs, the publication also has a Facebook page, Twitter feed, and other new media platforms for distribution of content. Interestingly, in 2014, the conservative *Telegraph* was called out by its more liberal competitor *The Guardian* for "continuing to publish Russian propaganda," i.e., RBTH, even as it condemned the Kremlin for complicity in events in eastern Ukraine (see Greenslade 2010).

Regarding the latter, this mercurial status was dramatically underscored in 2014 when the Sochi Winter Olympics provided the country with an awesome global platform for projecting a controlled and consumable image of the Russian Federation, just as the crisis in Crimea came to a head. Without a doubt, Sochi ranks as the single most important branding effort for the Russian Federation since it came into being (Ostapenko 2010; Basulto 2013; McPhee 2014). The opening ceremonies, deployed under the such sloganeering as "Russia—Great, New, Open!" (Müller 2014, 629), allowed a resurgent Russia to rewrite its history (particularly via the grandiosely staged Russian Alphabet/*Azbuka* ceremony), while projecting a contemporary country that is confident, capable, and in control of its own destiny. By taking control of the Debordian spectant/spected relationship with all eyes on southern Russia, the country's nation branders were in a most enviable position (see Makarychev and Yatsyk 2013). However, in the months following, Russia saw the branding dividends of the winter games go up in smoke as its "little green men" (Rosenberg 2014) marched into the Crimean Peninsula, abetting a secession-cum-annexation by Moscow. Perceptions of the Russian state darkened even more in the summer as separatist rebels in Ukraine's Donetsk region purportedly brought shot down a commercial jet, Malaysian Airlines Flight 17, killing 298 people in one of the worst aviation disasters since the Cold War. As Moscow seeks to balance its larger geopolitical aims with its convoluted notions of "soft power," the events of the Ukrainian crisis have prompted the Kremlin to come to understand that altering its brand image on the world stage is something akin to moving a mountain and any positive change will be slow in coming. This is particularly true given the Kremlin's continued "hypersensitivity to any criticism from Western countries" (Feklyunina 2008, 612) and "oscillation between aggressive rhetoric and more conciliatory behavior" (Avgerinos 2009, 116).

However, the real challenge for Russia—and the same can be said for all the world's great powers—is that it has a hard time differentiating between nation branding and the related concepts discussed in Chapter 2, namely soft power and public diplomacy. In fact, nearly every academic analysis of Russia's nation branding over the past 15 years has similarly conflated multiple concepts (see, as examples, Lebedenko 2004, 2008; Semeneko, Lapkin, and Pantin 2007; Feklyunina 2008; Ostapenko 2008; Simons 2011, 2013). Likewise, when one meets a reclusive Russian nation brander, they are almost always charged with a wide variety of other portfolios, from managing trade deals to solving political imbroglios. Rarely, if ever, does the Russian Federation see fit to engage in the often tedious, grassroots sorts of nation branding campaigns detailed in the second chapter of this text. Perhaps the constantly juggling of Great Power politics, transshipment geo-economics, post-imperial relationships, and neoliberal pressures on the state make it nigh impossible to do so, a fact consistently recognized in various international brand rankings and reports. While the Kremlin and others may be confused about how to disentangle soft power and public diplomacy from nation branding efforts, this is not the case with the tiny post-Soviet republics of the Baltic Rim.

The Baltic *Wunderkinder*: Estonia, Latvia, and Lithuania

In conjunction with their embrace of Western-style capitalism and integration into Europe, the three Baltic States have all embraced the notion of the "brand state" since gaining independence. Besides geo-economic considerations, the Baltic countries have geopolitical reasons for engaging in sophisticated nation branding campaigns. As small nations with a shared history of Soviet domination, all three countries are anxious to protect their sovereignty as new members of the EU and NATO, while simultaneously differentiating their nations from Russia and other members of the Commonwealth of Independent States. Given its technical prowess, Tallinn sought to brand the country as "E-stonia," while its southern neighbors have been somewhat more cultural and historical in their branding, with Riga choosing to label the country "Latvia: The Land that Sings" (referencing the importance of the folk music tradition in the independence struggle) and Vilnius adopting the mantle "Lithuania the Brave" (citing the country's medieval heritage and bold resistance to Soviet rule in the late 1980s). While the long-term success of the Baltics' branding efforts remains unclear, the three countries represent clear models for their former partners in Soviet bondage, the Eastern European republics of Ukraine, Belarus, and Moldova. Mixing tourism investment, economic reform, and high-priced image makeovers (including quirky logos, sappy videos, Facebook pages, and other tools of the trade),[6] the Baltic States crafted attractive and marketable images in Europe and the Anglophone West, eventually winning admission to NATO and the European Union in 2004. The countries have in fact benefited from *complementary* though not necessarily *common* strategies, and—ironically, given the problem of differentiating Eastern European "-ia" countries (see Saunders 2012)—the tendency to be lumped together.[7] Simply "being Baltic" seems to have provided the necessary fundaments for success in the field, according to a number of nation branding analysts (Andersson 2007; Anholt 2008; Moilanen and Rainisto 2008).[8] However, there have often been negative and unintended consequences of the Baltics' sometimes overly zealous conversion into brand states.

Estonia, which is frequently included in international assessments of nation branding successes, represents the paradigm of the brand state within post-Soviet Eurasia. Small, clean, prosperous, and open to the outside world, the tiny republic is a short ferry ride from Helsinki and possesses direct flights to London, as well as host of Western European cities. With a well-educated, tech-savvy population, the highest level of English proficiency in the former USSR, and leadership bent on adopting most of the Anglo-Saxon economic model, the country seemed pre-destined for a rapid transition to neoliberalism (Curry Jansen 2008). Importantly, the small size of the country factored into the transition to a brand state, as decisions taken at government level tended to produce rapid and wide-ranging impacts across Tallinn, the primary focus of the manipulation of Estonian identity post-1991. In order to brand the country, Estonia contracted the international brand consultancy Interbrand to help it "punch above its weight" (see Interbrand 2008). The ultimate outcome of the collaboration is generally recognized as highly successful, particularly given that Interbrand worked closely with Enterprise Estonia, part of the Ministry of Economic Affairs. While Estonia was able to convey a variety of image-points to the outside world including its fresh perspective, self-starting capabilities, and Nordic temperament (Interbrand 2008), one of the sticking points in creating a new national image for the country —as one of the Interbrand team attested in an interview with the author—was a dogged obsession with the past.

According to Jeremy Faro (2009), a brand consultant who worked on the country's "Welcome to Estonia" and "Positively Transforming" campaigns, the Estonians wanted to go back nine centuries in explaining their country to Western European investors and tourists.[9] The obsession with the primordial past remains palpable even today, evidenced by the online video tutorials produced by Brand Estonia (http://brand.estonia.eu/en/), including "Welcome to Estonia" (Brand Estonia 2015a), which reminds Estonians that they have "been here thousands of years" as a cartoon screen behind the narrator flashes the acronym "DNA," or another video entitled "Rootedness—Estonians Have Deep Roots" (Brand Estonia 2015b), which claims that all this history lies "in Jüri's blood" ("Jüri" is the representational strawman meant to "teach" Estonians about their past). While Interbrand has marketed its efforts at creating national cohesion through positive-image production, other scholars have suggested that Estonia's overly ethni-cized approach to its brand has actually made relations with the resident Russian population worse (see Budnitskiy 2012).

Regardless, the efforts paid off, with Estonia emerging from the dissolution of the USSR as a "Nordic country with a twist" and a "model pupil" for EU admission (Aronczyk 2013). Blessed with a highly concentrated population in the capital and benefiting from decades of functioning as the Soviet Union's "window to the

West," telecom and IT were soon linked to country. The fact that the worldwide phenomenon Skype started in the country did a great deal to embed the linkage with information and communications technologies (see Saunders 2009). The shift towards e-government and the provision of Wi-Fi across the whole of the country allowed for fairly effective branding of the small state as "E-stonia," a strength that was used against it in the 2007 cyber-war, though even in the chaos of the attacks Estonia was lauded as a model for its extremely developed Internet infrastructure (Mansel 2013). While its northern neighbor Finland is also a tech giant, Estonia has eschewed Eurosocialism and high taxation and was thus able to present itself as a Baltic Ireland or Luxembourg. In terms of its historical and place-based identity, there was little of the "Baltic Tiger" patina; instead, Estonia successfully situated itself as quintessentially "European," fulgent in its medieval Germanic architecture, quasi-pagan traditions, and unspoiled natural environments (see Figure 8.2). While now often overshadowed by successes in both (eco-friendly) tourism and (market-ready) neoliberalism, the importance of Estonia's 2001 victory and subsequent hosting of the Eurovision Song Contest were also key elements in making the country known to masses west and south of the country (see Jordan 2014), as is—at least to Estonia's own branders—the global recognition of sumo wrestler Baruto Kaito, born Kaido Höövelson (Collier 2008).

While not being able to lay claim to the "case study in nation branding" status enjoyed by its northern neighbor, Latvia has also been quite successful in shedding

Figure 8.2 The Old Town in Tallinn, Estonia (CIA Factbook)

its post-Soviet baggage and defining itself as a modern, Western-oriented state with a wide array of attractive qualities. As one of the most desirable retirement areas for Soviet *nomenklatura* and military officers, Latvia presents an interesting case study in the transition from a closed (Eastern Bloc) geopolitical space to an open one (to Europe and the world), or, put another way, a nation that adapted to the "cultural logic of late capitalism" (Jameson 1991). With its comparatively large, cosmopolitan, and picturesque Art Nouveau capital of Riga, Latvia functions as the keystone of the Baltic States, linking its northern and southern neighbors, as well as Russia and Belarus, to the rest of the Baltic Rim. Due to historical connections and the aforementioned influx of Russians during the post-WWII era, Latvia also carries the brand of the European nation that "best understands" Russia (Frasher et al. 2003). With pristine wildernesses, attractive beaches, and commitment to environmental preservation, the country provides visitors with a welcome reminder of an older, more bucolic Europe—a view which is only underscored by the branding of the Latvians as the "last remaining genuine peasant nation in Europe" (Dzenovska 2005, 115). Working in conjunction with the government agency tasked with national-image management, the Latvia Institute (www.latvia.eu/), the Saïd Business School at Oxford University (Frasher et al. 2003) helped launch the country's state branding efforts in the year prior to EU admission, focused on enlightening "misinformed, unaware, and mistaken" audiences in the West, with specific attention being paid to Germany, Sweden, and the United Kingdom (according to its nation branders, Latvia is consistently thought to be "Balkan" rather than a Baltic country).[10] In 2007, the Latvian Institute contracted with nation-brander extraordinaire Simon Anholt to build on its brand post-accession. In a blog post made after returning from Latvia, Anholt (2007b) lamented the historical ignorance of the West about the lost histories of the former socialist states, but waxed hopeful that the beauties of Riga would give his client a leg up in the future.

Over the past decade, Latvia has seen its brand rise, fall, and rise again, reflecting the country's painful passage through the 2008–2009 global financial meltdown, which resulted in positive associations with the way the nation emerged as a "champion of curbing crisis" (Pētersone qtd. in Markessinis 2011). Now one of the fast-growing markets in Europe, Latvia has emerged on the other side of its economic woes with a fairly well-known national image in most of Europe. The Baltic state also enjoys an increasing level of brand recognition farther afield, including in China as the result of an extremely innovative presentation at the 2010 Shanghai Expo (Armin 2010). Despite measurable success, Latvia often still ends up the joke even in erudite circles, as in a *Smithsonian Magazine* article on nation branding which lampooned the country for marketing its high-quality medical services when it possessed six UNESCO World Heritage sites in a space the size of South Carolina, suggesting instead that the country motto should be: "Latvia: Now With More Bacon Buns" (Conniff 2011). All banter aside, Latvia has been steady and cool in its branding efforts, with an enduring commitment to its brand-cum-logo "Latvia: The Land That Sings," a historically-resonant reference to the Singing Revolutions, as well as the nation's connection to a lyrical tradition that predates Christianity's arrival in the country (see Strmiska 2005). However, the country has switched to a new motto, "Latvia: Best Enjoyed Slowly," reflecting a shift from historical associations to touristic ones (see Figure 8.3). Reflecting its strong connections to Scandinavia, Latvia has most recently crafted an image based on folk fashion, poetry and literature, performing arts, and architecture/ design, a sort of Denmark of the eastern Baltic Rim (see Latvia Institute 2014). More recently, the country's success on the ice has made hockey an important tool in promoting "positive emotional relationships between Latvia and other nations" (Brencis and Ikkala 2013, 251). As the country moves towards its centennial in 2018, Latvia seems to have found a comfortable position in Nordic Europe, with a slow-growing but nonetheless healthy *Fremdbild* in Western Europe.

Unlike Estonia and Latvia, Lithuania commands a historical past with deep roots, though ones which requires a bit of excavation to be operationalized for the neoliberal world system and the global supermarket of brands.[11] Once part of the great Polish-Lithuanian Commonwealth, Lithuania's return to the community of free and independent nations has been a long time coming. However, attachment to the past often functions as a countervailing force against the mobility necessary

Figure 8.3 "Latvia: Best Enjoyed Slowly" (Latvian Tourism Development Agency)

to adapt to the present. In the late 2000s, the country embarked on a new phase in its branding efforts, focusing on the notion of bravery. Speaking of this image makeover, government spokesman Laurynas Bučalis noted, "Bravery marks our history—from being the last pagan nation in Europe to a nation which sparked the Soviet Union's downfall" (qtd. in Vaiga 2015). Further reflecting this backward-looking obsession, the man behind the newly proposed strategy of "Lithuania the Brave,"[12] self-declared identity developer Paulius Senuta, defiantly declared: "For those who speak Lithuanian, there already is no such place as Lithuania—Lietuva is what the country is called inside our borders. That hasn't changed in a thousand years and we are not suggesting it change now" (qtd. in Potter 2008). As Dalia Bankauskaitė (2009), Executive Director of the Vilnius—European Capital of Culture 2009 program, told me, "There is an over concentration on the past. Lithuanians are more confident about the past than the present. Even the current slogan of 'Reality that exceeds your expectations' reflects this." As a representative of PR company European House/Europos Namai, the organization in charge of managing the country's image-formation strategy, she criticized the various stakeholders for their lack of ability to collaborate and identify common goals across the spectrum, suggesting that "turf wars" had led to the Lithuanian brand withering on the vine despite positive traction in the UK, Germany, and Ireland (the latter being a common destination for economic migrants post-EU accession).

Undoubtedly, sport plays a key role in Lithuania's nation brand—or, to be more specific, the sport of basketball. Despite being 27th in terms of the average height of males between 20–30 years of age in the country, Lithuania is a basketball powerhouse. In her essay on Lithuania's nation brand, Ausra Park writes, "If basketball were the most popular sport in the world, Lithuania would have no problem with its national image" (Park 2009, 67); however, basketball is not the world's sporting pastime, and so Park and others who have studied Lithuania's nation branding efforts remark that despite the efforts of the country, there has been a failure to achieve much success since the early 1990s. Unlike Estonia and Latvia, Lithuania is neither capable of nor particularly interested in becoming a "Nordic country,"[13] an important factor that has weighed heavily on the small country's national image, as it has been lumped in with a large number of other post-socialist "Eastern European" states, from Belarus to Albania. However, at the behest of the late Wally Olins (Olins and Hildreth 2009), Lithuania has made some progress in branding itself as a paragon of the "lively" and "romantic" countries of "Northeastern Europe" (see Vilimaviciute 2009), a rather novel geopolitical category which may over time shift from rarified to reified. Lithuania's geographical ambiguity is not its only issue; the country was constrained by its image as a redoubt of Communism in the 1990s, a haphazard branding strategy, and pessimistic estimations of the country on the part of its own residents and disconnect between reality and representation (though this began to change in the 2010s [see Lionikaitė 2012]). There have also been a few high-profile instances of popular-geopolitical assaults (see Aronczyk 2008; Park 2009). One of these came from the actor Mel Gibson, who in a rather gauche attempt to market his film *The Patriot* (2000) offhandedly opined, "Imagine if you woke up one

Figure 8.4 Lithuania's logo (Lithuanian State Department of Tourism)

morning here in Los Angeles and found Lithuanians with sharp teeth crawling up the beach with golf clubs to beat your brains out. What would you do? That's a reason to defend yourself" (qtd. in Bonin 2000). Another came from Jonathan Franzen, whose Oprah book club novel *The Corrections* (2001) depicted Vilnius as a place of "economic chaos, gangsterism, and a diet reliant on horse meat," prompting a swift rebuke from Vygaudas Ušackas, the Lithuanian ambassador to the US (Flamm 2001). Despite the troubles the country has encountered on its way to becoming a brand state, Lithuania the Brave perseveres. Today, the country has a logo, a well-liked Facebook page, and the hope that being an "old country" somewhere between Scandinavia and Central Europe provides it with something particularly special (see Figure 8.4). Even with the shortcomings of its ad hoc branding campaigns, Lithuania still commands a brand that remains comparatively unsullied by the post-Soviet frame. The same cannot be said of those countries farther east, which, despite their best efforts, remain ensconced in the Eurasian realm.

Out of the shadows: The Commonwealth of Independent States

Unlike the Baltic States, a group of undeniably "European" states that were dragged into the USSR against their will, the remaining post-Soviet republics have faced a host of complications in their respective efforts at becoming brand states, particularly the Central Asian Republics.[14] For these countries, as well as Azerbaijan, brand differentiation is a major problem. For others such as Belarus and Armenia, the greatest hurdles lie in the government's refusal or inability to adapt to the changing nature of international relations and the dissolution of the USSR. Georgia and Ukraine have seen burgeoning brands besmirched by conflicts with Russia, while Moldova faces the gravest of challenges in terms of creating a cogent national identity, thus making nation branding almost impossible. Regardless, all 11 of these countries (all of which were members of the CIS for at least some part of the past two decades) have sought to project a vibrant, unique, and compelling national image abroad, seeking tourists, FDI, international partnerships, and recognition of their place in the twenty-first century world. A variety of tools and techniques have been employed in these respective pursuits; however, a few trends have emerged in recent years: Eurovision, sport, mega-architecture, and "exotic" tourism. While not all of these countries have employed every one of these tools of the trade, nearly every CIS state has sought to use one or more to get noticed in the global supermarket of nation brands. In addition to common strategies, the states of the CIS also share a suite of problematic attributes which interrupt nation branding efforts, most notably the issue of corruption, the "Just Russia" frame, and generational resentment in the West, i.e., lingering from Cold War orientations towards the post-Soviet republics (Myroshnychenko 2009). While most of these Newly Independent States have not gained much traction as brand states, Ukraine, Azerbaijan, and Kazakhstan have seen a precipitous—though not uncomplicated—rise in international profiles since 1991. Consequently, the subsequent analysis will highlight each of these states as "market leaders" within their respective regions (see Dinnie 2008), i.e., the Western Republics, the South Caucasus, and Central Asia, respectively; however, the efforts of all post-Soviet republics will also be considered herein.

The Western Republics

While the 1990s saw few if any positives in the realm of Ukrainian nation branding under the Kravchuk (1991–1994) and Kuchma (1994–2005) presidencies, the Orange Revolution provided Ukraine with a global platform to makes its national image known and valued. Under the administration of Viktor Yushchenko, the pro-Western face of the Orange Revolution, Ukraine positioned itself as a champion of democratic reform across post-Soviet space. In the wake of political uprising, the Ukrainian brand also became highly personalized, manifesting in the visage of Yushchenko (particularly his scarred face, purportedly the outcome of a botched FSB-abetted assassination attempt using the poison dioxin) and that of then-prime minister Yulia Tymoshenko (whose fiery rhetoric, compelling aesthetics, and trademark fairy-tale braid made her an international celebrity). However, changing politics have negated this element of Ukrainian national image since the election of Viktor Yanukovych and the marginalization of Yushchenko and Tymoshenko (Sussman 2012), particularly after the latter was thrown in jail in 2011 for abuse of power. Reflecting Yanukovych's neo-Soviet political moorings, he sought to brand Ukraine as a country that is "making the world safer" through responsible provision of nuclear energy and as a fast-follower in the global space race, eschewing the "touchy-feely" elements of most brand campaigns while reinforcing late-Cold War stereotypes.[15]

In a curious repetition of recent history, the status of Ukraine's national image once again rose to prominence following the toppling of Yanukovych in the Euromaidan uprising, which began in November 2013. Rejecting government corruption and seeking to ensure the country's commitment to European integration, young people took to the streets of Kyiv facing down snipers and wintery conditions, ultimately producing a genuine political revolution. Unfortunately, the unintended consequences of this political change included the annexation of Crimea by Russia and a separatist revolt in the east of the country. Consequently, Brand Ukraine—once seen through the prism of "Little Russia" (Saunders 2008b)—is now weighed down by conflict with its eastern neighbor. Russia is deeply involved in the ongoing political crisis and is using its formidable diplomatic power to besmirch the country and its new government, as well as promote the existence of a "New Russia" in the Donbass (Gronlund 2014).

Besides the country's political brand, Ukraine has sought to market itself to the world via its attractive geography, music, and sport. A strong push was made for tourism under the banner of "Ukraine: Beautifully Yours," which included numerous advertising spots featuring a variety of female singers (Myroshnychenko 2014). Other recent efforts to attenuate the association with Communism and brand the nation's "clean cities, happy people, a healthy lifestyle, elegant architectural landmarks, and beautiful nature" include the slogans "Switch on Ukraine" or "High Time to See Ukraine" (Bagramian, Üçok Hughes, and Viscont 2012), as well as the "Ukraine: All about U" campaign on CNN, which focused on Ukraine as an "open country" in terms of lifestyle, business, and investment (Myroshnychenko 2014). In 2004, Ukraine, represented by the singer Ruslana, won the Eurovision Song contest, thus securing its capital Kyiv as the site for the 2005 event, which brought an enormous boost in the country's profile. In 2007, the participation of famous cross-dresser Andrei Danilko (via his persona Verka Serduchka) built on this pop-culture wellspring of positive association (Miazhevich 2012).[16] The country also received a significant boost in its international profile by co-hosting (with Poland) the 2012 UEFA European Cup, the premier event in European football—an event which provided a "historic opportunity" for nation-image management (Lebedenko 2008, 124). Other recent initiatives include positioning Ukraine as a site for filming movies, a hothouse for fashion, and a contemporary art mecca as the ethnic homeland of pop-art dynamo Andy Warhol (Myroshnychenko 2014). In terms of economic development, the country adopted the slogan "Ukraine: Moving in the Fast Lane." Looking beyond a focus on EU trade, Ukraine has sought to develop long-standing ties to the African continent in recent years, with some success, while also building new relationships with the wealthy Gulf States where significant numbers of Ukrainian expats live and work.

Turning to Belarus, often labeled "Europe's last dictatorship," President Aleksandr Lukashenko's (1994–) commitment to a command-and-control economy and maintenance of Soviet-era norms makes branding the country difficult given the neoliberal structures that scaffold the process. Nonetheless, Minsk hired world-class PR consultancy Bell Pottinger to help burnish its image in 2008, hoping to present the country through its "Hospitality beyond Borders" campaign; however, little if nothing ever emerged from the relationship.[17] More recently, Belarus has become a trend-setter in Eurasian new media nation branding, a strategy that reflects both the country's slavish mimicry of Russian mores (specifically the shift to political technology) and a propensity for outdoing "Big Brother" in the application of postmodern propaganda. Lukashenko's presidential page (http://president.gov.by/en/), carrying the neo-Soviet tagline "The State for the People!" shows a beaming, mustachioed Lukashenko with all signs of aging magically photoshopped away. While an executive-centric page is the new media branding norm for more authoritarian regimes in post-soviet Eurasia, Belarus also has a well-developed tourism portal (www.belarus.by/en/), which highlights reasons to visit the country, including hospitable people, opportunities for adventure, a unique cuisine, and a variety of architectural gems, thus situating the repressive republic closer to Croatia than Tajikistan on the spectrum of online outreach (Saunders 2014b). In an interesting elision between popular geopolitics and nation branding, President Lukashenko and French actor Gerard Depardieu appeared in a obviously choreographed "harvest" video in late July 2015, with the two men wielding scythes and then riding a tractor across Belarusian grass fields, thereby expanding Depardieu's Eurasianist credentials beyond Russia.

A country that exists in "relative obscurity on the fringe of Europe's outer-limits" (ICD 2011), Moldova is paralyzed by its poverty and internal divisions, thus lacking the resources to even engage with the necessary audiences to which a brand state must speak. Hitherto, the nation has mostly been known in the Anglophone West for sending much of its population abroad for employment in agriculture and sex work (IOM 2008) or as the locus of a convenient URL for US doctors' offices due to its .md top-level domain name (Franda 2002). Nonetheless, the country is promoting its viniculture through the "Wine Road in Moldova" campaign, bringing the role of "flagship exports" in nation branding into clear focus (Florek and Conejo 2007). However, of all the former Soviet republics, Moldova seems to have done the least to make itself known and lays bare the truism that "if you do not brand your nation, someone will do it for you" (in this case travel writer Tony Hawks or filmmaker Guillermo del Toro).

Transcaucasia, or the South Caucasus

Despite the problems that Ukraine, Belarus, and Moldova face in terms of branding, they do benefit from their unambiguous "European" geography, which guarantees these states a place in the continent's future and the geopolitical imaginary of "Europe." The same does not necessarily hold true for the countries of the South Caucasus. Regarding the national brands of these three small republics, there are clear weaknesses, but also significant room for improvement. Unlike the various 'Stans of Central Asia (discussed below), the Caucasian republics do not generally suffer from conflation and confusion with one another. As proud Christian nations with centuries of history, Armenia and Georgia possess a number of traits (monasteries, artwork, pilgrimage sites, etc.) that have served them well in terms of religious tourism (Armenia's tourism development agency brags that, as the mythical home of the Garden of Eden, the country has been a "Favorite Destination since Noah's Time"). Armenia's large diaspora in the US, France, and other Western countries serves as a natural bridge for building people-to-people contacts and strong international networks of knowledge and communication, as does the country's reputation for ecological sustainability, though the country has difficulty in overcoming the "decadence and dangers posed by the old [read Soviet] economy's rubble and waste that are witnessed in Armenia" (Pant 2005, 274). However, Simon Anholt (2015) has decided to call out the impoverished republic in his own work, chiding Yerevan that "Armenia has to do something for humanity" if it wants to be known and valued, suggesting that the country has some attributes in the field of education but allows these remain fallow on the global scale.

Putting aside the obvious cyberspatial issues that come from sharing its name with the US "Peach State," Georgia—with its decidedly pro-American, pro-European political orientation—has witnessed a dramatic rise in brand recognition in the past decade through positive media coverage, increasing ties to the EU, and well-developed educational programs linked to the Anglophone countries. Wallowing in post-Soviet obscurity and internecine conflict under the Gamsakhurdia (1991–1992) and Shevardnadze (1995–2003) regimes, the country's 2003 "Rose Revolution" political transformations (Djakeli 2013) and the "underdog war" with Russia increased awareness, making it the most recognizable state among the former Soviet bloc outside the Baltics (Myroshnychenko 2009). By way of example, in *Foreign Policy*'s first issue of 2009, the magazine featured a brief collection of current international reputation-management strategies entitled "Branded," including one on war-torn Georgia. In the blurb, which reminds the reader that "running a country is a lot like managing a business," *FP* leads with the M&C Saatchi tagline for the country—"The Winner is Georgia"—before going on to report that in a number of areas, including tourism and business, the Caucasian republic outperforms France, China, and Australia despite having lost on the battlefield. Shifting away from the geopolitical, the Georgian brand has steadily become linked to hospitality, seaside resorts, delectable wines, healing mineral waters, and a genuinely unique cuisine.

As a small Muslim country abutting a region where Islam is dominant, Azerbaijan lacks the quiddity possessed by its neighbors; however, nonetheless, the state's ample petrodollars have allowed it to increase its international profile from 2008 onwards, winning it a ranking as one of the "Top Ten Movers" for both People and Skills (#4) and Tourism (#5) in Brand Finance's (2014) annual report on nation brands, and prompting one commentator to note that "few countries have come as far in mastering the art of geopolitics as Azerbaijan" (Savdonik 2013). Most recently, the capital Baku hosted the Eurovision Song Contest in 2012,[18] attempting to showcase its economic development and "hip" music culture, which includes an indigenous form of rap music, and the 2015 European Games, which prompted sport marketing executive Michael R. Payne (2015) to tweet: "#Azerbaijan #Baku really understanding use of sport as nation brand builder—European Games, F1 now Euros. Destination will surprise people." Massive building and island-terraforming projects in Baku have made the Caspian city an authoritarian-architectural icon (Lacayo 2009), while the nation's growing reputation as a safe, freewheeling, comparatively liberal "Muslim state" with a penchant for partying has allowed the country to assume the role once played by Lebanon (Savdonik 2013). However, negative media reports about governance, from the smoldering Nagorno-Karabakh conflict to the repression of free speech, continue to problematize these nascent gains.

The Central Asian Republics (CARs)

Due to Soviet-imposed isolation, the Newly Independent States of Central Asia are—generally speaking—unknown quantities in large parts of the West, and as a result strive to use nation branding to achieve some level of international recognition (see Marat 2009). However, these states possess a rich cultural heritage in the form of the Silk Road. Once known as the "land of a thousand cities," the vast expanse between the Caspian Sea and Gobi Desert was home to many of the world greatest thinkers during the Islamic Golden Age (Starr 2009). For a millennium, the cities of Merv, Bukhara, Samarkand, and Kokand connected the Mediterranean to the Pacific Rim, serving a zone of transit for goods, people, and ideas, sustaining a realm of cultural diversity like no other in the world. Today, the various countries of the region are capitalizing on this robust commercial, cultural, and confessional legacy by branding their new nations as scions of ancient traditions associated with the Silk Road and the centrality of the "heart(land) of Eurasia" (Blinnikov 2011). Various efforts at Silk Road marketing are underway across the area, including various forms of "nomadic" eco-tourism, invocation of the Silk Road in support of trade relations, geopolitical positioning of Central Asia within the "Silk Road frame" (Ki-moon 2011), cooperation with international museum exhibitions touting Central Asia's contributions to world culture, and important international educational and cultural outreach programs. However, Central Asia continues to suffer from lingering Cold War prejudices among academics and journalists, as well as the withering effects of contemporary popular culture which frames the region as backward, unstable, and hostile, not least of which being Sacha Baron Cohen's long-running Borat parody. However, there is a certain iron at work here, the only country without any claims to Silk Road-era urbanism or ties to the Islamic Golden Age—Kazakhstan—is the CAR that has most closely approached the status of brand state.

When Kazakhstan was confronted with a challenge to its brand (see Saunders 2008a), one of the tactics it used to generate a more accurate view of the country was to employ YouTube to deliver short, informative videos about its best attributes. While President Nursultan Nazarbayev's (1991–) government did this alongside more traditional advertising supplements and other traditional forms of nation branding, the decision to "go viral," as it were, marked an important shift in the level at which nation branding takes place. While a full-page advertisement in

The New York Times will have a certain impact, it is comparatively ephemeral in the long term. Conversely, the videos "Kazakhstan Unveiled" and "Kazakhstan: Reaching for the Future," posted by Kazakhstan's former spokesperson for the US Embassy Roman Vassilenko (now Chairman of the Committee for International Information of Kazakhstan's Foreign Ministry), remain available for all to see. These short films—one is just a minute long while the other runs for about 10 minutes—rings all the right bells in terms of nation branding.

We must remember that the first requirement of a successful (Anholtian) branding campaign is a positive perception of law-abiding elites, stability, and good governance (these videos stress the "remarkable leadership of Nursultan Nazarbayev" and Kazakhstan as a "respected and reliable partner in the international community"). The second factor is the attitudes and orientation of the people. Kazakhstan, as we learn, was recognized by the late Pope John Paul II for its ethnic and religious tolerance in a region of the world where this is not the norm. Likewise, the videos offer proof of a well-educated, entrepreneurial population full of optimism for the future. Third, nation branding must highlight suitability for foreign direct investment. These videos position Kazakhstan as the "economic engine of Eurasia," discussing how the country transitioned from one of the "worst-off fragments of the Soviet empire into an economically strong and dynamically developing democratic country" that is now "one of the fastest growing economies in the world." While the videos do not specifically address the metric of personal safety and security of property, the overall imagery, tone, and content of the films suggests a country that is law-abiding, safe, and prosperous. In terms of reliable and desirable export products, we learn that Kazakhstan will produce over 3 billion barrels of oil a year by 2015 (in fact, the country only reached about 700 million in 2015). In terms of its unique cultural heritage, Vassilenko's clips inform us of the country's "deep cultural traditions" and vast array of nationalities and religions. With regards to an attractive geography, the viewer is educated on the "stunning natural beauty" of "a country four times the size of Texas which links Europe and Asia," and its national treasures and tourism sites that include parts of the old Silk Road, the attractive new capital of Astana, and mountain and wilderness recreation. Kazakhstan makes a strong case for environmental and natural resource protection by declaring its "courage" in the face of the Kremlin's anger over ending nuclear testing and its "voluntary nuclear disarmament" as a model to the rest of the world. And, lastly, Kazakhstan underlines its globalization-readiness and receptiveness, with references to its prospective WTO membership, OSCE leadership, hosting of the Asian Winter Games, "globally minded workforce," and global leadership in the commercial space flight and satellite industry.

Remembering to keep it light, Vassilenko even engaged with commenters on YouTube. In one exchange, Alexander721 asks "Hey! Where's all the potassium?" (a reference to Borat's spurious claim that Kazakhstan is world's largest exporter) to which Vassilenko replied:

Because of the huge demand in potassium from Kazakhstan, spiked by the Borat movie, we have exported all of it and are no longer able to supply it to the needy Americans. If you do require some, Alexander721, I will do my best to try to find whatever remains.

Elsewhere, the discussion becomes more serious as a viewer critiques Kazakhstan's record on freedom of the press, to which Vassilenko offers a balanced response and promises to share more information upon request. Putting aside issues of human rights (which dogged Almaty's bid for the 2022 Winter Olympic games up until 31 July 2015, when the city lost to Beijing in a two-way race),[19] such playful banter suggests that Kazakhstan has learned the (sometimes painful) pedagogy of neoliberalism, unlike its southern neighbors, who continue to follow (post-)Sovietesque pathways to national image production. By way of evidence, Kazakhstan was the only post-Soviet republic to be included in Brand Finance's "Best Performers" category in that year, witnessing a 37 percent increase to achieve an estimated brand value of $164 billion (BF 2014).[20] In addition to the above-mentioned efforts at nation branding, the Nazarbayev regime invest heavily in a variety of other "soft" initiatives, such as sponsoring one of the top-performing cycling teams, Astana Pro Team, subsidizing cinema that showcases Kazakh national identity, convening inter-faith conventions, hosting diplomatic summits, sponsoring fashion shoots, and laying out untold sums of money to turn Astana into one of the world's most striking "new capitals." Without qualification, "Brand Kazakhstan" weathered the Borat storm, sailing forth with flying colors.

Uzbekistan, the largest country in the region by population, has not fared as well. Engaged in a domestic-focused branding campaign to convince the populace of the greatness of their forebear Tamerlane (see Bell 1999; Adams 2010), President Islam Karimov (1990–) has doubled down on presenting his nation as a paragon of torture, religious strife, and duplicitous diplomacy, all characteristics of the "Father of the Modern Uzbek Nation," Timur-I Lang (see Marozzi 2006). Perhaps modeling a new nation on a personage who has come to exemplify the darkest and most threatening aspects of the Islamic East (beginning with Christopher Marlowe's play *Tamburlaine the Great* [1590] and continuing through to today, with a recent episode of the CBS geopolitical drama *Madame Secretary* titling an episode about an Iranian coup attempt "Tamerlane") might have been poor planning on the part of Tashkent. Nonetheless, Uzbekistan has sought to engage in some form of nation branding in the West, though as Sally Cummings (2013) points out, this presents a paradox alongside the adoration of Timur, given that much of the representational strategy relies on selling Soviet-era infrastructure (the Soviets decried Tamerlane as an imperialist despot). Uzbekistan, like its neighbors, suffers from an inability to positively project a helpful narrative about the country due to the fact that "images are developed and circulated by ruling elites and diplomats under the government's strict supervision" (Marat 2009, 1124). While Tashkent has sought to market its Silk Road cities as evidence of Uzbekistan's claimed sobriquet "Crossroads of Civilizations," most efforts have failed to connect, particularly after the Andijan Massacre in 2005.

After enjoying great success in its national-image management in the 1990s, Kyrgyzstan has become somewhat of comi-tragedy in the past decade. Once labeled the "Switzerland of Central Asia," for both its homespun democracy and its stunning alpine geography, the ousting of President Askar Akayev (1990–2005) led to a sustained period of political instability, with ethnic clashes and governmental turnover becoming the norm. Today, Kyrgyzstan hangs its hat on being the home of Santa Claus, assuming that St. Nick runs his operation like FedEx. Stemming from a study by a Swedish logistics company that suggested that the mountainous republic would be the best staging-ground for global gift deliveries on Christmas Eve (Marat 2008), the government staged a large festival with hundreds of National Guard in Santa costumes in 2008. While being picked up in the international media, the strategy became somewhat of a running joke, particularly given that Bishkek was going head-to-head with nation-brand behemoths like Finland, Norway, Canada, and the US to lay claim to being the permanent residence of the "world's strongest brand," Santa Claus (Hall 2008). Kyrgyzstan has found more success in marketing itself as a ecofriendly-adventure tourism destination, given the allure of Issyk Kul and its ability to market the country as "Land of the Tien Shan"; however, considering that *The New York Times* recently rendered the country's name as "Kyrzbekistan" (Scheib 2015), there remains much work to be done. Nonetheless, with a strong focus on the German market, the Central Asian Republic has gained traction in branding itself as a purveyor of authentic "Eurasian" culture, high-end fabrics (including merino wool and organic cotton), and bio-pharma and holistic medicines.

Turkmenistan offers a study in contrasts. Universally lambasted for its repressive policies, particularly towards Internet use (Annasoltan 2010), the fossil fuel-rich desert country has made some rather surprising efforts to burnish its image in cyberspace and attract tourists and investors. In the wake of Türkmenbaşy's death, there was hope that the country would democratize and undergo other reforms; however, his successor (and reputed body-double) Gurbanguly Berdimuhamedow has done little to bring his country in line with Western expectations of governance. While the country's human rights record remains abysmal, Ashgabat has undertaken a number of initiatives that are branding Turkmenistan in the global supermarket. One is the new Caspian coast mega-city named after the late leader. Rising out of the old railroad junction Krasnovodsk, this architectural showcase represents a latter-day Potemkin village; replete with casinos and high-end shops, the city of Türkmenbaşy represents a rare free economic zone in a state that still tightly controls the market. Plans for this building project were announced shortly after the death of the long-running president, with the slogan "A new era, a new Turkmenistan" (Antelava 2015). While the project has foundered, the capital is emerging as an architectural colossus, with dozens of awe-inspiring buildings and monuments, including the laughably grotesque statue of the *Ruhnama*, or "Book of the Soul," required reading for every Turkmen (see Figure 8.5). Sweeping domestic issues under the carpet, Berdimuhamedow has launched various branding campaigns, including announcing a new "Golden Age" in Turkmenistan, a country which is witnessing a "Great Era of Rebirth" and entering an "Epoch of Happiness" (Omelicheva 2014). With a sleek, interactive web portal for the country, Turkmenistan stands out among the authoritarian-inclined states of the former USSR, perhaps counting on the West's almost complete ignorance of the country some 25 years after independence, possibly explaining why it was so easy for investigative reporter Ken Silverstein (2008) to misrepresent himself as a K-Street lobbyist and shill for the neo-Stalinist regime in the halls of American power.[21]

Rounding out the Newly Independent States, we now turn our attention to Tajikistan, the country with perhaps the least-known brand in the whole of post-Soviet Eurasia. In 2011, I had the opportunity to speak with Imomudin Sattarow, Ambassador of Tajikistan to Germany, at a conference in Berlin on the topic of nation branding. While his proxy Lochin Fayzullaev, First Vice-Chairman of the Committee of Youth Affairs, touched on all the issues a branding state would want to address in his lecture "Tourism and Sustainable Development in Tajikistan" (2011), the ambassador waxed realistic in his private conversations with the attendees (including me). Two years later to the day, he would give his own speech, entitled "Nation Brand Tajikistan" (2013), lauding the diplomatic and economic connections between Germany and Tajikistan, making the case that the relationship opens the door to greater interaction with the West and the world. Despite the hope expressed by Sattarow, and a long list of national development projects, economic liberalization, and new tourism infrastructure, Tajikistan continues to be a victim of its geographical location and generalized perceptions about the region as a whole (e.g., Presidents for Life, clan-based corruption, Islamist threats, etc.). Beset by a host of problems—from the lingering problems associated with the Civil War (1992–1997) to its restive 1,400–kilometer border with Afghanistan to drug-trafficking—the small, mountainous republic faces an uphill battle to make

Figure 8.5 The Ruhnama Monument in Ashgabat (Arto Halonen Art Films)

itself known (and has yet to release a slogan to help do so), but being valued seems to be off the table as the best the country can hope for is becoming another 'Stan in a sea of them, despite a strong record on promoting women's rights and other factors that differentiate it from Afghanistan and Pakistan. Eco-tourism and mountain vacations are the mainstay of the country's strategy, but with safety issues, logistical difficulties, and poor off-season delivery of light and power to the provinces, only the most intrepid Western traveler is likely to board a flight for Dushanbe (although *renminbi*-laden Chinese investors seem to offer an alternate route to Tajikistan's goal of achieving a semblance of brand quality).

While few scholars are likely to conflate nation branding with popular geopolitics, the two are twinned processes in a contemporary world saturated by digital flows of foreign images. As I have argued, both practices are consumption-centric, geographically obsessed, ideologically driven, and dependent on "imagineers" of one sort or another. The driving engine behind nation branding is marketing; in practice, the focus is on international outreach, tourism development, positive messaging, showcasing greatness, and achievement of long-term resonance with various audiences. Pop-culture geopolitics, on the other hand, trades in tropes, stereotypes, and the ethnocentrism of domestic audiences, relying on metaphors and cultural prisms to achieve various ends (some political, some not). However, much links these two praxes: centrality of the visual, sound-bite language, neo-liberal worldviews, proselytizing orientations, etc. Rather than suggesting that we should keep these two fields of activity separated by artificial barriers, the concluding section will argue that much can be gained by bringing popular geopolitics and nation branding into conversation with one another.

Notes

1 In 2013 the protocols were updated, with many analysts suggesting a shift towards greater isolation in international affairs. For more information, see Monaghan 2013.

2 Lebedenko describes this process as the use of the "aggregate of technologies drawing resources from the depths of humanitarian knowledge and aimed at creating and changing the rules of human interaction in order to meet the challenges of the external environment" (2004, 71), although this rather fluid definition might as easily be applied to "state propaganda."

3 Political scientist Andrew Wilson describes political technologists as something much more than "spin doctors," claiming instead that they use all the tools and techniques of postmodern media and society in their "construction of politics as a whole" (2005, 49).

4 In the West, Russia is widely seen as having won the ground war, but lost the media battle in the South Ossetian War (see Bookman 2008; Pitter 2008). With an English-speaking president who was willing to be interviewed by CNN multiple times per day, as well as a host of new media allies in the world of Facebook and other social media sites, Georgia was able to market itself as a "David" taking on a Russian "Goliath," a frame that played with popular-cultural depictions of imperial Rus.

5 In late 2014, RT UK was called before Ofcom, the British television regulator, for violations of impartiality in its coverage of the crisis in Ukraine.

6 Of all the former Soviet republics, only the Baltic States have managed Facebook pages with unique branding-style content: "Visit Estonia" (132,000 likes); "If you like

Latvia, Latvia likes you" (82,000 likes); and "We love Lithuania" (190,000 likes); see Saunders 2014b.

7 Both Estonia and Lithuania have seriously considered changing their country names since independence, with Estland (the German name for Estonia) and Lietuva (the Lithuanian rendering of the country's name) being the most popular options, though Anholt (2008) publicly condemned such ideas as "daft."

8 The various states of the Baltic Rim regularly engage in conversations about common elements of their identity with the aim of increasing links between the various countries of the region. Most recently, this was evidenced by the 2013 conference on "The Baltic Sea Region: Identity, Branding and Communication," held in Helsinki, Finland.

9 Follow-on campaigns would include "Positively Surprising" and "Introduce Estonia" (see Mändmets 2010), as well as the tech-focused "EST_IT@2018" (see Curry Jansen 2012)

10 A problem in image that is less often spoken of is the pro-Nazi sheen that characterizes some Western views of the country. Many Latvians still see the Nazis as liberators who freed their nation from Soviet annexation, and a recent rise in anti-Russian sentiment with strong fascist overtones (including commemorations of the Latvians who served in Waffen-SS divisions) only reinforces this "image nightmare" for Latvia's nation branders (see Kaža 2015).

11 If one accepts that souvenirs are a form of nation branding, then we can identify a curious commonality across the Baltic States (as well as Ukraine): a Viking fetish. In recent trips to these three countries, I was astounded by the popularity of Viking-themed tchotchkes for purchase in touristic areas, from medieval weaponry and armor to mead horns and mini-Viking statues.

12 The tagline was never fully deployed, as it did not register with the citizenry and ultimately became somewhat of a laughingstock, though it tested well abroad (Bankauskaitė 2009).

13 Although the country is committed to "being Baltic" (much like its neighbour Poland), thus linking it to the group of countries that make up Norden, an increasingly cohesive geopolitical region within the larger framework of Europe.

14 A number of scholars have taken up the theme that it is almost "unfair" to compare the nation branding efforts of the Baltic States (particularly Estonia) to other post-Soviet republics, thus reinforcing the geopolitical diversity that exists within Eurasia (see, for instance, Gardner and Standaert 2003).

15 In a full-page 22 February 2012 *New York Times* branding advert, Yanukovych appears in a Sovietseque stock photo underwritten by the caption: "The energy future of Ukraine is impossible without atomic energy," eerily situating the country within the popular-geopolitical frame discussed in Chapter 7.

16 However, as Miazhevich points out, such linkages of national image to "Eurotrash" culture is problematic in terms of branding, as it reinforces Western stereotypes of post-Soviet space such as those discussed in Chapter 6.

17 For several months in 2009, I was in communication with Paul Baverstock, who was managing the contract, with the goal of discussing strategy and tactics. Ultimately, he declined to speak on the topic. Lord Bell later commented that all he seemed to do for Belarus was attempt to "explain" Lukashenko's behavior to the British diplomats (Mance 2014).

18 The contest, however, proved problematic as many activists used the global media attention to bring attention to human rights issues in the country, proving that alternative narrations of the nation and national "culture jamming" are major issues in contemporary state branding efforts (see Saunders 2014b).

19 While Kazakhstan's record on media freedom and its treatment of the opposition remains poor, the country has not ranked as low in international reports on human rights abuses as either China (its sole competitor for the Games) or many of its peers in Eurasia, including Russia, Turkmenistan, and Uzbekistan, all of which were targeted for opprobrium by Freedom House in its 2014 Freedom in the World report (with the latter two being labeled the "Worst of the Worst").

20 This was in stark opposition to Ukraine, which saw its brand drop by 37 percent in 2014, a rather distressing finding for Kyiv which is often in competition with Astana for the status of the "second most important" state to emerge from the dissolution of the USSR.
21 Peter van Ham, author of the influential essay on "Brand States" (2001), once quipped that Sacha Baron Cohen would have been wiser to have used Turkmenistan, given that it is almost completely unknown in Europe and North America (see Metahaven 2008).

References

Adams, Laura L. 2010. *The Spectacular State: Culture and National Identity in Uzbekistan.* Durham, NC: Duke University Press.

Andersson, Marcus. 2007. "Region Branding: The Case of the Baltic Sea Region." *Place Branding and Public Diplomacy* (3):120–130.

Andreev, Pavel. 2014. "Convincing a Seven-Year-Old: National Identity in Russia's Soft Power." *Valdai Discussion Club*, available at http://valdaiclub.com/politics/67024.html [last accessed 18 July 2015].

Anholt, Simon. 2006. "What is a Nation Brand?" *Superbrands*, available at www.superbrands.com/turkeysb/trcopy/files/Anholt_3939.pdf [last accessed 2 January 2007].

——. 2007a. *Competitive Identity: The New Brand Management for Nations, Cities and Regions.* Houndsmills, UK: Palgrave Macmillan.

——. 2007b. "Latvia and the Legacy of Communism " *Simon Anholt's Placeblog*, available at http://simonanholt.blogspot.com/2007/11/latvia-and-legacy-of-communism.html [last accessed 24 July 2015].

——. 2008. "Rules of Attraction." *City Paper*, available at www.citypaper.ee/rules_of_attraction/ [last accessed 1 May 2009].

——. 2013. "Russia's International Image, and Why It Matters." *Valdai Discussion Club*, available at http://valdaiclub.com/politics/62321.html [last accessed 1 July 2011].

——. 2015. "Armenia Has To Do Something for Humanity." *MediaMax*, available at www.mediamax.am/en/news/interviews/13053/?utm_source=mediamax.am&utm_medium=widget_300x300&utm_campaign=partnership#sthash.ccXQp5QF.dpuf [last accessed 27 July 2015].

Annasoltan. 2010. "State of Ambivalence: Turkmenistan in the Digital Age." *Digital Icons: Studies in Russian, Eurasian and Central European New Media* no. 3:1–13.

Antelava, Natalia. 2015. "Turkmenistan Starts Tourist Drive." *BBC*, available at http://news.bbc.co.uk/2/hi/asia-pacific/6911661.stm [last accessed 31 July 2013].

Armin. 2010. "Manufacturing Happiness." *Brand New*, available at www.underconsideration.com/brandnew/archives/manufacturing_happiness.php#.VbJ3PPkfPNg [last accessed 24 July 2015].

Aronczyk, Melissa. 2008. "'Living the Brand': Nationality, Globality and the Identity Strategies of Nation Branding Consultants." *International Journal of Communication* no. 2:41–65.

——. 2013. *Branding the Nation: The Global Business of National Identity.* Oxford and New York: Oxford University Press.

Auchter, Jessica. 2014. *The Politics of Haunting and Memory in International Relations.* London: Routledge.

Avgerinos, Katherine P. 2009. "Russia's Public Diplomacy Effort: What the Kremlin is Doing and Why It's Not Working." *Journal of Public and International Affairs* no. 20:115–132.

Babich, Dmitry. 2007. "The Rediscovery of the 'Russian World'." *The School of Russian and Asian Studies*, available at www.sras.org/rediscovery_of_the_russian_world [last accessed 29 December 2009].

Bagramian, Ruben, Mine Üçok Hughes, and Luca M. Viscont. 2012. "Bringing the Nation to the Nation Branding Debate: Evidence From Ukraine." In *Proceedings of the Academy of Marketing Science Cultural Perspectives in Marketing Conference*, available at www.academia.edu/16460703/Bringing_the_nation_to_the_nation_branding_debate_Evidence_from_Ukraine [last accessed 4 April 2016].

Bankauskaitė, Dalia. 2009. Author interview, 29 May.

Basulto, Dominic. 2013. "The Sochi Olympics and the Re-Branding of Russia." *Russia Direct*, available at www.russia-direct.org/analysis/sochi-olympics-and-re-branding-russia [last accessed 20 July 2015].

Bell, James. 1999. "Redefining National Identity in Uzbekistan: Symbolic Tensions in Tashkent's Official Public Landscape." *Cultural Geographies* no. 6 (2):183–213.

Beumers, Birgit. 2012. "National Identity through Visions of the Past: Contemporary Russian Cinema." In *Soviet and Post-Soviet Identities*, edited by Mark Bassin and Catriona Kelly, 55–71. Cambridge: Cambridge University Press.

Blinnikov, Mikhail S. 2011. *A Geography of Russia and Its Neighbors*. New York: The Guilford Press.

Bogomolov, Alexander, and Oleksandr Lytvynenko. 2012. "A Ghost in the Mirror: Russian Soft Power in Ukraine." *The Aims and Means of Russian Influence Abroad Series*. London: Chatham House, available at www.chathamhouse.org/publications/papers/view/181667 [last accessed 28 March 2016].

Bonin, Liane. 2000. "Mel Gibson Explains When War is Right." *Entertainment Weekly*, available at www.ew.com/article/2000/07/05/mel-gibson-explains-when-war-right [last accessed 24 July 2015].

Bookman, Jay. 2008. "Russian "Brand" Takes a Hit after Losing the Media Wars." *Atlanta Journal-Constitution*, available at www.ajc.com/search/content/shared/news/stories/2008/09/BOOKMAN_COLUMN_0915_COX.html [last accessed 1 July 2009].

Borenstein, Eliot. 2008. *Overkill: Sex and Violence in Contemporary Russian Popular Culture*. Ithaca, NY: Cornell University Press.

Brand Estonia. 2015a. "Brand Estonia—Welcome to Estonia." *Brand Estonia—Teeme Eesti tuntuks—Estonia.eu*, available at http://photos.visitestonia.com/eng/videos/oid-1577/?#id=1577 [last accessed 23 July 2015].

———. 2015b. "Rootedness—Estonians Have Deep Roots." *Brand Estonia—Teeme Eesti tuntuks—Estonia.eu*, available at http://photos.visitestonia.com/eng/videos/oid-1580/?#id=1580 [last accessed 23 July 2015].

Brand Finance. 2014. "Annual Report on Nation Brands." *Brand Finance*, available at www.brandfinance.com/knowledge_centre/reports/brand-finance-nation-brands-2014 [last accessed 27 July 2014].

Brencis, Ainars, and Jacob Ikkala. 2013. "Sports as a Component of Nation Branding Initiatives: The Case of Latvia." *Marketing Review* no. 13 (3):241–254.

Buckley, Neil. 2006. "Moscow Hopes PR Group Will Improve Its G8 Image." *Financial Times*, available at www.ft.com/cms/s/0/7fa75188-d8ae-11da-9715-0000779e2340.html#axzz3gGXAvO00 [last accessed 2 May 2006].

Budnitskiy, Stanislav. 2012. *Lost Tribes of Nation Branding? Representations of Russian Minority in Estonia's Nation Branding Efforts*. Budapest: Central European University.

Burt, Julian. 2007. "A World Beyond Spin—Re-branding Nations." *Pop Matters*, available at www.popmatters.com/post/a-world-beyond-spin-re-branding-nations/ [last accessed 13 July 2015].

Cassiday, Julie A., and Emily D. Johnson. 2010. "Putin, Putiniana and the Question of a Post-Soviet Cult of Personality." *Slavonic & East European Review* no. 88 (4):681–707.

Churchill, Winston. 1939. "Russia: A Riddle, Wrapped in a Mystery, Inside an Enigma." Radio broadcast, London.

Collier, Mike. 2008. "The Image Makers: Estonia." *Baltic Times*, available at www.baltictimes. com/news/articles/so839/ [last accessed 3 March 2009].

Conniff, Richard. 2011. "Strike Up the Band." *Smithsonian Magazine*, available at www. smithsonianmag.com/arts-culture/strike-up-the-brand-50059089/?no-ist [last accessed 24 July 2015].

Cromwell, Thomas. 2013. "National Identity and National Image: A Purpose-Driven Approach to Nation Branding." *East West Communications*, available at www. eastwestcoms.com/res_national.htm [last accessed 5 January 2014].

Cummings, Sally N. 2013. *Symbolism and Power in Central Asia: Politics of the Spectacular*. London: Routledge.

Curry Jansen, Sue 2008. "Designer Nations: Neo-liberal Nation Branding—Brand Estonia." *Social Identities* no. 14 (1):121–142.

———. 2012. "Redesigning a Nation: Welcome to E-stonia, 2001–2018." In *Branding Post-Communist Nations: Marketizing National Identities in the "New" Europe*, edited by Nadia Kaneva, 79–98. New York and London: Routledge.

Čvoro, Uroš. 2014. *Turbo-folk Music and Cultural Representations of National Identity in Former Yugoslavia*. Surrey, UK and Burlington, VT: Ashgate Publishing, Ltd.

Dinnie, Keith. 2008. *Nation Branding: Concepts, Issues, Practice*. Oxford: Butterworth-Heinemann.

Djakeli, Kakhaber. 2013. "Country's Branding for More Attractive Image of Georgia." *IBSU Journal of Business* no. 2 (1):15–20.

Dunlop, John B. 2000. "Sifting Through the Rubble of the Yeltsin Years." *Problems of Post-Communism* no. 47 (1):3–15.

Dzenovska, Dace. 2005. "Remaking the Nation of Latvia: Anthropological Perspectives on Nation Branding." *Place Branding* no. 1 (2):173–186.

Faro, Jeremy. 2009. Author interview, 11 May.

Fayzullaev, Lochin. 2011. "Tourism and Sustainable Development in Tajikistan." In *The Berlin International Economics Congress 2011*. Berlin: Institute for Cultural Diplomacy.

Feklyunina, Valentina. 2008. "Battle for Perceptions: Projecting Russia in the West." *Europe-Asia Studies* no. 60 (4):605–629.

Flamm, Matthew. 2001. "Jonathan Franzen's Latest Faux Pas: Lithuania Delegates Say 'The Corrections' Maligns Their Country." *Entertainment Weekly*, available at www.ew.com/article/2001/11/30/jonathan-franzens-latest-faux-pas [last accessed 24 July 2015].

Florek, Magdalena, and Francisco Conejo. 2007. "Export Flagships in Branding Small Developing Countries: The Cases of Costa Rica and Moldova." *Place Branding and Public Diplomacy* no. 3 (1):53–72.

Foreign Policy. 2009. "Branded." *Foreign Policy* (170):28–28.

Foxhall, Andrew. 2013. "Photographing Vladimir Putin: Masculinity, Nationalism and Visuality in Russian Political Culture." *Geopolitics* no. 18 (1):132–156.

———. 2015. "The War at Home: How Russia Is Winning the Battle for Hearts and Minds." *New Statesman*, available at www.newstatesman.com/politics/2015/03/war-home-how-russia-winning-battle-hearts-and-minds [last accessed 2 May 2006].

Franda, Marcus F. 2002. *Launching Into Cyberspace: Internet Development and Politics in Five World Regions*. Boulder, CO: Lynne Rienner Publishers.

Frasher, Spencer, Michael Hall, Jeremy Hildreth, and Mia Sorgi. 2003. *A Brand for the Nation of Latvia*. Oxford: Oxford Saïd Business School.

Freedom House. 2014. *Freedom in the World 2014*. Washington, DC: Freedom House Press.

Gardner, Stephen, and Mike Standaert. 2003. "Estonia and Belarus: Branding the Old Bloc." *Brand Channel*, available at www.brandchannel.com/print_page.asp?ar_id=146&section=main [last accessed 5 January 2009].

Gorham, Michael. 2011. "Virtual Rusophonia: Language Policy as 'Soft Power' in the New Media Age." *Digital Icons: Studies in Russian, Eurasian and Central European New Media* no. 5:23–48.

———. 2014. *After Newspeak: Language Culture and Politics in Russia from Gorbachev to Putin*. Ithaca, NY: Cornell University Press.

Goscilo, Helena. 2013. *Putin as Celebrity and Cultural Icon*. London: Routledge.

Greenslade, Roy. 2010. "Telegraph to Continue Publishing Russian Propaganda Supplement." *The Guardian*, available at www.theguardian.com/media/greenslade/2014/jul/29/dailytelegraph-russia [last accessed 13 July 2015].

Gronlund, Jay. 2014. "Re-brand Ukraine? How to Create a New, Trusted Nation Brand." *Biznology*, available at http://biznology.com/2014/05/re-brand-ukraine-create-new-trusted-nation-band/ [last accessed 26 July 2015].

Hall, C. Michael. 2008. "Santa Claus, Place Branding and Competition." *Fennia* no. 186 (1):59–67.

Havens, Timothy, and Kati Lustyik. 2013. *Popular Television in Eastern Europe During and Since Socialism*. London: Routledge.

Hudson, Victoria. 2013. "The *Russkiy Mir*: Easing Russia's Way to Post-Colonialism". Paper presented at Global Russian: Exploring New Research Perspectives, The Princess Dashkova Centre, University of Edinburgh (24 January).

ICD. 2011. *Cultural Diplomacy Outlook Report 2011*. Berlin: Institute for Cultural Diplomacy.

Interbrand. 2008. "Country Case Insight—Estonia." In *Nation Branding: Concepts, Issues, Practice*, edited by Keith Dinnie, 229–235. Oxford: Butterworth-Heinemann.

IOM. 2008. *Migration in Moldova: A Country Profile*. Geneva: International Organization for Migration.

ITAR-TASS. 2012. *Rossotrudnichestvo Will Work on Russia's Image. ITAR-TASS (Russian Press Review)*, 4 September, available from http://tass.ru/en/russianpress/681373 [last accessed 4 April 2016].

IWS. 2011. "Top Ten Internet Languages." *Internet World Stats*, available at www.internetworldstats.com/stats7.htm [last accessed 3 March 2013].

Jameson, Frederic. 1991. *Postmodernism or the Cultural Logic of Late Capitalism*. Durham, NC: Duke University Press.

Jordan, Paul. 2014. *The Modern Fairy Tale: Nation Branding, National Identity and the Eurovision Song Contest in Estonia*. Tartu, Estonia: University of Tartu Press.

Kaneva, Nadia. 2012. "Nation Branding in Post-Communist Europe: Identities, Market, and Democracy." In *Branding Post-Communist Nations: Marketizing National Identities in the "New" Europe*, edited by Nadia Kaneva, 3–22. New York and London: Routledge.

Kaža, Juris. 2015. "Latvian March Remembers Veterans Who Fought Alongside Nazis." *Wall Street Journal*, available at www.wsj.com/articles/latvian-march-remembers-veterans-who-fought-with-nazis-1426520842 [last accessed 24 July 2015].

Ki-moon, Ban. 2011. Opening address by the Secretary General of the United Nations delivered on day one of the Silk Road Summit, Almaty, Kazakhstan (30 November), available at www.economistinsights.com/event/silk-road-summit/tab/0 [last accessed 30 March 2016].

Kottasova, Ivana. 2015. "Putin Drops His American PR Company." *CNN Money*, available at http://money.cnn.com/2015/03/12/media/russia-putin-pr-ketchum/ [last accessed 20 July 2015].

Lacayo, Richard. 2009. "The Architecture of Autocracy." *Foreign Policy*, available at http://foreignpolicy.com/2009/10/07/the-architecture-of-autocracy/ [last accessed 9 September 2010].

Latvia Institute. 2014. "Culture: the Key to Latvia." *Latvia.eu Resources*, available at www.latvia.eu/brochures [last accessed 24 July 2015].

Lebedenko, Vladimir. 2004. "Russia's National Identity and Image-Building." *International Affairs: A Russian Journal of World Politics, Diplomacy & International Relations* no. 50 (4):71–77.

——. 2008. "Country Case Insight—Russia: On National Identity and the Building of Russia's Image." In *Nation Branding: Concepts, Issues, Practice*, edited by Keith Dinnie, 107–129. Oxford: Butterworth-Heinemann.

Leone, Piero, and Alex Gorshkov. 2004. "Russia's New National Pride." *Brand Strategy* (179):54–55.

Lewis, M. Paul. 2009. *Ethnologue: Languages of the World, Sixteenth Edition*. Dallas, TX: SIL International.

Lionikaitė, Jūratė. 2012. Internal Perceptions as Groundings of Value for the Nation Branding: The Case of Lithuania. In *Berlin International Economics Congress*. Berlin: Institute for Cultural Diplomacy.

Lukyanov, Fyodor. 2014. "The Lost Dream of a Common European Home." *Russia Beyond the Headlines*, available at http://rbth.com/opinion/2014/10/27/the_lost_dream_of_a_common_european_home_40915.html [last accessed 9 July 2015].

Makarychev, Andrey, and Alexandra Yatsyk. 2013. "Sochi Olympics—The Dangers of Rebranding." *Open Democracy*, available at http://opendemocracy.net/od-russia/andrey-makarychev-and-alexandra-yatsyk/sochi-olympics-%E2%80%93-dangers-of-rebranding [last accessed 20 July 2015].

Mance, Henry. 2014. "Lord Bell's Textbook for Old School Public Relations." *Financial Times*, available at www.ft.com/intl/cms/s/0/3d39bf02–4ae2–11e4–839a-00144feab7de.html#axzz3h1t88ck7 [last accessed 26 July 2015].

Mändmets, Leitti. 2010. "The Story of Creating Brand Estonia." *Estonian Ministry of Foreign Affairs Yearbook* no. 1:71–76.

Mansel, Tim. 2013. "How Estonia Became E-stonia." *BBC News*, available at www.bbc.com/news/business-22317297 [last accessed 23 July 2015].

Marat, Erica. 2008. "World's Famous Santa Clauses United in Kyrgyzstan." *Central Asia-Caucasus Analyst* no. 10 (24):15–16.

——. 2009. "Nation Branding in Central Asia: A New Campaign to Present Ideas about the State and the Nation." *Europe-Asia Studies* no. 61 (7):1123–1136.

Markessinis, Andreas. 2011. "Latvia—State of the Nation Brand." *Nation Branding*, available at http://nation-branding.info/2011/04/27/latvia-state-of-the-nation-brand/ [last accessed 24 July 2015].

Marozzi, Justin. 2006. *Tamerlane: Sword of Islam, Conqueror of the World*. New York: Da Capo Press.

McPhee, Sarah. 2014. "Securing Sochi: Nation Branding and Mega-Events." *Ellison Center for Russian, East European and Central Asian Studies*, available at http://ellisoncenter.

washington.edu/community-spotlight/securing-sochi-nation-branding-mega-events/ [last accessed 20 July 2015].

Metahaven. 2008. "Brand States: Postmodern Power, Democratic Pluralism, and Design." *e-Flux*, available at www.e-flux.com/journal/brand-states-postmodern-power-democratic-pluralism-and-design/ [last accessed 20 July 2009].

Miazhevich, Galina. 2012. "Ukrainian Nation Branding Off-line and Online: Verka Serduchka at the Eurovision Song Contest." *Europe-Asia Studies* no. 64 (8):1505–1520.

Minenko, Dmitriy. 2012. "RuNet 2012: Unprecedented Online Market Growth In Russia Continues." *Multilingual Search*, available at www.multilingual-search.com/runet-2012–unprecedented-online-market-growth-in-russia-continues/13/07/2012/ [last accessed 8 March 2013].

Moilanen, Teemu, and Seppo Rainisto. 2008. *How to Brand Nations, Cities and Destinations: A Planning Book for Place Branding*. Houndsmills, UK: Palgrave Macmillan.

Monaghan, Andrew. 2013. *The New Russian Foreign Policy Concept: Evolving Continuity, Russia and Eurasia*. London: Chatham House.

Müller, Martin. 2014. "After Sochi 2014: Costs and Impacts of Russia's Olympic Games." *Eurasian Geography and Economics* no. 55 (6):628–655.

Myroshnychenko, Vasyl. 2009. Author interview, 9 January.

——. 2014. Author interview, 17 October.

Nagornykh, Irina, and Ivan Safronov. 2015. "The Brand Is Trending [*Brend popal v trend*]." *Russkiy Mir*, available at https://news.mail.ru/society/21282344/?idc=1 [last accessed 20 July 2015].

Nunn, Sam, and Adam N. Stulberg. 2000. "The Many Faces of Modern Russia." *Foreign Affairs* no. 79 (2):45–62.

Olins, Wally, and Jeremy Hildreth. 2009. *Selling Lithuania Smartly: A Guide to the Creative-Strategic Development of an Economic Image for the Country*. New York: Saffron Brand Consultants.

Omelicheva, Mariya Y. 2014. "Eye on International Image: Turkmenistan's Nation branding." In *Nationalism and Identity Construction in Central Asia: Dimensions, Dynamics, and Directions*, edited by Mariya Y. Omelicheva, 91–110. Lanham, MD: Lexington Books.

Orttung, Robert, Elizabeth Nelson, and Anthony Livshen. 2015. "How Russia Today Is Using YouTube." *Washington Post*, available at www.washingtonpost.com/blogs/monkey-cage/wp/2015/03/23/how-russia-today-is-using-youtube/ [last accessed 20 July 2015].

Ostapenko, Nikolai. 2008. "Sochi, Russia, 2014: What Should You Ask the Golden Fish For? Nation Branding through the Olympics." Paper presented at the Oxford Business and Economics Conference, Oxford, UK, 1–10.

——. 2010. "Nation Branding of Russia through the Sochi Olympic Games of 2014." *Journal of Management Policy and Practice* no. 11 (4):60–63.

Pant, Dipak R. 2005. "A Place Brand Strategy for the Republic of Armenia: 'Quality of Context' and 'Sustainability' as Competitive Advantage." *Place Branding* no. 1:273–282.

Park, Ausra. 2009. "'Selling' a Small State to the World: Lithuania's Struggle in Building Its National Image." *Place Branding and Public Diplomacy* no. 5 (1):67–84.

Payne, Michael R. 2015. "#Azerbaijan #Baku." *Twitter (@MichaelRPayne1)*, available at https://twitter.com/michaelrpayne1/status/512950370008834048 [last accessed 27 July 2015].

Pitter, Silvio. 2008. "Local Win, Global Loss." *Johnson's Russia List*, available at www.russialist.org/archives/2008-152-25.php [last accessed 1 July 2011].

Pomerantsev, Peter. 2014. "Yes, Russia Matters." *World Affairs* no. 177 (3):16–23.

Potter, Mitch. 2008. "The Man Who's Trying to Change Lithuania." *Star (Toronto)*, available at www.thestar.com/news/2008/02/16/the_man_whos_trying_to_change_lithuania.html [last accessed 24 July 2015].

Rainsford, Sarah. 2015. "Brand Putin: Russia's President Still in Fashion 15 Years On." *BBC News*, available at www.bbc.com/news/world-europe-32076836 [last accessed 20 July 2015].

Rizvi, Haider. 2015. "Foreign News Channels Drawing U.S. Viewers." *North America Inter Press Service*, available at http://ipsnorthamerica.net/news.php?idnews=2815 [last accessed 20 July 2015].

RMF. 2012. "Vyacheslav Nikonov and Lyudmila Verbitskaya Deliver Opening Address to the Russkiy Mir Assembly." *Russkiy Mir Foundation*, available at www.russkiymir.ru/en/news/130287/ [last accessed 3 March 2014].

———. 2015. "Russian Center—Definition and Mission." *Russkiy Mir*, available at http://russkiymir.ru/en/rucenter/what-is.php [last accessed 20 July 2015].

Rosenberg, Steven. 2014. "Ukraine Crisis: Meeting the Little Green Men." *BBC News Europe*, 30 April, available from www.bbc.com/news/world-europe-27231649 [last accessed 17 July 2014].

Rossotrudnichestvo. 2013. "Rossotrudnichestvo." available at http://rs.gov.ru/ [last accessed 29 December 2013].

RT. 2015. "About Us—Distribution." *RT*, available at www.rt.com/about-us/distribution/ [last accessed 20 July 2015].

Ryazanova-Clarke, Lara. 2014. "The Russian Language, Challenged by Globalisation." In *The Russian Language Outside the Nation*, edited by Lara Ryazanova-Clarke, 1–31. Edinburgh: Edinburgh University Press.

Salmin, A. M. 2000. "Russia, Europe, and the New World Order." *Russian Social Science Review* no. 41 (3):4–16.

Sattarow, Imomudin. 2013. "Nation Brand Tajikistan." In *The Berlin International Economics Congress 2011*. Berlin: Institute for Cultural Diplomacy. Berlin: Institute for Cultural Diplomacy.

Saunders, Robert A. 2008a. "Buying into Brand Borat: Kazakhstan's Cautious Embrace of Its Unwanted 'Son'." *Slavic Review* no. 67 (1):63–80.

———. 2008b. "Ukraine's Nation Brand—Why It Matters." *Zovnishni Spravy (UA "Foreign Affairs")* no. 7:38–39.

———. 2009. "Wiring the Second World: The Geopolitics of Information and Communications Technology in Post-Totalitarian Eurasia." *Digital Icons: Studies in Russian, Eurasian and Central European New Media* no. 1:1–24.

———. 2012. "Brand Interrupted: The Impact of Alternative Narrators on Nation Branding in the Former Second World." In *Branding Post-Communist Nations: Marketizing National Identities in the "New" Europe*, edited by Nadia Kaneva, 49–78. New York and London: Routledge.

———. 2014a. "The Geopolitics of Russophonia: The Problems and Prospects of Post-Soviet 'Global Russian'." *Globality Studies Journal* (40):1–22.

———. 2014b. "Mediating New Europe-Asia: Branding the Post-Socialist World via the Internet." In *New Media in New Europe-Asia*, edited by Jeremy Morris, Natalya Rulyova, and Vlad Strukov, 143–166. London: Routledge.

———. forthcoming. "Geopolitical Enemy #1? VVP, Anglophone 'Popaganda' and the Politics of Representation." In *Russian Culture in the Era of Globalisation*, edited by Vlad Strukov and Sarah Hudspith. London: Routledge.

Savdonik, Peter. 2013. "Azerbaijan Is Rich. Now It Wants to Be Famous." *The New York Times*, available at www.nytimes.com/2013/02/10/magazine/azerbaijan-is-rich-now-it-wants-to-be-famous.html [last accessed 9 September 2013].

Scheib, Katrin. 2015. "New York Times Blunder Creates New Country of Kyrzbekistan." *Moscow Times*, available at www.themoscowtimes.com/news/article/new-york-times-blunder-creates-new-country-of-kyrzbekistan/514145.html [last accessed 31 July 2015].

Semeneko, Irina, Vladimir Lapkin, and Vladimir Pantin. 2007. "Russia's Image in the West (Formulation of the Problem)." *Social Sciences* no. 38 (3):79–92.

Shafer, Jack. 2007. "Hail to the Return of Motherland-Protecting Propaganda!" *Slate*, available at www.slate.com/articles/news_and_politics/press_box/2007/08/hail_to_the_return_of_motherlandprotecting_propaganda.html [last accessed 16 January 2009].

Shleifer, Andrei, and Daniel Treisman. 2005. "A Normal Country: Russia After Communism." *Journal of Economic Perspectives* no. 19 (1):151–174.

Shuster, Simon, and Charlotte Mcdonald-Gibson. 2015. "Putin's On-Air Army." *Time* no. 185 (9):46–51.

Silverstein, Ken. 2008. *Turkmeniscam: How Washington Lobbyists Fought to Flack for a Stalinist Dictatorship*. New York: Random House.

Simons, Greg. 2011. "Attempting to Re-Brand the Branded: Russia's International Image in the 21st century." *Russian Journal of Communication* no. 4 (3/4):322–350.

——. 2013. "Nation Branding and Russian Foreign Policy." *Swedish Institute of International Affairs Occasional Papers* no. 21:1–19.

Starr, Frederick. 2009. "Rediscovering Central Asia." *The Wilson Quarterly* no. 33 (3):33–44.

Strmiska, Michael. 2005. "The Music of the Past in Modern Baltic Paganism." *Nova Religio: The Journal of Alternative and Emergent Religions* no. 8 (3):39–58.

Strukov, Vlad, and Vera Zvereva. 2014. *From Central to Digital: Television in Russia*. Voronezh: Voronezh State Pedagogical University Press.

Sussman, Gerald. 2012. "Systemic Propaganda and State Branding in Post-Soviet Eastern Europe." In *Branding Post-Communist Nations: Marketizing National Identities in the "New" Europe*, edited by Nadia Kaneva, 23–48. New York and London: Routledge.

Tchouikina, Sofia. 2010. "The Crisis in Russian Cultural Management: Western Influences and the Formation of New Professional Identities in the 1990s–2000s." *Journal of Arts Management, Law & Society* no. 40 (1):76–91.

Vaiga, Laima. 2015. "Lithuania Ponders Brave New Image . . . and Name." *Baltic Times*, available at www.baltictimes.com/news/articles/19740/ [last accessed 24 July 2015].

van Ham, Peter. 2001. "The Rise of the Brand State: The Postmodern Politics of Image and Reputation." *Foreign Affairs* no. 80 (5):2–7.

Vilimaviciute, Ieva. 2009. *Lithuania, the Brand: An Examination of Nation Branding Efforts and Nation Brand Image in Foreign Press*. Amsterdam: Uinversity of Amsterdam.

Volčič, Zala, and Mark Andrejevic. 2015. *Commercial Nationalism: Selling the Nation and Nationalizing the Sell*. London: Palgrave Macmillan.

Wilson, Andrew. 2005. *Virtual Politics: Faking Democracy in the Post-Soviet World*. New Haven, CT: Yale University Press.

Yagodin, Dmitry. 2012. "Blog Medvedev: Aiming for Public Consent." *Europe-Asia Studies* no. 64 (8):415–434.

Zwick, Jesse. 2012. "Pravda Lite." *New Republic* no. 243 (5):8–9.

Conclusion: Post-Soviet Eurasia

The once and future geopolitical imaginary

The world is an unimaginably vast space and an exceptionally complex place. Would it were that a medieval peasant were magically transported to Manhattan's Times Square or Piccadilly Circus in London, s/he would likely be confronted with a traumatic cascade of images, many being *so foreign* in nature that the most likely outcome would be the subject entering a fugue state from which they might never recover. Yet we, as postmodern denizens of the global imaginarium that is the twenty-first-century world, are inoculated against the swirling cacophony of informational steams, sights, sounds, and emotive triggers we encounter every day. Raised from birth to process these ideologically and geopolitically coded representations, renderings, likenesses, and simulacra—or what I have hitherto referred to as "false seeings"—the postmodern subject has a tumescent quiver of mental and affective tools to navigate this realm. That being stated, such praxes are invariably imbued with deep-seated ideological and prejudicial lenses, such that "seeing" becomes a political act in and of itself. For the Anglophone West, a centuries-long engagement with the outside world—via imperialism, overseas trade, immigration/emigration, and, most importantly, popular culture—has established thousands of prepared responses to encounters with the Other. Like the stylus touching the grooves of a long-playing record, when the Western audience is exposed to the images of far-away places, spaces, and peoples, a predictable "music" is produced. This symphonic interaction is rarely chaotic, despite its varying and often dissonant "notes"; instead, it is fairly prosaic, assuming that the receptor avoids "scratching the surface."

Popular culture is a colossal if ambient force in contemporary international relations. While much of the rank and file of academia have been slow in recognizing this fact due to concerns over diluting the erudition of an already-suspect discipline with the social sciences, this does not make the importance of pop culture any less trenchant. For a wide swathe of the English-speaking world, all they will ever know about Nicaragua, Namibia, or Nauru will come from the images they see/hear/feel in films, comic books, TV series, videogames, and pop music, and such knowledge may never (need to) be verified. This is important given that "the control of the imaginary" where such images live and breathe is "fundamental" to the reproduction of power in the world system (Stocchetti 2011, 19). In short, popular culture is power, and those with greater capacities to produce and distribute popular culture possess more power. As Roland Bleiker informs us:

> Images shape what can and cannot be seen and, indirectly, what can and can-
> not be thought. They influence not only what can be said legitimately in pub-
> lic but also what cannot be said. They help prevent some political positions
> from being established while leaving open discursive space that can be occu-
> pied by others. (Bleiker, Campbell, Hutchinson, and Nicholson 2013, 400)

Postmodernity has seen a dramatic rise in the consumption of popular culture, graphic and spectacular media coverage, and an overall rise in "image satura-tion" that results in an entirely novel form of "truth-telling" and "reality con-struction" (Debrix 2004, 153). Only one-third of all US citizens holds a passport, with a sizeable percentage of those individuals never traveling beyond their near-backyard (Canada, Mexico, and the Caribbean Basin).[1] Only about 10 percent of the US population actually travels overseas—that is, to destinations where they might encounter a culture significantly different from their own, as the Anglophone United Kingdom remains the perennial top destination (Chalmers 2012). Yet even among the comparatively worldly British, travel tends to focus on zones where English is spoken, the sun shines most days, and there is a quick-and-EasyJet connection. Even when the odd, intrepid Anglophone sets out into the unknown and alien world, the power of geopolitical imagination results in an environment where one "knows" a place before one "sees" it, so when one encounters data that does not fit with the meta-structures of understanding, it is often disregarded or discarded.

This result is, in large part, due to the combined (though often conflicting) forces of geopolitical staging via the branding of a country vis-à-vis other mem-bers of the international community and "geo-graphed" pop-cultural production of place/space, Self/Other, and home/away. While this study has hitherto treated *popular geopolitics* and *nation branding* as opposing and/or countervailing forces, I do recognize that both phenomena rely on similar tools, techniques, and approaches vis-à-vis the scripting of reality, including the production of consump-tion-centric geographies manufactured and refined by cultural intermediaries, be they brand consultants, political technologists, videogame designers, journalists, or studio executives. Furthermore, both are exercises in ideology (Kaneva 2012) and propaganda (Sussman 2012), which reinforce neoliberal hegemony, while also being mired in the ideological fundaments of Cold War propaganda and other decades-old information wars. Consequently, both nation branding and geopo-litical imagineering ultimately serve the interests of Western economic elites and more firmly root the non-Western spaces within the field of power (Bourdieu 1993) and the system of structural production (Debord 1983 [1967]) that currently dominate global affairs. As a result, the work of cultural producers is the work of the state, and should not be ignored or dismissed by those scholars operating in the field of international relations.

For the vast and diverse post-Soviet realm, a veritable universe of its own during its Cold War existence, the so-called "end of history" (Fukuyama 1992) has triggered a cascade of geopolitical, economic, cultural, and social change. Once easily lumped together as single country, and later as post-Soviet Eurasia, the countries that make up the Newly Independent States are now emerging as textbook examples of difference and distinction within a single (academically constructed) world region, with any comparison of Estonia and Tajikistan proving this theorem. Yet, as discussed in the chapter on national image, the history of Russia continues to "haunt" (Auchter 2014) the space that is Eurasia. Arguably, this is the result of the power and resonance of the ideation of an "immutable Rus." For the West, and particularly the Anglophone West (and the US and UK specifically) with its set of peculiar interactions with Russia since the 1800s, Russia represents the unflagging Other, a convenient foil that is knowable but recondite, near but distant, somewhat like "us" but more often like a "them." Moreover, all this geopolitical and ideological positioning is often done subconsciously. Iver Neumann condemns much of the literature of foreign policy analysis because of a critical blind spot: perceptions of the Other (particularly, enemies). Neumann argues that IR tends to fail to "socially situate" the self, and therefore is ineffective in understanding the nexus between the Self/Other in terms of national image (Neumann 1999, 13). While it is difficult to counter this argument, it does not seem to address the normative fundament of IR, i.e., the pursuit of the national interest (however that slippery concept is constructed or construed). Regardless, national image is and remains a seminal if amorphous element of foreign policy, international relations, and global affairs despite the paucity of scholarly literature on the subject.

In the early days of the Cold War, Kenneth Boulding called for a Copernican Revolution in the realm of national image, one that would transform our "unsophisticated" image production and perception, which "sees the world only from the viewpoint of the viewer," into something more nuanced, a true *Weltanschauung* that would perceive the world from "many imagined viewpoints, as a system in which the viewer is only a part" (1959, 130). Perhaps his wish has come true with the end of the Cold War and its concomitant effects on the myriad flows of ideas, people, products, and money abetted by "globalization, deterritorialization, cultural fragmentation, and the proliferation of information" (Debrix 2004, 159). However, to extend the Copernican metaphor, there may have been paradigm shift in national-image production and consumption since the days of Eisenhower and Khrushchev, but we are still dealing with loads of Ptolemaic baggage, and we have yet to pass through our Galilean and Newtonian stages. Old prejudices die hard, and the deeper the stereotype, the more difficult it is to unmake it: witness the slow death of the "African savage" as an acceptable trope in Western discourse, particular in the context of the recent Ebola epidemic (see Seay and Dionne 2014).

Despite its capacity for radicalization and revolution and a reputation for liberalism and transgressiveness, most forms of popular-cultural production are conservative and imitative in nature (Nachbar and Lausé 1992), with the overwhelming majority of cultural output ensconcing stereotypes, scaffolding patriarchy, regurgitating hoary myths, defending traditionalism, and tending to evolve slowly over time. By producing the "golden cages of illusion," cultural producers, intentionally or not, create intellectual culs-de-sac that hamper change and support the "hegemony of the present" (Stocchetti 2011, 34). At the most basic level of popular-cultural exchange, we find the *joke*, a simple verbal delivery system designed to provoke laughter. However, there are no jokes without cultural knowledge, and each and every joke relies on the receiver to be able to navigate (or at least hark back to) a shared repository of ideas for it to be "funny." Consequently, humor is boxed into a comparatively fixed space wherein the jokester and the audience are bound by common life experiences, histories, belief systems, and social mores. Taking this structuration up the pop-culture food chain—from political cartoons to comic books to pop songs to radio programs to television series to film to novels to travelogues to news reports—the dynamic of the schema holds. As Juha Ridanpää reminds us, "Popular culture in general occupies a fundamental role in the social processes of how our 'geo' is politically 'graphed'" (2009, 731). Thus, the business of popular culture is the production of worlds, and no cultural producer ever got famous unless they were able to do likewise. From Scott Adams' mundane office-space in *Dilbert* to J. R. R. Tolkien's fantastical Middle-earth, imagined and imaginative geographies are the basic building blocks of any pop-cultural construct. While videogame-worlds, warped and wefted with affective landscapes that exist nowhere else in the universe, represent the paragon of this trend, even a child's toy like G.I. Joe's Cobra Commander action-figure or an 1980s hit like Men at Work's "Down Under" prove that novel-yet-familiar (or at least intelligible) space must be constructed for popular culture to gain purchase amongst its intended audience(s).

Bringing together the power of entrenched stereotypes, operationalized national images, and popular-cultural production, we saw in the chapters on the post-Soviet bogeyman, the laughable nations of the former USSR, and the mapping of "Trashcanistan" that Eurasia faces steep challenges in the current neoliberal world system. Ingrained antipathy associated with the Soviet Union's military-industrial complex (espionage, WMDs, gulags, etc.) has melded with new fears about the effects of "opening up" and "unleashing" Russia to/on the world (mafi-aization, mercenarism, transnational terrorism, radiological contagion, etc.) to produce a new form of popular geopolitics for the post-Cold War era, manifesting most visibly in geopolitical-thriller series like the Bond, Bourne, and *Mission: Impossible* franchises, as well as topical television series like *24*. As a jocose corollary to the fearful popular geopolitics of the 1990s and beyond, there was also a new geopolitical construct which emerged in the post-1989 era: the post-Soviet buffoon.[2] Taking cues from the bumbling cartoon spies Boris and Natasha, and the wackiness of the Leningrad Cowboys, the new Second World creature, beset with "consumerist angst" (Morozov 2011, 59) and incapacitated by the "collapse" of the USSR (Borenstein 2008, 49), became almost as recognizable as the rogue Russian general or the deranged separatist from some "unknown corner" of the Caucasus. While Borat might have been the most famous iteration of this trope, he stood on the shoulders of countless goofy *Homo post-Sovietici*, faced and faceless rubes from Europe's eastern backyard that were meant to provoke haughty bouts of laughter while assuaging any guilt the West might have for "winning" the Cold War. However, as these fanciful human geographies are incontestably parodic in nature, they cannot be effectively countered by their victims as they are masked as ridiculous simulacra rather genuine knowledge (see Wallace 2008). This curious paradox presents a pernicious problem for those seeking to improve the national image of their respective countries, as messages embedded in popular culture are often superimposed or linked recursively to "official" discourse (Boulton 2008). Indubitably, states seeking to improve or even simply maintain their national images must mount innovative defensive campaigns in this growing "battle of the texts" (typically via the tactic of nation branding).

Beyond geopolitical cinema and spoofs, we must also consider the affective power of everyday post-Soviet landscapes and how representations of place and space construct imagined worlds where threats abound. Keeping in mind the impossibility of disentangling popular culture from politics (Bourdieu 1978), the various streams of anxiety, narcissism, and fetishism that characterize the inter-twining pedagogical and disciplining discourses that present themselves in these texts come into focus. Thus, we can identify aspects of what Dalby and Ó Tuathail (1996) refer to as the "constellation of power"—that is, the West's mediatized relationship with its various Others, particularly the post-Second World. In doing so, we are provided with a glimpse of the role of the neoliberal ordering and struc-turing of the world through information flows, mediatized representations, and the visual-textual graphing of space. Add to this mix the discursive legitimization of us/them security metaphors (Palu 2011) and the mental mapping of spaces and places (Brown 2009) of the "new" Eurasia: frigid taiga underlain by long-dormant evils; colorless constructivist dystopias of plutonian ennui; urban warzones of thuggish violence; and toxic wastelands of preternatural horrors. Rooted in cen-turies of travel writing, jaundiced depictions of Eurasia's physical geography play as important a role in shaping political assumptions about the former Soviet republics as any spy-thriller or sitcom. Arguably, they may even do more damage as they tap one's deepest and darkest emotions (fear, loathing, sadness, guilt, etc.), and are seen as relatively immutable when contrasted against human characteris-tics which can change over time (albeit slowly).

Taken together, these three strains of representation (bogeys, buffoons, and badlands) have, on many levels, ineluctably besmirched the image of the post-Soviet East, creating a powerful feedback loop that seeps into quotidian geopol-itics and increasingly informs Western elites' political communication about the former USSR. Regardless, the conditions of late capitalism demand that states respond to the here-and-now realities of international relations. From São Tomé to the People's Republic of China, nation branding has come to be a necessary evil and often a common good for states large and small. For political, eco-nomic, social, cultural, and intellectual elites in the post-Soviet republics, there is a particularly acute perceptual need for country branding, given the lingering power of the Soviet frame and national image issues associated with the Cold War, as well as the problems of isolation and submersion of separate identi-ties within the USSR. From Facebook pages to presidential tweets, from adver-tising supplements in *The Telegraph* to YouTube videos, from the Olympics to Eurovision, the new states of the old USSR have endeavored to make themselves known and valued in the global supermarket of brands. While the results have been mixed, it is clear that nation branding in Eurasia is established, evolving, and occasionally entropic. Moreover, it seems that latent tendencies associated with the Soviet experience continue to bubble to the surface as portions of the region gravitate towards innovative forms of neo-authoritarianism, with its strange mix of openness, anti-Westernism, and cultural distinctiveness based on established postmodern paradigms (Umland 2012), just as other states in the region embrace the "transatlantic consensus" on what it means to be a responsible, neoliberal country in the twenty-first century (Plehwe, Walpen, and Neunhöffer 2007).

Holistically speaking, post-Soviet Eurasia—as a zone of the third wave of decolonization, an arena of traumatic politico-economic transition, and a complex yet cogent geopolitical entity that is neither wholly European nor completely non-Western—represents a hothouse for experimentation and innovation when it comes to geopolitical staging, in both indigenous (nation branding) and external (popular culture) forms. This study has attempted to address a small sliver of this protean nexus where the state, media, and ideology meet to produce geographies of the mind's eye as it relates to the Anglophone West's relationship with the former Soviet Union. The once and future geopolitical imaginary of Eurasia—a zone burdened by centuries of ideated baggage, from Ivan the Terrible throwing puppies from the windows of the Kremlin to President Putin shooting whales with crossbows in the northern reaches of the Pacific Ocean—is indeed a complicated entity, and one deserving of further study.

Notes

1 These numbers represent a precipitous increase compared to pre-9/11 figures. Following the 11 September 2001 attacks, the US instituted passport requirements for its citizens when traveling to a large number of destinations that previously only required a US state-issued identification card.
2 In a late June 2015 report on the impending opening of Patriot Park in Kubinka, a "military Disneyland" where "children can play with grenade launchers and clamber over heavy weaponry" (Walker 2015), the Anglo-American comedian John Oliver neatly linked the fearsome and quirky aspects of such depictions, stating: "Now, on to Russia—a country that will continue to be funny, until it suddenly isn't."

References

Auchter, Jessica. 2014. *The Politics of Haunting and Memory in International Relations.* London: Routledge.

Bleiker, Roland, David Campbell, Emma Hutchison, and Xzarina Nicholson. 2013. "The Visual Dehumanisation of Refugees." *Australian Journal of Political Science* no. 48 (4):398–416.

Borenstein, Eliot. 2008. *Overkill: Sex and Violence in Contemporary Russian Popular Culture.* Ithaca, NY: Cornell University Press.

Boulding, Kenneth E. 1959. "National Images and International Systems." *Journal of Conflict Resolution* no. 3 (2):120–131.

Boulton, Andrew 2008. "The Popular Geopolitical Wor(l)ds of: Post-9/11 Country Music." *Popular Music and Society* no. 31 (3):373–387.

Bourdieu, Pierre. 1978. "La sociologie de la culture populaire." In *Le Handicap socio-culturel en question*, edited by Centre de recherche de l'éducation spécialisée et de l'adaptation scolaire, 117–120. Paris: ESF.

———. 1993. *The Field of Cultural Production: Essays on Art and Literature*. New York: Columbia University Press.

Brown, Neville. 2009. *The Geography of Human Conflict: Approaches to Survival*. Eastbourne, UK: Sussex Academic Press.

Chalmers, William D. 2012. "The Great American Passport Myth: Why Just 3.5% Of Us Travel Overseas!" *Huffington Post*, available at www.huffingtonpost.com/william-d-chalmers/the-great-american-passpo_b_1920287.html [last accessed 20 July 2015].

Dalby, Simon, and Gearóid Ó Tuathail. 1996. "The Critical Geopolitics Constellation: Problematizing Fusions of Geographical Knowledge and Power." *Political Geography* no. 15 (6–7):451–456.

Debord, Guy. 1983 [1967]. *Society of the Spectacle*. Detroit, MI: Black & Red.

Debrix, François. 2004. "Tabloid Realism and the Revival of American Security Culture." In *11 September and Its Aftermath: The Geopolitics of Terror*, edited by Stanley D. Brunn, 151–190. London and Portland, OR: Frank Cass.

Fukuyama, Francis. 1992. *The End of History and the Last Man*. New York: Free Press.

Kaneva, Nadia. 2012. "Nation Branding in Post-Communist Europe: Identities, Market, and Democracy." In *Branding Post-Communist Nations: Marketizing National Identities in the "New" Europe*, edited by Nadia Kaneva, 3–22. New York and London: Routledge.

Morozov, Evgeny. 2011. *The Net Delusion: How Not to Liberate the World*. London: Allen Lane.

Nachbar, Jack, and Kevin Lausé. 1992. *Popular Culture: An Introductory Text*. Madison, WI and London: Popular Press.

Neumann, Iver B. 1999. *Uses of the Other. The "East" in European Identity Formation*. Minneapolis: University of Minnesota Press.

Palu, Helle. 2011. "The Politics of Visual Representations: Security, the US and the 'War on Terrorism'." In *Images in Use: Towards the Critical Analysis of Visual Communication*, edited by Matteo Stocchetti and Karin Kukkonen, 151–180. Amsterdam and Philadelphia, PA: John Benjamin Publishing Co.

Plehwe, Dieter, Bernhard J. A. Walpen, and Gisela Neunhöffer. 2007. *Neoliberal Hegemony: A Global Critique*. London: Routledge.

Ridanpää, Juha. 2009. "Geopolitics of Humour: The Muhammad Cartoon Crisis and the *Kaltio* Comic Strip Episode in Finland." *Geopolitics* no. 14 (4):729–749.

Seay, Laura, and Kim Yi Dionne. 2014. "The Long and Ugly Tradition of Treating Africa as a Dirty, Diseased Place." *Washington Post*, available at www.washingtonpost.com/news/monkey-cage/wp/2014/08/25/othering-ebola-and-the-history-and-politics-of-pointing-at-immigrants-as-potential-disease-vectors/ [last accessed 31 December 2015].

Stocchetti, Matteo. 2011. "Images: Who Gets What, When, and How?" In *Images in Use: Towards the Critical Analysis of Visual Communication*, edited by Matteo Stocchetti and Karin Kukkonen, 11–38. Amsterdam and Philadelphia, PA: John Benjamin Publishing Co.

Sussman, Gerald. 2012. "Systemic Propaganda and State Branding in Post-Soviet Eastern Europe." In *Branding Post-Communist Nations: Marketizing National Identities in the "New" Europe*, edited by Nadia Kaneva, 23–48. New York and London: Routledge.

Umland, Andreas. 2012. "The Claim of Russian Distinctiveness as Justification for Putin's Neo-Authoritarian Regime." *Russian Politics & Law* no. 50 (5):3–6.

Walker, Shaun. 2015. "Vladimir Putin Opens Russian 'Military Disneyland' Patriot Park." *The Guardian*, available at www.theguardian.com/world/2015/jun/16/vladimir-putin-opens-russian-military-disneyland-patriot-park [last accessed 6 October 2015].

Wallace, Dickie. 2008. "Hyperrealizing 'Borat' with the Map of the European Other." *Slavic Review* no. 67 (1):36–49.

Index

Taylor & Francis eBooks

Helping you to choose the right eBooks for your Library

Add Routledge titles to your library's digital collection today. Taylor and Francis ebooks contains over 50,000 titles in the Humanities, Social Sciences, Behavioural Sciences, Built Environment and Law.

Choose from a range of subject packages or create your own!

Benefits for you

» Free MARC records
» COUNTER-compliant usage statistics
» Flexible purchase and pricing options
» All titles DRM-free.

REQUEST YOUR FREE INSTITUTIONAL TRIAL TODAY

Free Trials Available
We offer free trials to qualifying academic, corporate and government customers.

Benefits for your user

» Off-site, anytime access via Athens or referring URL
» Print or copy pages or chapters
» Full content search
» Bookmark, highlight and annotate text
» Access to thousands of pages of quality research at the click of a button.

eCollections – Choose from over 30 subject eCollections, including:

Archaeology	Language Learning
Architecture	Law
Asian Studies	Literature
Business & Management	Media & Communication
Classical Studies	Middle East Studies
Construction	Music
Creative & Media Arts	Philosophy
Criminology & Criminal Justice	Planning
Economics	Politics
Education	Psychology & Mental Health
Energy	Religion
Engineering	Security
English Language & Linguistics	Social Work
Environment & Sustainability	Sociology
Geography	Sport
Health Studies	Theatre & Performance
History	Tourism, Hospitality & Events

For more information, pricing enquiries or to order a free trial, please contact your local sales team:
www.tandfebooks.com/page/sales

Routledge
Taylor & Francis Group

The home of
Routledge books

www.tandfebooks.com

Printed and bound by CPI Group (UK) Ltd, Croydon, CR0 4YY

24/10/2024

01778279-0008